Air Quality Monitoring, Assessment and Management

Air Quality Monitoring, Assessment and Management

Editor

Henri Duff

Air Quality Monitoring, Assessment and Management

Edited by **Henri Duff**

ISBN: 978-1-68117-239-2

Library of Congress Control Number: 2016934779

© 2017 by
SCITUS Academics LLC,
www.scitusacademics.com
Box No. 4766, 616 Corporate Way,
Suite 2, Valley Cottage,
NY 10989

Preface

Human beings need to breathe oxygen diluted in certain quantity of inert gas for living. The atmosphere is a complex natural gaseous system that is essential to support life on planet Earth. Stratospheric ozone depletion due to air pollution has been recognized as a threat to human health as well as to the Earth's ecosystems. Air pollution has become a problem of major concern in the last few decades as it has caused negative effects on human health, nature and properties. Air is considered to be polluted when it contains certain substances in concentrations high enough and for durations long enough to cause harm or undesirable effects. The atmosphere is susceptible to pollution from natural sources as well as from human activities. To improve air quality, It is important for governments, industry and the public to understand the interconnected components of an air quality management system. Air quality management (AQM) refers to all the activities to make sure that the air we breathe is safe, both outdoors and indoors. The AQM process is the system of understanding the sources that contribute to pollution in the air and the health and environmental effects of the pollutants, and then taking steps to reduce or control the sources to reach or maintain agreed upon target pollution levels in the air. These levels may vary from country to country, but the overall system for planning, assessing, characterizing, mitigating, and implementing control strategies is similar. While AQM is generally handled at the national government level, regional and local governments, industry, and the public all have important roles to play in this system. Each air quality management activity is related to the others. This book, entitled Air Quality Monitoring, Assessment and Management is invaluable tool for all those, whether as a student, researcher, entrepreneur, consultant, or government agency with obligation in this area.

Table of Contents

CHAPTER 1

Air Pollution Monitoring and Prediction

Sheikh Saeed Ahmad[1, 2], Rabail Urooj[1, 2] and Muhammad Nawaz[1, 2]

[1] *Department of Environmental Sciences, Fatima Jinnah Women University, Mall Road, Rawalpindi, Pakistan*
[2] *BZ University, Multan, Pakistan*

1. INTRODUCTION

One of the most important emerging environmental issues in Asian cities is air pollution. Air pollution is an atmospheric condition in which the concentration and duration of certain substances present in the air produce injurious and destructive effects on both man and the surrounding environment [1]. The most common pollutants in air are sulfur oxide, nitrogen dioxide, carbon monoxide and dioxide, and particulate matter.

Geographical Information Systems (GISs) are computer-based applications used for mapping and analyzing the earth and related spatially distributed phenomena. GIS applications integrate unique visualizations with common databases, which make it possible to capture, model, manipulate, retrieve, analyze, and present the geographically referenced data. Compared to other information systems, GIS systems have advantages, including the high power of analyzing spatial data and handling large spatial databases.

GIS applications can be used in air quality management and for controlling pollution, for handling and managing large amount of data. GIS systems manage spatial and statistical data, which facilitates depiction of the association between the frequency of human activities leading to bad environmental health and poor air quality. GIS modeling and statistical analysis also enables to examine and predict the impact of climatic variables on air pollution. In this way, GIS systems help in monitoring air pollution and emissions of pollutants from different sources.

Air pollution mapping is a helpful method for determining the concentration of pollutants. As the result of air pollution mapping, overviews of pollution in cities can be created and their sources of pollution emission can be identified, which help in controlling emissions. Different studies have been executed on air pollution in conjunction with GIS [2-11]. Consequently, GIS applications in air monitoring are necessary to determine air quality to reduce pollution to such a level at which harmful impacts on human health and the environment is reduced.

With the help of GIS applications, an output report of pollutants in Air Quality Management Systems (AQMSs) can be achieved in the form of three-dimensional (spatial) records. In AQMS emission time, concentration and place of air pollutants are regulated in order to achieve the predefined air quality standards of ambient air. It encompasses the estimation of the pollutants' emission schedule in a way to determine the consequences to air quality and the design of alternative programs for emission control in order to meet air quality standards, which are subject to some limitations, for example, technological viability and lowest charges. For environmental modeling with GIS applications, AQMSs are considered to locate monitoring stations, for development of geospatial model for air quality, and for spatial decision-support systems. However, the most significant step in an AQMS is data mining. The data mining method is a skill, which is used to analyze the data, uncover hidden patterns, and find interesting information from large amounts of data or huge databases. The most commonly used technique in data mining is artificial neural networks [12].

The human brain consists of a large number of neurons connected to each other by synapses to make networks, and these networks of neurons are called neural networks, or natural neural networks. Similarly, the artificial neural network (ANN) is basically a mathematical model of a natural neural network. The ANN uses a mathematical or computational model based on connectionist approach for solving the given problem. The concept of ANN is derived from biological neural network systems. The key applications of neural networks are control systems, classification systems, and prediction and vision systems.

Three basic components are important in order to make functional model, like: synapses of neuron; an added that sum all input in form of weights; and activation function. In Figure 1, synapses are shown by weights. Basically, a strong connection between input and neuron is noted by synapses or value of weight. Negative values reflect inhibitory connections, whereas excitatory connections are shown by positive values. Activation functions regulate the output of neurons within an acceptable range from -1 to 1.

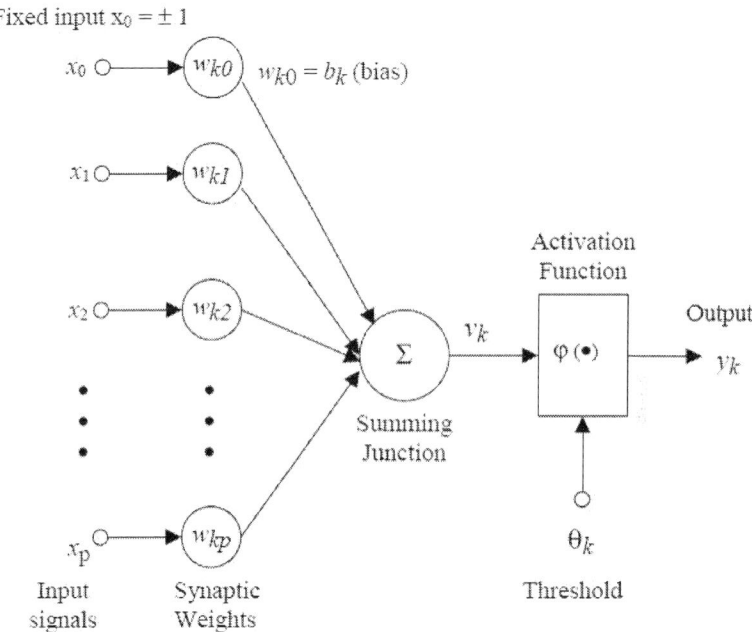

Figure 1. Model of a neuron

1.1. Sources

Air pollution takes place due to natural and anthropogenic activities. But air pollution as the result of man-made activities like fossil fuel combustion, construction, mining, agriculture, and warfare are the most significant and cause problems in the atmosphere [13].

Basically, two types of pollution sources have been categorized, i.e., Stationary and Mobile. The stationary source is a type of source that is fixed or is a preset pollutant emitter, for example, fossil fuel burning power plants and refineries. The mobile source is a nonstationary type of pollutant emitter, for example, vehicles. The most emerging and leading cause of air pollution is the motor vehicle [14]. Pollutants that are emitted directly from the source into the air are known as primary pollutants, for example, carbon dioxide, carbon monoxide, sulfur dioxide, etc. When these primary pollutants react in atmosphere with each other to form another type of pollutants, they are called secondary pollutants, which are not directly emitted but formed as a result of primary pollutants' reaction in the atmosphere. For example, ozone forms when nitrogen oxides react with hydrocarbons in the presence of sunlight, and the resulting nitrogen dioxide reacts further with oxygen and forms ozone as pollutant.

1.2. Health Effects

Air pollution and its effects in rural and urban areas are directly related to the ongoing activities. For example, in cities, pollution is related to the products of combustion in industries and vehicles. Many large cities all over the world exhibit excessive levels of air pollutants. Among all dangerous pollutants, nitrogen dioxide (NO_2) is important due to its capacity of causing dangerous effects on humans and the environment, which results in photochemical oxidation and acid rain.

The effects of air pollution cannot be ignored even within homes. Many air pollutants can cause cancer and other diseases among inhabitants. In 1985, it was reported that indoor toxic chemicals are three times more potent in causing cancer than outdoor air pollutants [15]. In America, health issues caused by buildings are called "sick building syndrome"[16].

1.3. Case Study

In Pakistan, air pollution is emerging as a serious problem in its mega cities, which needs to be monitored and addressed at the root level in order to reduce the lethal impacts of pollutants on man and environmental health. The present study of Pakistan focuses on the most important twin cities of Pakistan, which are Rawalpindi and Islamabad. Both cities are commonly viewed as one unit and are 15 km apart. The study area with 135 sampling locations is shown in Figure 2. The climatic condition of Rawalpindi and Islamabad is sub-humid to tropical, with hot and long summers (May to August) accompanied by a monsoon season (July to August) followed by short and mild winters (October to March). The average low temperature is 12.05 °C in January and average high temperature is 31.13 °C in July.

For the monitoring campaign, the maximum area (135 sampling sites) was covered in order to represent different traffic intensity and congestion levels in the urban area of Rawalpindi and Islamabad, for sampling. These sites included dual carriageways, major, linking, and small roads, healthcare centers, educational institutes, commercial areas, old residential areas, modern residential areas, recreational spots and semi-rural areas.

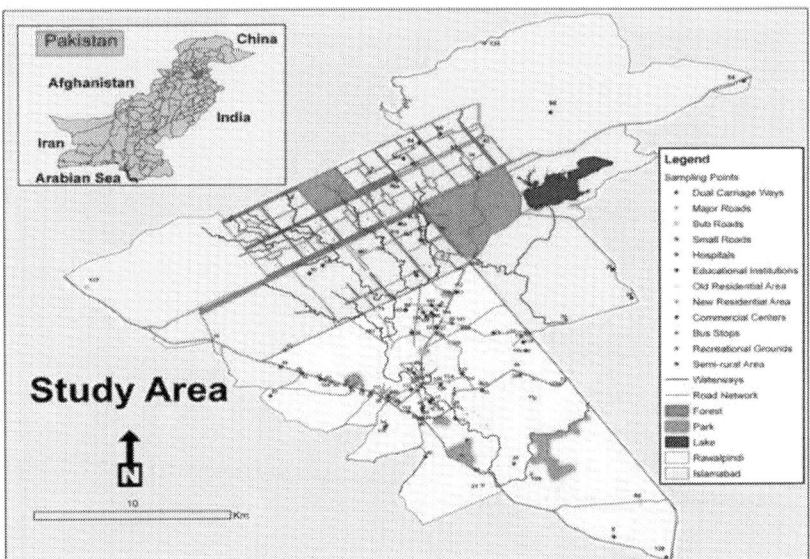

Figure 2. Base map

Research was carried out in order to monitor the NO$_2$ concentration in the ambient air of Rawalpindi city. Passive samplers were used within the city from January to December in 2008. The average concentration found was 27.46±0.32 ppb. The highest concentration was recorded near the main roads and in the vicinity of schools and colleges due to the large number of transport vehicles, which exceeded the set limit concentration value given by the World Health Organization.

2. EXPERIMENTAL DESIGN

2.1. Passive Sampling Of No$_2$

The most frequent method in monitoring studies for passive sampling of NO$_2$ is using diffusion tubes described by Atkins [17]. This method for NO$_2$ measurement is reliable, easy to handle, and it is an inexpensive method for screening air quality. Moreover, passive samplers are preferably appropriate for extensive spatial measurement of NO$_2$, and they have been reported in many studies of NO_2 monitoring of air in many countries like the United Kingdom, USA, France, Turkey, Argentina, and China [18].

Basically, passive samplers are designed on the principle of air diffusion having an efficient absorber at one end of the tube, and the flow rate (sampling rate) at constant temperature can be measured by using Flick's Law [19]. For that, the length and diameter of diffusion tubes are known, whereas sampling by using diffusion tubes is independent of air pressure.

2.2. Neural Network Design

From different sampling sites covering the whole study area, data was collected for neural network analysis. Collected data was fed to the neural network that has area_id, season_id, temperature, humidity, rainfall, and the respective concentrations as columns. For the neural network, the marked value was set to predict concentrations and rests were used as input to the neural network.

Neural network has two phases: training and testing. In the first phase (training), the network is trained by providing the complete information about the characteristics of data and observable outcomes to perform a particular task.

A neural network can develop a model that learns the relationship between input data and the desired outcome in the training phase. In the testing phase, testing data are provided as input. The performance of the testing phase depends upon the training phase (it depends on the number of samples that are provided during the training phase and also on the number of times that the network is accurately trained. However, it is impossible that the output is 100% precise for any network input. MS Access was used as the database engine because it is easy to use for all.

For testing the neural network, the cross validation method is used by using holdout method in which data was divided into testing and training data. The database consisted of two tables: training_ data and testing_data. The function of training_data is to train the ANN by adjusting weights in order to maximize the predictive ability of ANN and minimize error during forecasting. Testing data was used to test the prediction accuracy of ANN on new data. The structure of training data and testing data is given in Table 1.

In Table 1, the first key "id" is primary key, which contains the number that indicates row number and the second key "loc_id" contains the number that indicates location from where data is gathered, loc_name indicates the name of location and the next six fields indicate position of location with respect to north and east. The next two indicate temperature and humidity levels.

The 13th and 14th fields indicate concentration of NO_2 and level of concentration value. The last field of dataset contains week number, which indicates the number of weeks in which data is gathered from particular location. The attribute for testing data are the same in the testing data structure.

Table 1. Structure of training data

Field Name	Data type	Primary key	Field size
Id	Number	Yes	Long Integer
loc_id	Number		Long Integer
loc_name	Text		50
map_id	Number		Long Integer
north_d	Number		Long Integer
north_m	Number		Long Integer
north_s	Number		Long Integer
east_d	Number		Long Integer
east_m	Number		Long Integer
east-s	Number		Long Integer
Temp	Text		50
Humidity	Text		50
Concentration	Number		Long Integer
con_level	Number		Long Integer
Week	Number		Long Integer

For designing a network, we need to specify the architecture of a neural network by designing a number of hidden layers and units in each layer along properties of network that describe error function and network activation.

For optimal generalization of collected data, two types of architectures: the rtNEAT (real-time neuro evolution of augmented topologies) architecture with evolution algorithm and the feed forward architecture with back propagation algorithm of ANN are used in order to ensure high accuracy of ANN prediction about impacts of NO_2 concentration achieved in future. This rtNEAT architecture is used to train neural network with evolutionary algorithm, which has three steps, i.e., selection, mutation, and reinsertion. But before the training of neural network, the topology has to be created in the design of the neural network. A neural network is a connection of neurons, which contains three types of nodes: input, output, and hidden node. All nodes are randomly created during its execution.

Table 2 describes the properties of network, which contains an error function and network activation parameters. These properties are functional to all tested networks by the architecture search method and manually selected network.

Table 2. Network properties

Parameter	Value
Input activation FX	Logistic
Output name	Concentration
Output error FX	Sum-of-squares
Output activation FX	Logistic

The logistics function has a sigmoid curve and sum of squares. The sum of squares is the most frequent function error, which is used for the classification problem. The error is the sum of the square differences between the real input value and neural network target value.

2.3. Architecture Search

A heuristic search is used to search the dataset for the best networks. Heuristic methods are used to speed up the process of finding a satisfactory solution. The architecture search for the designed neural network NO_2 is given in Table 3.

Table 3. Heuristic architecture search for NO2

ID	Architecture	# of Weights	Fitness	Train Error	Validation ErrorError	Test ErrorError	AIC	Correlation	R-Squared
1	[8-1-1]	11	0.079965	11.371084	12.151164	12.505404	-4220.122343	0.652911	0.424949
2	[8-20-1]	201	0.080369	11.295746	12.093373	12.442678	-3846.484024	0.653839	0.427147
3	[8-12-1]	121	0.080841	11.017718	12.044772	12.37002	-4030.333807	0.668927	0.446734
4	[8-7-1]	71	0.080593	11.182193	12.108417	12.407984	-4116.153129	0.662147	0.438331
5	[8-16-1]	161	0.081507	10.941978	11.986108	12.26894	-3956.935279	0.670637	0.448777
6	[8-18-1]	181	0.080474	10.917611	12.026946	12.4264	-3919.068903	0.676823	0.457557
7	[8-14-1]	141	0.080839	11.105445	12.044827	12.370266	-3982.744071	0.666178	0.44366

2.4. Training Of Neural Network

The next step is to train the neural network for the NO_2 dataset by using the propagation algorithm. Weight change is calculated by the quick propagation algorithm by utilizing the quadratic function $f(x) = x^2$. In neural networks, several layers contain neurons in each layer that are connected with each other like neurons in the input layer connected to one or more neurons of the hidden layer, which are further connected to the output layer's neuron. With each presentation in neural network, error is computed as the difference between network output and observable output. The combination of randomly assigned weight (giving low error) replaces weights that are at the first location. This is called training to adjust the connection weights to enable the network to produce the expected output. Two different weights having two different error values are two points of a secant. Relating this secant to a quadratic function, it is possible to calculate its minimum $f(x) = 0$. The x-coordinate of the minimum point is the new weight value.

$$S(t) = \frac{\partial E}{\mathbb{R}w_i(t)}$$

$$\Delta w_i(t) = \alpha \cdot \frac{\partial E}{\partial w_i(t)} \text{(Normal back propagation)}$$

$$\frac{\Delta w_i(t)}{\alpha} = \frac{\partial E}{\partial w_i(t)}$$

$$S(t) = \frac{\partial E}{\partial w_i(t)} = \frac{\Delta w_i(t)}{\alpha}$$

$$\Delta w_i(t) = \frac{S(t)}{S(t-1) - S(t)} \cdot \Delta w_i(t-1) \text{(Quick propagation)}$$

Here w =weight, i =neuron, E =error function, t =time (training step), α= learning rate, and μ= maximal weight change factor

The quick propagation coefficient was set to 1.75, learning rate was 0.1, and iterations were 500. The training graph for dataset errors for NO_2 is shown in Figure 3.

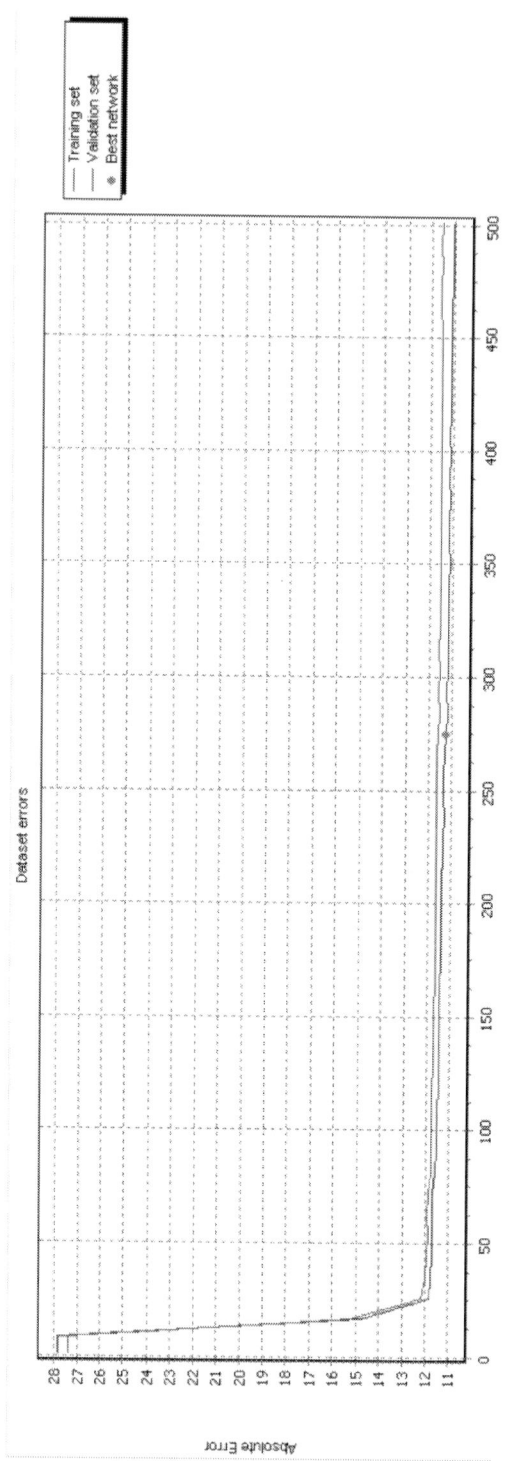

Figure 3. Dataset errors for the NO$_2$ dataset

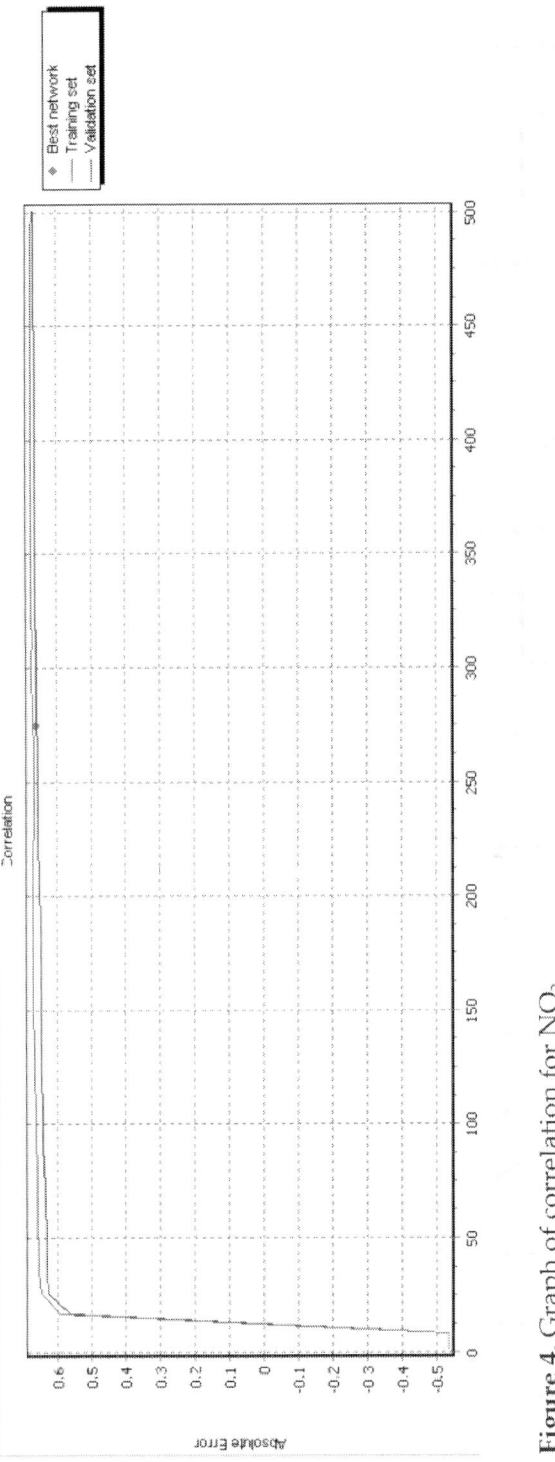

Figure 4. Graph of correlation for NO_2

The training graph of correlation for NO_2 is shown in Figure 4.

The graph of error improvement – network errors for NO_2 is shown in Figure 5.

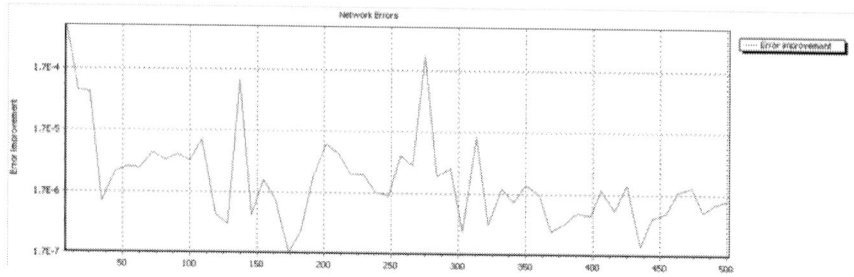

Figure 5. Network errors for NO_2

The error distribution of network statistics obtained after training of neural network is shown in (Figure 6).

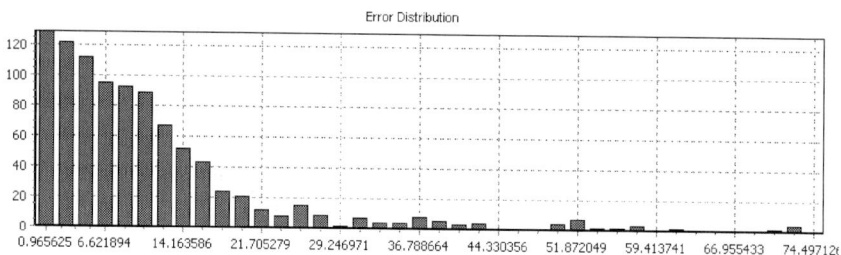

Figure 6. Error distribution for NO_2

3. DATA ANALYSIS

In order to determine the seasonal variation and statistical significance, results are presented in tabular format. Tables 4 a and 4 b show the average concentration level of NO_2, season-wise, along standard deviation (SD) values measured at different sampling sites of study.

Table 4 a shows average values of NO_2 concentration in different seasons of 12 major sampling categories in urban Rawalpindi and Islamabad from November 2009 to July 2010.

Table 4 b shows the seasonal average concentration of NO_2 of 12 major sampling categories in urban Rawalpindi and Islamabad from September 2010 to March 2011.

Table 5 presents NO_2 concentration for each selected category, as described in study area profile, to understand the general trends of NO_2 concentration levels among different categories during the course of experimental period.

Table 4. (a): Seasonal mean values of NO_2 from November 2009 to July 2010 (b): Seasonal mean values of NO_2 from September 2010 to March 2011

	Sampling Categories	Mild Winter (Nov)	Winter (Dec to Jan)	Early Spring (Feb)	Spring (Mar)	Mild Summer (April)	Summer (Pre-Monsoon) (May to June)	Monsoon (July to August)
	NO₂ Conc. (weekly basis)	(ppb)	(ppb)	(ppb)	(ppb)	(ppb)	(ppb)	(ppb)
Rawalpindi	Dual Carriage Ways (5)	87±19.78	98±26.87	63±12.29	53±6.49	44±10.64	22±4.22	18±1.91
	Major Roads (10)	60±12.19	68±9.56	52±13.52	45±10.23	36±8.97	26±5.88	19±4.74
	Sub-roads (6)	74±20.50	86±24.47	60±16.49	50±11.05	38±12.65	33±13.01	21±4.39
	Small Roads (3)	55±9.78	63±4.89	47±5.57	40±8.24	31±3.40	25±4.68	18±4.81
	Public Hospital (5)	48±18.71	63±18.40	37±0.74	29±2.29	22±2.24	18±0.79	14±0.96
	Private Hospitals (8)	61±14.47	75±14.19	38±1.16	32±2.03	25±2.29	20±5.57	14±3.98
	Public EI (11)	85±30.58	95±32.94	75±23.75	63±17.94	47±17.37	31±10.14	20±1.94
	Private EI (17)	55±9.71	66±9.54	45±4.56	43±9.65	38±10.89	26±4.61	18±3.18
	Old Residential Areas (5)	83±15.24	95±16.09	55±13.32	51±6.66	37±6.44	26±2.54	19±1.05
	Modern Residential Areas (5)	65±20.07	73±14.89	69±24.49	59±12.55	36±7.13	28±5.08	21±2.61
	Commercial Area (2)	75±0.83	82±17	61±6.69	51±7.11	36±4.29	21±6.20	18±4.78
	Bus Stops (9)	74±20.26	83±31.47	69±33.78	58±17	39±17.32	28±8.41	20±5.25
	Recreational Spots (9)	75±38.40	87±40.76	62±36.39	56±21.88	43±19.97	31±11.12	19±2.37

Table 4 (continued)

Sampling Categories	Mild Winter (Nov)	Winter (Dec to Jan)	Early Spring (Feb)	Spring (Mar)	Mild Summer (April)	Summer (Pre-Monsoon) (May to June)	Monsoon (July to August)
NO$_2$ Conc. (weekly basis)	(ppb)	(ppb)	(ppb)	(ppb)	(ppb)	(ppb)	(ppb)
Islamabad							
Dual Carriage Ways (3)	84±28.73	95±33.64	66±23.78	57±12.31	45±16.69	24±5.98	19±4.16
Major Roads (3)	50±3.72	60±2.04	40±0.81	32 ±2	26±4.42	21±2.97	15±2.16
Sub-roads (4)	54±6.06	67±6.39	49±6.49	43±7.24	38±12.79	25±1.38	18±1.26
Small Roads (3)	59±12.65	64±6.33	51±9.60	44±8.93	35±4.53	26±3.66	20±3.08
Public Hospitals (3)	44±0.58	57±0.29	39±0.29	32±0.58	23±1.47	19±0.51	15±1.71
Private Hospitals (1)	42	56	38	30	24	19	14
Public EI (5)	53±13.34	64±9.32	46±9.30	39±10.76	34±14.19	25±6.01	18±1.28
Private EI (6)	58±11.23	63±7.18	49±7.72	39±9.93	31±5.85	24±2.66	17±1.77
Commercial Area (1)	61	68	57	50	35	25	16
Bus Stops (12)	72±14.25	78±16.23	65±7.51	55±5.23	34±6.22	25±3.21	19±2.56
Recreational Spots (2)	62±5.97	69±4.58	57±2.45	48±1.59	38±2.48	25±3.15	17±1.56
Semi-Rural Areas (7)	46±8.98	59±5.64	42±6.41	33±7.87	31±5.29	24±3.22	18±3.19

(a)

Sampling Categories	Mild Winter (Nov)	Winter (Dec to Jan)	Early Spring (Feb)	Spring (Mar)	Mild Summer (April)	Summer (Pre-Monsoon) (May to June)	Monsoon (July to August)
Rawalpindi							
Dual Carriage Ways (5)	30±4.00	51±7.52	88±22.28	100±26.42	63±18.52	50±20.32	
Major Roads (10)	27±6.57	49±9.72	61±10.33	68±9.34	48±3.76	37±3.79	
Sub-roads (6)	32±7.31	53±12.97	74±23.42	87±26.60	58±13.69	39±6.36	
Small Roads (3)	28±3.65	37±2.53	53±6.30	62±2.58	54±12.39	43±11.13	
Public Hospital (5)	20±0.98	32±5.46	48±18.01	64±18.03	40±6.13	29±3.94	
Private Hospitals (8)	23±3.90	38±7.19	60±15.37	73±14.03	40±4.05	31±1.95	
Public EI (11)	44±16.98	81±36.87	86±31.78	96±34.20	73±21.26	63±18.23	
Private EI (17)	31±4.85	42±6.10	55±8.91	66±9.82	45±7.08	35±5.90	

Table 4 (continued)

	Sampling Categories	Mild Winter (Nov)	Winter (Dec to Jan)	Early Spring (Feb)	Spring (Mar)	Mild Summer (April)	Summer (Pre-Monsoon) (May to June)	Monsoon (July to August)
	NO₂ Conc. (weekly basis)	(ppb)	(ppb)	(ppb)	(ppb)	(ppb)	(ppb)	(ppb)
Islamabad	Dual Carriage Ways (3)	31±5.72	50±11.40	82±21.11	99±32.70	67±19.78	49±12.84	
	Major Roads (3)	22±0.80	37±1.93	53±4.33	65±0.30	44±2.30	33±2.00	
	Sub-Roads (4)	26±2.48	39±4.60	54±5.74	65±4.08	46±3.02	35±9.17	
	Small Roads (3)	30±5.94	41±4.12	54±4.24	63±6.67	47±2.60	38±2.79	
	Public Hospitals (3)	22±2.14	34±2.66	45±0.80	60±1.41	45±0.22	34±0.82	
	Private Hospitals (1)	22	31	40	55	38	30	
	Public EI (5)	31±7.41	40±3.60	52±7.94	64±8.39	46±9.34	37±9.62	
	Private EI (6)	29±11.58	41±8.65	54±10.14	63±7.03	47±7.93	36±9.17	
Twin Cities	Old Residential Areas (5)	27±2.97	61±14.74	84±14.18	95±16.51	58±12.41	48±10.06	
	Modern Residential Areas (5)	32±7.86	49±11.70	66±20.07	75±16.16	60±19.16	48±16.53	
	Commercial Area (3)	32±1.23	46±6.09	63±1.00	71±3.57	56±7.02	48±8.41	
	Bus Stops (11)	32±9.11	53±20.30	76±20.07	87±32.40	69±31.34	54±19.54	
	Recreational Spots (10)	37±18.55	52±25.23	71±37.63	84±39.83	57±29.71	46±24.78	
	Semi-Rural Areas (7)	31±9.47	41±7.44	53±6.51	62±6.21	44±7.50	36±6.99	

(b)

Table 5. Average NO₂ concentration levels in twin cities from November 2009 to March 2011

Sampling Categories	No. of Sites	Average NO₂ Conc. (ppb)
Dual Carriage Ways	8	55.23
Major Roads	13	53.56
Sub-roads	10	51.78
Bus Stops	11	51.62
Educational Institutions	39	51.26
Recreational Spots	10	51.18
Old Residential Area	5	48.97
Small Roads	6	48.23
Commercial Area	3	47.59
Hospitals	17	47.44
Modern Residential Area	5	46.25
Semi-Rural Area	7	37.65

In Table 5 most of the sampling sites of study area showed nearly similar average concentration from month of November 2009 to March 2011. Maximum concentration of NO_2 shown on dual carriage ways.

The possible cause of such elevated levels of NO_2 concentration is extensive increase in number of vehicles, increase in population, busy roads, fuel inefficient vehicles, driving ways, and traffic jams. Gilbert reported that NO_2 is considerably related to both the distance from the nearest highway and the traffic count on the nearest highway [20].

The rest of the categories showed nearly the same average concentration. Major roads and sub-roads showed average NO_2 concentration levels of 53.56 ppb and 51.78 ppb, respectively. Sub-roads, bus stops, recreational spots, and educational institutions showed similar concentration levels of approx. 51 ppb.

Educational institutions and recreational spots, being present close to the dual carriage ways, also experience elevated concentration levels. Old residential areas (48.97 ppb) showed slightly higher NO_2 concentration levels as compared to modern residential areas (47.59 ppb).

Narrow road, enclosing architecture, and congestion among the old residential areas result in traffic emission being trapped and buildup leading to higher NO_2 concentration levels, whereas in modern residential areas increased vehicular number is the major cause of elevated NO_2 levels. The minimum NO_2 concentration levels were indicated in semi-rural areas, that is 37.65 ppb. A study in Vilnius commented the same phenomena; NO_2 average rates depend upon traffic and are highest in cross roads and lowest at the background suburban areas [21].

For annual average concentration level of nitrogen dioxide, a spatial interpolation map has been developed by using inverse distance weighted (IDW). IDW in Figure 7 is clearly depicted as the areas of higher and lower concentration level of NO_2 in Rawalpindi and Islamabad.

Higher concentration levels are represented by darker shades while the lower concentration levels are shown with lighter shades. The maximum NO_2 values were found at the center of the city, where they reached the concentration of 83–110 ppb. Values were low on the outskirts of the city, with the lowest concentration in north (31–44 ppb).

A study in Vilnius commented the same phenomena; NO_2 average rates depend upon traffic and are highest in cross roads and lowest at the background suburban areas. Dual carriage ways, sub roads, major roads, commercial areas, old residential areas, and areas where schools and colleges are existing have higher concentration levels of NO_2. Intense traffic flow and congestion were the major reasons for these elevated levels of nitrogen dioxide concentration in those areas as vehicular emission is the predominant source of NO_2.

Vehicle growth rate in twin cities is extensively high. Load of traffic is continuously increasing with growing population rate and demand of motor producing industry. Due to this, traffic congestion is also increasing day by day with growing vehicle population, resulting in highest emission rates per vehicle.

The higher emission rate of NO_2 can also be attributed to the type of fuel and quality of fuel [22]. InFigure 7 Rawalpindi showed more concentration levels than Islamabad due their building patterns.

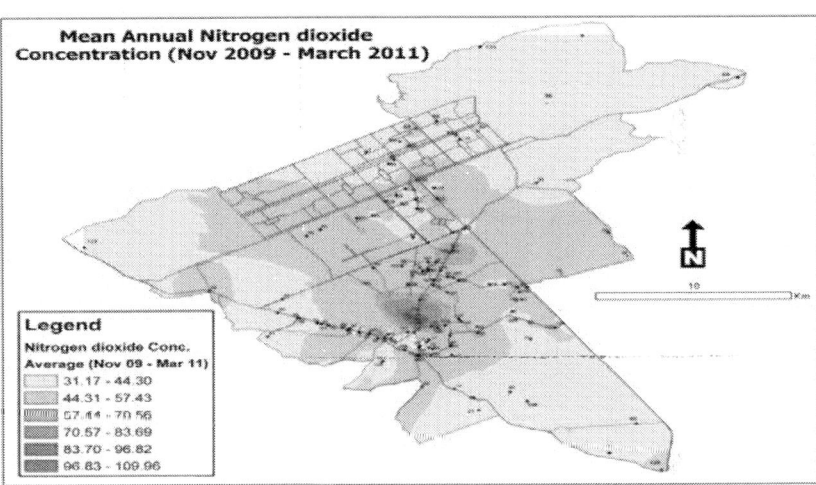

Figure 7. Spatial distribution of NO_2 concentration

3.1. Neural Network Data Analysis

Based on the design of neural network, with the neural architecture and properties discussed, the data space is searched by using heuristic search method with 500 iterations and fitness criteria is set to Inverse Test error. The best top 5 networks explored from the space by the heuristic search are graphically shown (Figure 8).

Heuristic search is a problem-solving method that analytically searches a space of problem states. The best network is obtained when the absolute error gets minimum in the initial iterations so the best network out of the 5 best networks is shown (Figure 9).

Results for all data sets produced after training and testing data. Real vs. target graph represented a line graph of real- and network-predicted target values for record displayed in Table 6. X-axis shows the selected input column values and Y-axis represents network-predicted output values. Table 6 presents the summary of the real vs. output table after training.

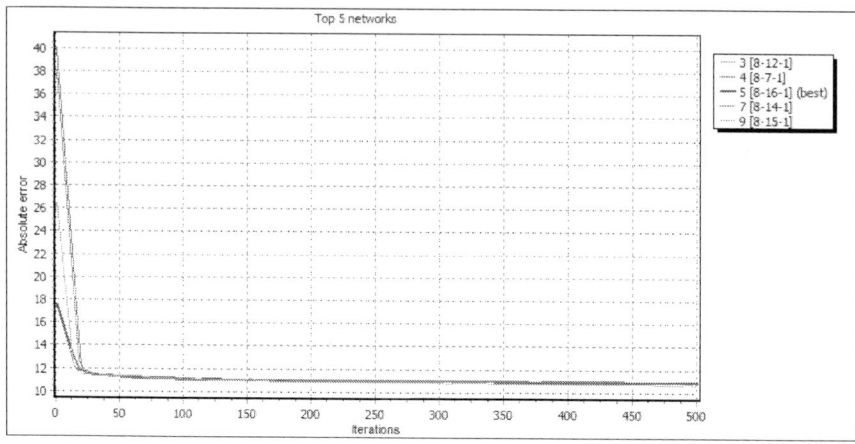

Figure 8. The top five networks explored by heuristic search approach

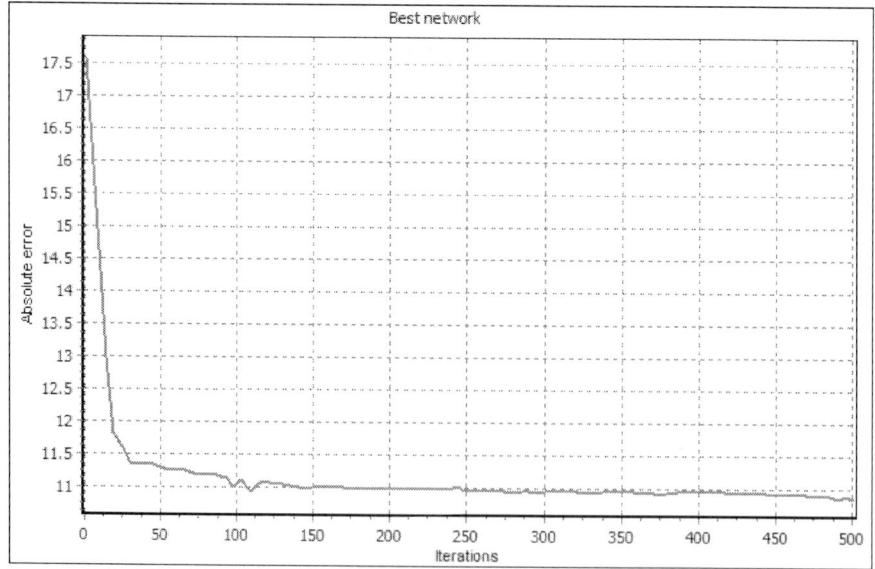

Figure 9. Network explored by heuristic search

Table 6. Summary of real vs. target

	Target	Output	AE	ARE
Mean:	45.237265	45.09091	11.341221	0.250292
StdDev:	20.98552	13.97879	11.112997	0.180871
Min:	11.3	20.16986	0.004569	0.000171
Max:	132.72	63.353673	73.986765	1.096446

Correlation, 0.653989; R-squared, -0.290243

The visualization for real vs. output with row number on x-axis and target/output (area_id) on y-axis is shown (Figure 10).

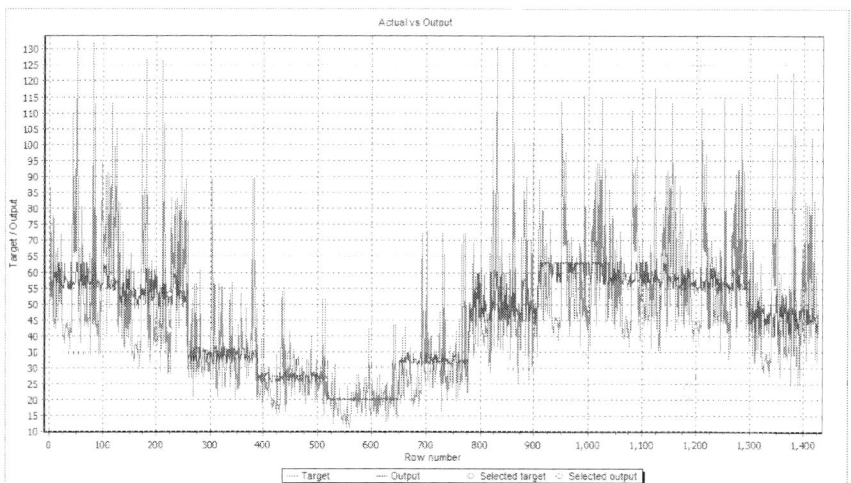

Figure 10. Real vs. network output

Figure 11 shows a scatter plot of the real and forecasted output values. X-axis presents the real values and Y-axis shows predicted network values.

Graph in Figure 12 shows the Network Error Dependence on values, which are numerically input in columns of data sheet. Through graph of Error Dependence, the ranges of the selected input column that can produce network error can be identified.

The last phase after the neural network is trained and tested is to query the network. The concentration is the output value for the neural network. So the input queries are subjected area_id, season_id, temperature, relative humidity, and rainfall (Figure 13).

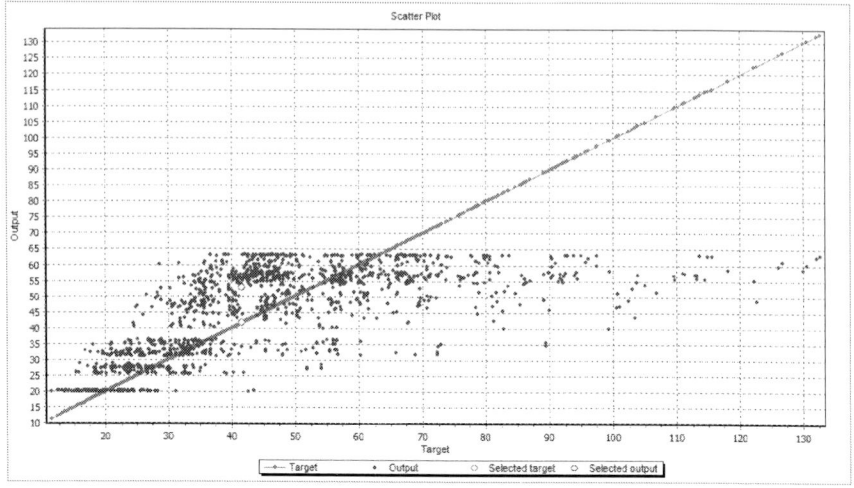

Figure 11. Scattered plot of real and output network values

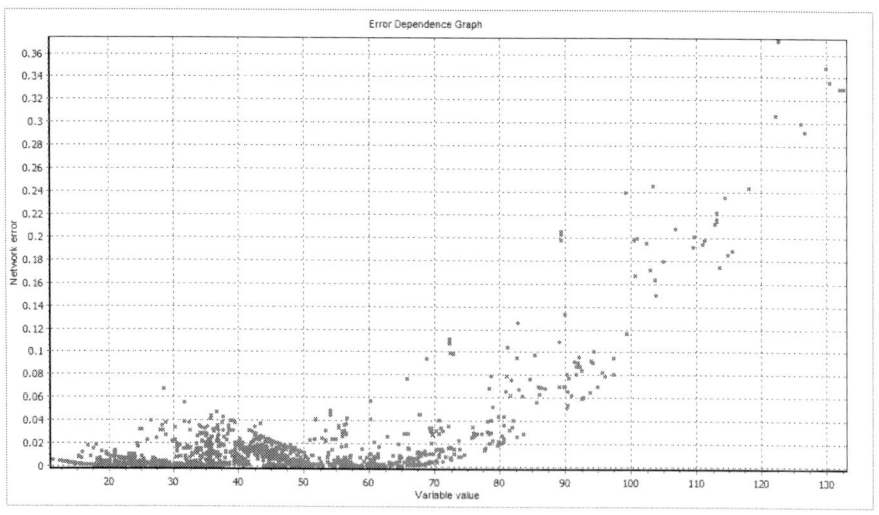

Figure 12. Graph of error dependence

The input Excel sheets are prepared for the GIS mapping. Sheets include area_id, their latitude, longitude, and their concentrations. With the help of interpolation, maps are created for the service.

season	temprature	humidity	rainfall	area_id
1	22	43	134	55
max: n/a	max: 30	max: 76	max: 284	max: 126
min: n/a	min: 11	min: 42	min: 0	min: 0

concentration

49.557664

Figure 13. Excel sheet presenting manual query

Temporal variation can be explained through meteorological recorded conditions. However, most of the variations on a local scale are due to the impact of air pollutants.

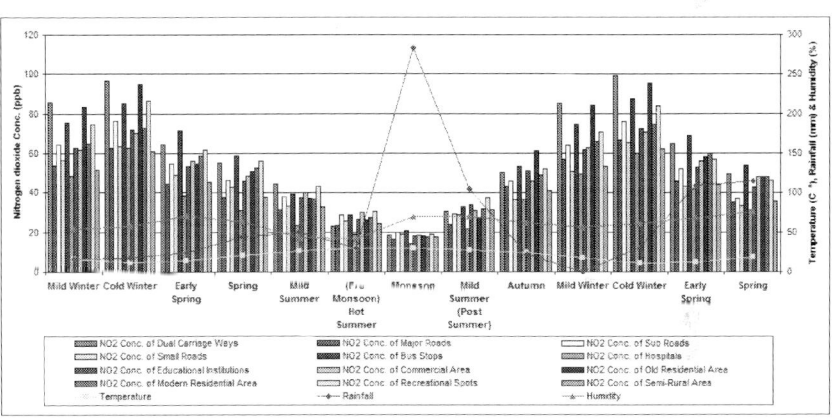

Figure 14. Relationship of rainfall, temperature, and humidity with NO$_2$ concentration (November 2009–March 2011)

Figure 14 indicates the positive association of NO$_2$ concentration level with humidity (RH in %) and negative association with the temperature. Figure 15 shows the concentration of NO$_2$ during summer when recorded temperature, rainfall, humidity are 31^{0}C, 67, and 17mm, respectively.

Figure 16 shows the concentration of NO$_2$ during the winter season at 11 ^{0}C, 68% humidity, and 9mm rainfall.

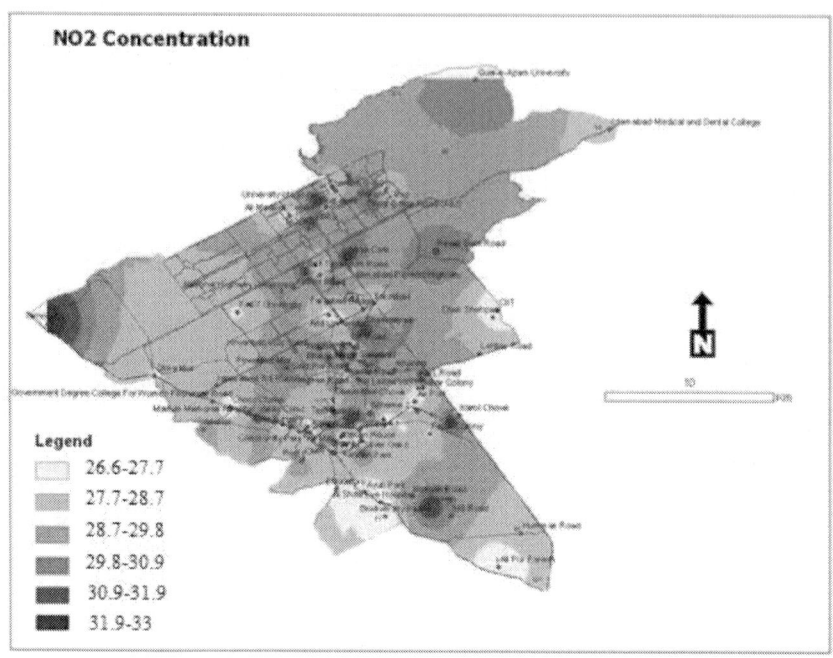

Figure 15. NO$_2$ concentration in summer

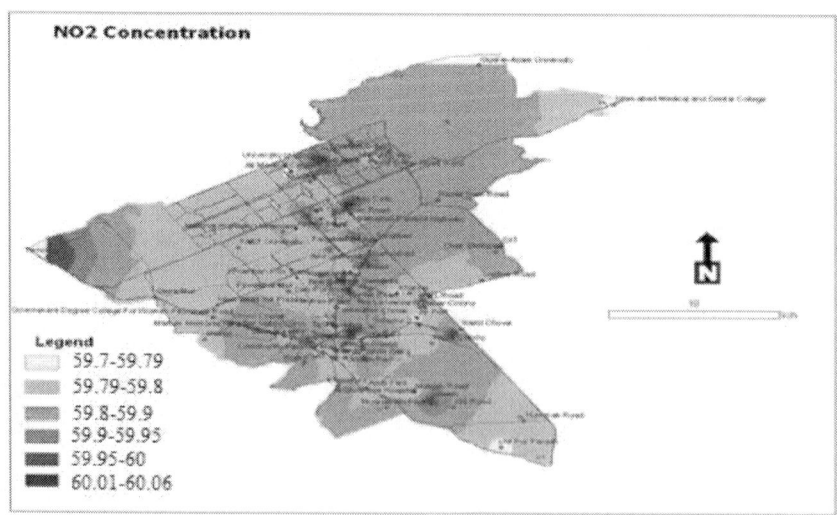

Figure 16. NO$_2$ concentration in winter

Concentration of NO_2 during the spring season, shown in Figure 17, when recorded temperature is 35°C, humidity is 58%, and rainfall is 60 mm.

Figure 17. NO_2 concentration in spring

Figure 18 shows predicted concentration of NO_2 in autumn season when recorded temperature, humidity, and rainfall are 29 °C, 69, and 22 mm, respectively.

Figure 19 shows that concentration of NO2 varies in different seasons. The months from May to August were months in which the minimum value of NO2 was recorded, and the maximum concentration was measured in the winter season from December to January.

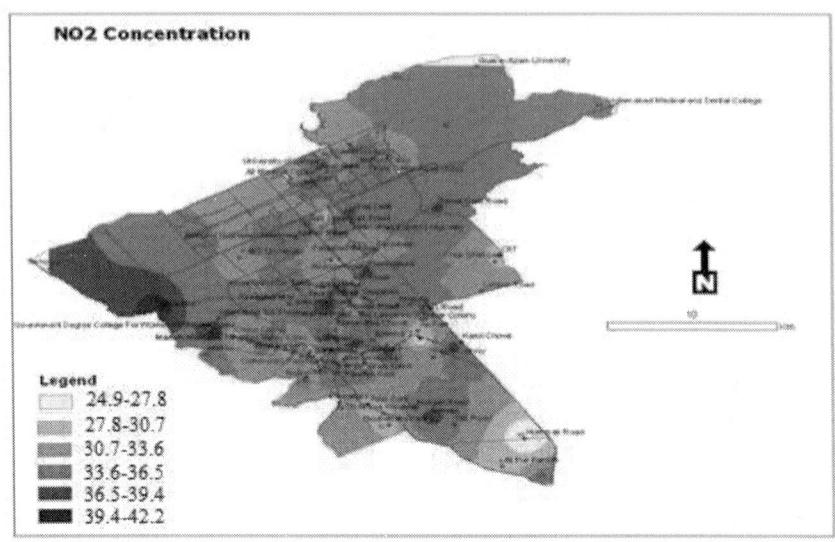

Figure 18. NO₂ concentration in autumn

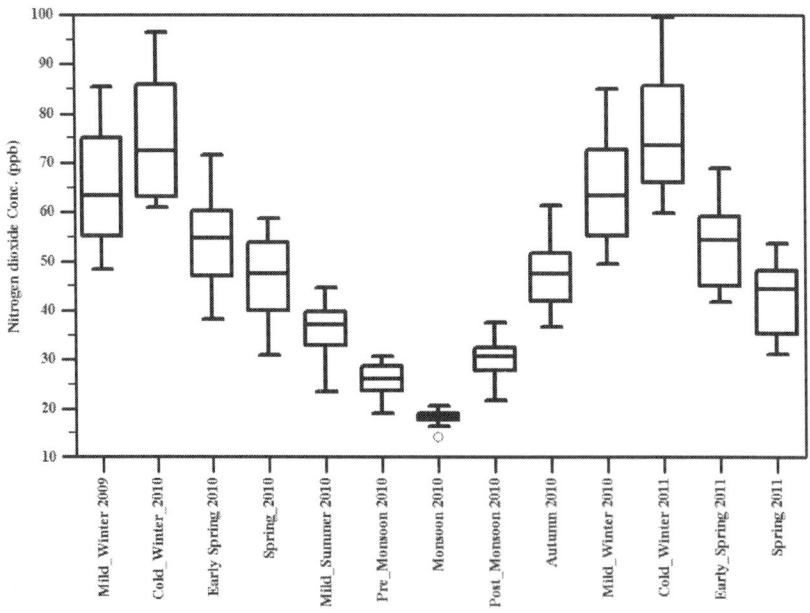

Figure 19. Seasonal variation in NO₂ concentration levels (November 2009 – March 2011)

4. CONCLUSION

NO_2 concentration levels were recorded on hourly and weekly basis in Rawalpindi and Islamabad city by using diffusion tubes. Artificial neural networks were trained to generalize the process of air pollutant spread over three dimensions. Prediction capabilities of ANN were analyzed through generalization by using hold-out evaluation method of classification. Results showed the advantage of using rtNEAT-like architecture of ANN where a neural network can modify its architecture to reduce the error up to the maximum possible limit. Results showed that annual average concentration of NO_2 concentration was 44 ± 6 ppb. However, the highest concentration was recorded in winter season near the dual carriage ways, schools, and colleges because of the higher number of transport vehicles on the road. This endorsed the fact that the reduced photolysis leads to the accumulation of NO_2 during winter due to less solar radiation. This is again attributed by the results of correlation, which reveal the negative correlation of nitrogen dioxide concentration levels with rainfall and temperature and the positive correlation with humidity. Moreover, the results of correlation reveal that the measured NO_2 concentration levels at different sampling areas exceeded the set limit of concentration value of the World Health Organization and Pak-EPA standard policy. This type of investigative study of artificial neural networks in the area of air pollution modeling shows promising applications for advanced machine learning algorithms in the emerging area of research called eco-informatics.

REFERENCES

1. Mulaku C. Mapping and analysis of air pollution in Nairobi, Kenya. International conference on spatial information for sustainable development, Kenya, 2001.

2. Gualtieri G and Tartaglia M. Predicting urban traffic air pollution: a GIS framework. Transportation Research – D; 1998; 3(5): 329–336.

3. Pummakarnchana O, Tripathi N and Dutta J. Air pollution monitoring and GIS modeling: a new use of nanotechnology based solid state gas sensors. Science and Technology of Advanced Materials 2005; 6: 251–255.

4. Afshar H and Delavar MR. GIS-based air pollution modeling in Tehran. Environmental Informatics 2007; 5: 557–566.

5. Barnes J, Parsons B and Salter L. GIS Mapping of nitrogen dioxide diffusion tube monitoring in Cornwall, UK. Air Pollution 2005; 13: 157–166.

6. Elbir T, Mangir N, Kara M, Simsir S, Eren T and Ozdemir S. Development of a GIS-based decision support system for urban air quality management in city of Istanbul. Atmospheric Environment 2010; 44: 441–454.

7. Veen AVD, Briggs DJ, Collins S, Elliott S, Fischer P, Kingham S, Lebret E, Pryl K, Reeuwijk HV and Smallbone K. Mapping urban air pollution using GIS: a regression-based approach. International Journal of Geographical Information Science 2010; 11(7): 699–718.

8. Vienneau D, de Hoogh K and Briggs D. A GIS-based method for modeling air pollution exposures across Europe. Science of the Total Environment 2009; 408: 255–266.

9. Banja M, Como E, Murtaj B and Zotaj A. Mapping air pollution in urban Tirana area using GIS. International Conference SDI, Skopje, 15–17 September 2010, 105–114.

10. Jensen SS. Mapping human exposure to traffic air pollution using GIS. Journal of Hazardous Material 1998; 61(3): 385–392.

11. Kim JJ, Smorodinsky S, Lipsett M, Singer BC, Hodgson AT and Ostro B. Traffic-related air pollution near busy roads. American Journal of Respiratory and Critical Care Medicine 2004; 170(5): 520–526.

12. Alexander SM. Data mining 2005. hhp//www.eco.utexas.edu/~norman/ BUS, FOR/course.mat/Alex/ (Accessed 12th February 2008).

13. United Nation. Conference on the Human Environment, Sewden, 1972.

14. United Nation. Environmental Performance Annual Report, New York, 2001.

15. US-EPA. Health affects of different air quality index (AQI) levels caused by nitrogen dioxide, 2008.

16. Miller D. Potential hazards of future volcanic eruptions. California, 1989.

17. Atkins DHF, Sandallas J, Law DV, Hough AM and Stevenson K. The measurement of nitrogen dioxide in the outdoor environment using passive diffusion tube samplers. AEA Technology, 1986.

18. Varshney CK and Singh AP. Passive samplers for NOx monitoring: a critical review. The Environmentalist 2003; 23: 127–136.

19. Palmes ED, Gunnison AF, Dimattio J, and Tomczyk C. Personal sampler for nitrogen dioxide. American Industrial Hygiene Association Journal 1976; 37: 570–577.

20. Gilbert NL, Goldberg MS, Beckerman B, Brook JR and Jerrett M. Assessing spatial variability of ambient nitrogen dioxide in Montreal, Canada, with a land-use regression model. Journal of the Air & Waste Management Association 2005; 65: 1059–1063.

21. Lozano A, Usero J, Vanderlinden E, Raez J, Contreras J, Navarrete B and Bakouri HEI. Air quality monitoring network design to control nitrogen dioxide and ozone applied in Granada Spain. Ozone: Science & Engineering 2011; 33(1): 80–89.

22. Heywood JB. Internal combustion engine fundamentals. McGraw-Hill, New York, 1998.

CHAPTER 2

Air Pollution in Welding Processes — Assessment and Control Methods

Farideh Golbabaei[1] and Monireh Khadem[1]

[1] Department of Occupational Health Engineering, School of Public Health, Tehran University of Medical Sciences, Tehran, Iran

1. INTRODUCTION

Welding is a very common operation in many industries and workplaces [1, 2]. According to American Welding Society, it is defined as "a metal joining process wherein coalescence is produced by heating to suitable temperature with or without the use of filler metal" [3]. There is a variety of welding processes that are used in different working conditions. According to some reports, from 0.2 to 2.0% of the working population in industrialized countries are engaged in welding activities [4]. Worldwide, over five million workers perform welding as a full time or part time duty [5, 6]. These welders, depending on conditions, work in outdoor or indoor workplaces, in open or confined spaces, underwater, and above construction sites. Welding operators face various hazards resulting in different injuries, adverse health effects, discomfort and even death. Furthermore, air pollution due to welding leads to certain consequents on humans and environment. Therefore, there are strong reasons to deal with the welding processes and the working environment of the welder from different aspects. A large number of welders experience some type of adverse health effects. Other workers near the place where welding process is done may be affected by the risks generated by it [1, 7]. Totally, welding risks can be classified as risks deriving from physical agents and risks related to the chemical components. The main hazards related to welding include electricity, radiation, heat, flames, fire, explosion, noise, welding fumes, fuel gases, inert gases, gas mixtures and solvents. Welders may be exposed to other hazards not directly related to welding, such as manual handling, working at height, in confined spaces, or in wet, hot or humid situations, and working with moving equipment, machinery and vehicles. Welding in a static awkward or horizontal posture may result in musculoskeletal injuries, such as strains and sprains. Prolonged use of a hard hat and a helmet can cause strain on the neck. Furthermore, long-term exposure, repetitive motions with arms and hands, and tasks inducing high force may lead to cumulative effects, increasing risk of injury. The main components of welding emissions are oxides of metals due to contact between the oxygen in the air and the vaporized metals. Common chemical hazards include particulates (lead, nickel, zinc, iron oxide, copper, cadmium, fluorides, manganese, and chromium) and gases (carbon monoxide, oxides of nitrogen, and ozone). Recently, nanoparticles emitted by welding operations are considered as an important group of air pollutants and there is a need to assess particle sizes and size distributions when risk assessment is done. Each welding technique produces a distinctive range of particulate composition and morphology. Different and complex profiles of exposures may be related to various welding environments [8-10].

Table 1. The hazards associated with welding Processes

HAZARD	WELDING PROCESS			
	PAW/PAC Carbon Arc Processes	SMAW GTAW GMAW FCAW	SAW	Oxyfuel
Ergonomic	+	+	+	+
Electric Shock	+	+	+	x
Bright light	+	+	-	+
Ultraviolet radiation	+	+	-	x
Toxic fumes and gases	+	+	-	+
Heat, Fire, and Burns	+	+	+	+
Noise	+	x	x	x

x No hazards, + Hazard present, - Hazard present if SAW flux is absent [11]

2. WELDING TECHNOLOGY

2.1. APPLICATIONS

Welding is used extensively in various manufacturing industries including shipyards, automobile factories, machines, home appliances, computer components, bridge building and other constructions. Welding is used for manufacturing pressure vessels, heat exchangers, tanks, sheet metal, prefabricated metal buildings and architectural work. Also, welding is an applicable technique in maintenance operations and repair shops. It is used in mining, oil and gas transmission companies, piping systems, heavy equipment manufacturing, aerospace, electronics, medical products, precision instruments, electric power, and petrochemical industries. Perhaps artists and sculptors are the smallest group who use welding techniques to create artworks. Therefore, many things that people use in daily lives are welded or made by welded parts [12].

2.2. Workplace Conditions

Welders, depending on conditions, work in outdoor or indoor workplaces, in open or confined spaces, underwater, and above construction sites. In some conditions, welding processes are carried out in confined spaces where the welding work area is surrounded on most sides by walls and there is no sufficient space for the installation of a conventional exhaust hood [1, 7].

Working in indoor environments includes all works which are done in buildings like workshops, repairing shops, storages, office, and any closed area in industries, factories, and other places. Welders may work in indoor areas to do welding tasks full time or part time. An important benefit of indoor workplaces is the protection against environmental factors such as rain, wind and sunshine. Outdoor workers spend long periods of time working in open areas. They are exposed to different hazards depending on their type of work, as well as geographic region, season, and the period of time they are outside. Outdoor works include agriculture, construction, mining, oil and gas transmission through pipelines, transportation, warehousing, utilities, and service sectors. Sometimes welders should work in such workplaces to do their tasks. Some workplace hazards related to outdoor areas include unpredictable weather conditions, bugs and wild animals, extreme heat, extreme cold, and ultraviolet (UV) radiation.

Many workplaces contain spaces that are considered "confined" because their configurations hinder the activities of employees who must enter, work in, and exit them. A confined space has limited or restricted means for entry or exit. Confined spaces include underground vaults, tanks, storage bins, manholes, reactor vessels, silos, process vessels, and pipelines. Confined spaces have the following characteristics: limited space, entry, or exit; poor ventilation and lack of safe breathing air. Welders may experience various hazards when welding in confined spaces, such as fire, explosion, electric shock, asphyxiation, and exposure to hazardous air contaminants [13-16].

2.3. Types Of Welding Processes

There are different welding processes (over 50 types) that differ greatly in some parameters such as heat, pressure, and the type of equipment used. Welding process can be classified into various types based on different literatures. Some common types of welding are listed in five categories each of which includes some subcategories (Figure 1). The most common and known types of welding include:

Shielded Metal Arc Welding: (SMAW) also is known as Manual Metal Arc welding (MMA) or stick electrode welding. It is one of the oldest, simplest, and most versatile arc welding processes used for carbon steel welding and low alloy welding. In SMAW, the electrode is held manually, and the electric arc flows between the electrode and the base metal. The electrode is covered with a flux material which provides a shielding gas for the weld to help minimize impurities. A wide range of metals, welding positions and electrodes are available based on intended requirements. This type of welding is especially suitable for jobs such as the erection of structures, construction, shipbuilding, and pipeline work. Contrary to the other methods requiring shielding gas which are unsuitable in wind, SMAW can be used outdoors in different weather conditions. However, owing to the time required for removing the slag after welding and changing the electrodes, its arc time factor is relatively low. As a disadvantage, forming fumes in SMAW makes the process control difficult.

Gas Metal Arc Welding: (GMAW) or metal inert gas (MIG) welding is used for most types of metal and is faster than SMAW. It may be applied to weld vehicles, pressure vessels, cranes, bridges and others. This process involves the flow of an electric arc between the base metal and a continuous and consumable wire electrode. Shielding gas (usually an argon and carbon dioxide mixture) is supplied externally; therefore, the electrode has no flux coating or core. MIG welding is used for mild steel, low alloyed and stainless steel, for aluminum, for copper, nickel, and their alloys. Some parameters can affect MIG welding process, such as:

- Electrode diameter

- Voltage

- Wire feed speed and current

- Welding speed

- Shielding gas and gas flow rate

- Torch and joint position

To perform an optimum welding, most of the mentioned parameters should be matched to each other. In addition to affecting the quality of welding, some of these parameters can influence the fumes and gases emitted from the process. However, the fume produced by MIG welding is less than that of SMAW. Unlike the SMAW that is discontinuous due to limited length of the electrodes, GMAW is a continuous welding process. There is no slag and no need for high level of operators' skill. Nevertheless, expensive and non-portable equipment is required, and also outdoor applications are limited because of the negative effects of weather conditions like wind on the shielding gas [17, 18].

Gas Tungsten Arc Welding: (GTAW) is also known as tungsten inert gas (TIG) welding. GTAW is used on metals such as aluminum, magnesium, carbon steel, stainless steel, brass, silver and copper-nickel alloys. This technique uses a permanent non-consumable tungsten electrode. The filler metal is fed manually, the weld pool and the electrode are protected by an inert gas (usually argon), and high electrical currents are used in this type. Welding of stainless steel, welding of light metals, such as aluminum and magnesium alloys, and the welding of copper are the main applications of TIG welding. GTAW welds are highly resistant to corrosion and cracking over long time periods. However, TIG welding is suitable to weld thin materials and produces a high quality weld of most of metals. There is no need for slag removal in GTAW process. The concentration of heat takes place in a small zone, resulting in the minimal thermal distortion of work piece. The TIG welding has some disadvantages including low welding rate, expensiveness, and need for high level of operators skill. Although during TIG welding operators are exposed to dangerous gases and fumes, the generation of these compounds is very little in comparison with other welding processes.

Submerged Arc Welding: (SAW) is a highly-productive welding method (4-10 times as much as the SMAW). SAW may be automatic or semi-automatic. It is used to weld thick plates of carbon steel and low alloy steels. In this welding process, the electric arc flows between the base metal and a consumable wire electrode; however, the arc is not visible since it is submerged under flux material. This welding process is usually used for large structures such as large tubes, cylindrical vessels, and plates in shipyards. Some parameters can affect SAW process such as welding arc voltage, arc current, the size and shape of the welding wire, and the number of welding wires. A low fume emission is produced during SAW process and there is a little ozone, nitric oxide and nitrogen dioxide generation because of the invisibility of the arc. Very high welding rate, suitability for automation, suitability for both indoor and outdoor works, and high weld quality are mentioned as advantages of SAW. Some limitations of this welding process include: slag inclusion, limited applications often for welding in a horizontal position, and need for precise parameter setting and positioning of the wire electrode.

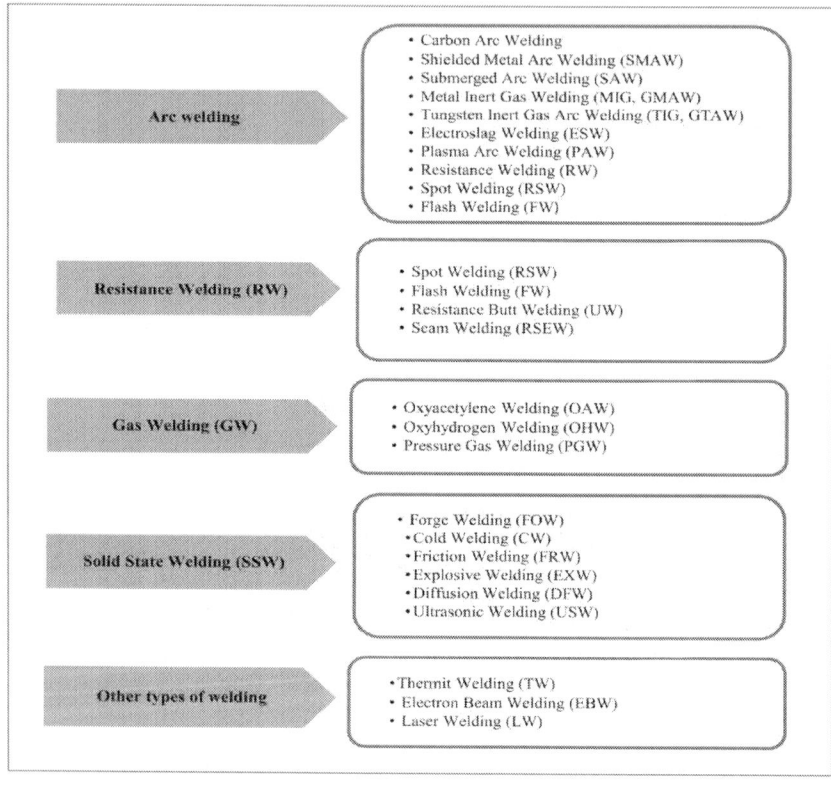

Figure 1. Classification of welding processes [18]

Plasma Arc Welding: (PAW) is an arc welding process in which arc is formed between an electrode and the workpiece. In PAW process, the plasma arc can be separated from the shielding gas cover by positioning the electrode within the body of the torch. It can be named as a key difference between GTAW and PAW. Two inert gases are used in the process, one forms the arc plasma and the second shields the arc plasma. Applying the plasma arc welding is being increased in industries, because it provides a high level of control and accuracy to produce high quality welds. Also, using the PAW leads to long electrode life for high production conditions. This welding process is suitable for both manual and automatic applications. It can be used for precise welding of surgical equipment, jet engine blades, and instruments required for food and dairy industry. There is a low level of fume generation during PAW, but welding gases especially ozone is often formed in this process. Need for less operator skill, high welding rate, high penetrating capacity, long electrode life, high accuracy and precision, and short weld time are considered as the advantages of PAW process. Its limitations include expensive process tools, needs for high power electrical equipment, more distortion and loss of mechanical properties due to the greater heat input.

Flux Core Arc Welding: (FCAW) is used for carbon steels, low alloy steels and stainless steels. This welding process has similarities to both SMAW and GMAW. This process is used in construction because of its high welding speed and portability. The consumable tubular electrode is continuously fed from a spool and an electric arc flows between the electrode and base metal. The electrode wire has a central core containing fluxing agents. There are a variety of cored wires; some of them require the use of shielding gas like carbon dioxide or the mixture of argon/carbon dioxide and the others (self-shielded flux cored wires) do not require additional shielding gas. The slag produced in FCAW process acts as an additional protection during cooling time but has to be chipped away after that. Like other welding process, FCAW has some advantages and limitations. No needs for skilled operators and pre-cleaning of metals, suitability for use in the outdoor or windy condition (it is true about self-shielded flux cored wires), suitability for use in all positions, and ease of varying the alloying constituents are mentioned as FCAW advantages. Its limitations include: emission of considerable amount of fumes in self-shielded wires, higher price of filler material and wire in comparison with GMAW, and needs for slag removal. Also, escaping of the shielding gas from the welded area leaves holes in welded metal, resulting in porosity in products [17, 18].

3. AIR POLLUTION OUT OF WELDING

According to Flagan and Seinfeld definition, "the phenomenon of air pollution involves a sequence of events: the generation of pollutants at and their release from a source; their transport and transformation in and removal from the atmosphere; and their effects on human beings, materials, and ecosystems" [19]. Air pollution is indoor or outdoor contamination by particulates, biological molecules, or other harmful materials that changes the natural characteristics of the Earth's atmosphere. Household combustion devices, motor vehicles, forest fires, and industrial processes are common sources of air pollution. Major industrial sources of particulate matter include the metals, mineral products, petroleum, and chemical industries. Air pollution is considered as a threat to human health as well as to the Earth's ecosystems. Based on WHO report, around 7 million people worldwide died due to the air pollution in 2012 [20]. Welding, as an important operation in most industries, can considerably cause air pollution. In all types of welding processes, fume and gases are formed as air pollutants. Due to high temperature during the welding process, different substances in the arc are vaporized. Then, the vapor condenses and oxidizes in contact with the air, leading to the formation of fumes. The fume particles are so small and they can reach the narrowest airways of respiratory system (respiratory bronchioles). Some parameters like the welding type and consumables (filler metal and surface coatings) determine the kind and amount of generated particles and gases.

The composition of welding fumes and their generation rate is a function of different parameters. Welding fume particles are in the fine (<2.5 μm) to ultrafine (<100 nm) respirable size and can penetrate into the alveolar regions of the lungs. The generation of fumes depends on:

- -Amperage, voltage, gas and arc temperatures and heat input in the welding process
- -Consumables like electrodes
- -Materials
- -Welding duration [9, 21].

The most common gases emitted during welding are ozone, nitrous gases and carbon monoxide. Phosphine and phosgene are the other gases that may be produced during welding. Gases are generated due to the high temperature and ultraviolet (UV) radiation from the arc. Like fumes, some factors can affect the emission of gases during welding processes. For instance, ozone formation during welding depends on process type, used material, and shielding gases. Welding gases can also be generated when surface coatings or contaminants contact with hot surfaces or UV radiation.

Along with harming human health, air pollution may lead to various environmental impacts. Air pollution can adversely cause critical impacts on the atmosphere and natural environment in many ways. Welding, as an industrial process, causes serious impacts on the environment depending on its operation mode and the technological equipment. Environmental pollution in welding process is the result of some parameters, such as high percentage of heat that is released into the environment and materials including large amount of gases and fumes. Some factors needed to carry out the welding operation include: energy, mineral or organic substances (protective gases, cooling water, oils, grease and protective substances etc.). These consumables can be harmful for the environment. Furthermore, produced waste during the welding processes results in undesirable impact on the work or natural environment. To protect the welding region and prevent oxidation, inert gases like carbon dioxide and argon are used because of their availability and low cost. They are used as shielding gases and have undesirable impacts on the environment. To protect the environment and keep the resource for future, energy conservation and reducing greenhouse gas emissions should be considered. In this respect, the average consumption rate, usage rate and the purity of products and consumables are important factors [22, 23].

The generation of fumes and gases is directly related to the welding process. Fumes emitted during manual metal arc welding (MMA) and MIG welding is the same. In some conditions, the level of fume generated during MIG welding (with solid wire) may be much lower in comparison with the fumes produced by MMA. In TIG welding, a lower level of fumes is emitted compared to MMA and MIG welding. The composition of fumes is directly associated with the composition of used wire. MMA welding causes adverse health effects because of forming the hexavalent chromium (Cr (VI)) in the process. In addition, high rates of emission of toxic compounds generate in MMA-stainless steel (MMA-SS) welding [24]. During TIG welding, very little fume are generated. Welding fumes may be composed of oxides of chromium, nickel and copper, with very low specific limit values. The individual elements and also their synergetic effect must be considered when assessing fume toxicity. Lower ozone and nitrogen oxides are emitted during TIG welding than those in MIG/MAG welding. The amount of mentioned gases during TIG welding is dependent on current, arc length and the flow and type of shielding gas. High electrical currents cause the significant levels of ozone, nitric oxide and nitrogen dioxide. During MIG welding, significant levels of ozone and nitrogen oxides are produced because of intense current levels.

There is a little information concerning emissions during plasma arc welding (PAW). Due to the similarity of TIG and PAW welding techniques, they may probably emit air pollutants with the same magnitude. MIG welding of aluminum produces larger quantities of ozone than TIG welding of aluminum. Forming more nitrogen oxides in the latter process will keep the emitted ozone levels down [25, 26]. A study by Schoonover et al. showed that welders performing MIG and SMAW are exposed to higher fume concentrations than welders performing TIG. According to mentioned study, exposure to manganese during MIG was nearly two and ten times higher than in SMAW and TIG, respectively. In fact, not using a consumable electrode during TIG welding results in lower exposures. The highest average exposures occur in SMAW, followed by GMAW, and GTAW [21]. K. Fuglsang et al. investigated the Fume

Generation Rates (FGR). This rate for MMA was 3-5 times higher than that found for MAG and MIG. The same FGR was found for TIG and MIG/MAG welding [27].

Various welding processes generate particles in different size distributions. Particles produced during MMAW, MAG, MIG, and laser welding are quite similar in size. Resistance Spot Welding (RSW) and TIG welding have a completely different structure for particle size distribution. These techniques produce particles smaller than 100 nm, in which, at least 90% are smaller than 50 nm. Particles generated during processes with high mass emission rates (MMAW, MAG, MIG, and Laser) have diameters about 100–200 nm and there are few nanoscaled particles between them. Processes with low mass emission rates (TIG and RSW) generate exclusively particles smaller than 50 nm; however, the number concentration of particles in these techniques is similar to the others. Although, welding types with low mass emission rates are called "clean techniques", their potential toxicological properties and health effects due to exposure to nanoscaled particles should be further studied [28].

A study by Keane M. introduced the pulsed axial spray method (from MIG process) as the best choice of the welding processes because of minimal fume generation (especially Cr (VI)) and cost per weld. The advantages of this method include usability in any position, high metal deposition rate, and simple learning and use. Totally, the highest amounts of fume are produced by the self-shielded cored wire electrodes. These electrodes are used without a shielding gas. Using solid wire electrodes results in emission of ozone and nitrogen oxides as in MAG welding [25, 29].

Airborne particles with diameter smaller than 100 nm are known as nanoparticles or ultrafine particles. According to researches, nanoparticles are more harmful to human health than larger particles. They can deeply penetrate inside the respiratory system and then enter the blood stream. The main character of nanoparticles is the high surface area, and their toxicity depends on the shape and penetration potential inside the respiratory system. In addition to the emission of fine particles with diameter less than 10 µm, nanoparticles may be emitted during welding operations. Some studies have indicated that the highest values of nanoparticles are related to MAG and TIG processes when applying the highest current intensities. Therefore, the higher amounts of nanoparticles are emitted by processes in which the higher energy intensities are used.

As it was stated, the emission of nanoparticles during welding operations increases with the increase of welding parameters like current intensity. Welding with short-circuit mode results in lower value of nanoparticles, because its low current intensity and tension causes an electric arc with lower temperature and thus emitting lower amounts of elements. Also, the high quantity of nanoparticles is generated by the stainless steel welding, which can be related to the presence of helium in the gas mixture of welding. Helium, due to high ionization energy, results in electric arc with high temperature that generates higher values of nanoparticles. Furthermore, the study of different base materials indicated that the higher quantity of nanosized particles is obtained for stainless steel compared to carbon steel. According to data from different investigations, the lowest level of ultrafine particles deposited in alveolar region of lungs was related to FSW, followed by TIG and MAG. Totally, all welding processes can result in deposition of a significant concentration of nanosized particles in lungs of exposed welders [30-32].

4. WELDING HEALTH EFFECTS

Fume and gases emitted during welding pose a threat to human health while welding. The exposures may be varied depending on where the welding is done (on the ship, in confined space, workshop, or in the open air). The welding process and metal welded affect the contents of welding fumes. On the other hand, physical and chemical properties of the fumes and individual worker factors are effective on deposition of inhaled particles. In this respect, particle size and density, shape and penetrability, surface area, electrostatic charge, and hygroscopicity are the important physical properties. Also, the acidity or alkalinity of the inhaled particles are the chemical properties that may influence the response of respiratory tract. Welding gases can be classified into two groups; some gases are used as a shielding gas and the others are generated by the process. Shielding gases are usually inert, therefore, they are not defined as hazardous to health but they may be asphyxiants. Gases generated by welding processes are different based on welding type and may cause various health effects if over-exposure occurs. Welding emissions depending on some factors like their concentration, their properties, and exposure duration can lead to health effects on different parts of human body.

Hazards on Respiratory System - The inhalation exposures may lead to acute or chronic respiratory diseases in all welding processes. In the occupational lung diseases, the various reactions produced in respiratory tract depend on some parameters such as the nature of the inhaled matter, size, shape and concentration of particles, duration of exposure, and the individual workers susceptibility. Chronic bronchitis, interstitial lung disease, asthma, pneumoconiosis, lung cancer, and lung functions abnormalities are some hazardous effects on respiratory systems. The pulmonary disorders are various based on the differences in welding metals and their concentrations. Ozone, at low concentrations, irritates the pulmonary system and can cause shortness of breath, wheezing, and tightness in the chest. More severe exposures to ozone can lead to pulmonary edema. Exposure to nitrogen dioxide may cause lung function disorders like decrements in the peak expiratory [33, 34]. Kim JY in a study showed the PM2.5 concentration for welders (1.66 mg/m^3) was significantly greater than that for controls (0.04 mg/m^3), and the exposure of healthy working population to high levels of welding fumes resulted in the acute systemic inflammation [35].

Hazards on Kidney- Substantial exposure to metals and solvents may be nephrocarcinogenic. Chromium can deteriorate renal function because of accumulation in the epithelial cells of the proximal renal tubules and induce tubular necrosis and interstitial changes in animals and humans. Tubular dysfunctions have been identified in subjects occupationally exposed to Cr (VI) [33, 36]. Welders exposed to heavy metals like cadmium and nickel have also experienced kidney damage [7]. Pesch et al. indicated that there was an excess nephrocarcinogenic risk involved with soldering, welding, milling in females. So, it can be considered an evidence for a gender-specific susceptibility of the kidneys [37].

Hazards on Skin - Erythema, pterygium, non-melanocytic skin cancer, and malignant melanoma are the adverse health effects of welding on the skin among which erythema is a common one. The intense UV as well as visible and infrared radiations are produced by welding arc machines. Exposure to UV can lead to short- and long-term injuries to the skin [33, 38-40]. Some metals like beryllium, chromium and cobalt can cause direct effects (irritation and allergic impacts) on the skin. Also, they may be absorbed through the skin and cause other health effects such as lung damage. When the particles are small and there are cuts or other damages to the skin, the absorption through the skin is raised [7, 36]. Chromium (VI) may cause irritating and ulcerating effects when contacting with skin. An allergic response including eczema and dermatitis may be induced in sensitized individuals exposed to Cr (VI) [34].

Hazards on the visual systems - Most welding processes emit intense ultraviolet as well as visible and infrared radiations. Adverse effects on the eyes may be induced by these optical radiations. In addition, Tenkak reported that, welding may cause photokeratitis and some types of cataract. Erhabor et al. showed the most frequent symptoms among the welders were eye irritation (95.43%). Exposure to UV radiation can lead to short- and long-term injures to the eyes. Acute overexposure to UV radiation can result in the photokeratitis and photoconjunctivitis that are the inflammation of the cornea and the conjunctiva, respectively. These responses of the human eye to UV radiation are commonly known as snow blindness or welder's flash [33, 38, 41].

Hazards on Reproductive System - In the past, some studies have indicated the increased risk for infertility and reduced fertility rate in mild steel welders. There are some evidences that reduced fecundity can be related to exposure to hexavalent chromium and nickel. According to new investigations, damages to male reproduction system have been reported less than before, probably because of decreasing the exposure levels in the developed countries. However, some special tasks like stainless steel welding may impair welders' reproduction system [42-44]. A study by Bonde showed that mild steel welding, but not stainless one, resulted in significant effects on the fertility during years [45]. Mortensen [46] observed a greater risk for poor sperm quality among welders compared to controls, especially welders who worked with stainless steel. Therefore, welding in general, and specifically with stainless steel, may cause the reduced sperm quality. According to Sheiner, impaired semen parameters can be associated with the exposures to lead and mercury [47].

Hazards on the nervous system - Memory loss, jerking, ataxia and neurofibrillary degeneration have been attributed to exposure to aluminum. The accumulation of aluminum in the brain may develop some neuropathological conditions, including amyotrophic lateral sclerosis, Parkinsonian dementia, dialysis encephalopathy and senile plaques of Alzheimer's disease [36]. A review of literatures by Iregren suggests that occupational exposure to manganese results in the central nervous system damage that is generally irreversible [48]. Although there are multiple toxic agents in welding, more literatures have dealt with manganese as an important agent of toxicity. Welders are also exposed to high concentrations of carbon monoxide and nitrogen dioxide. Carbon monoxide can cause the neurological impairments of memory, attention, and visual evoked potentials. Both central and peripheral nervous system damages may be induced by exposure to welding fumes [49]. Some neurobehavioral impairments associated with exposure to lead and manganese have been indicated by Wang [50]. A study by Bowler (2003) showed there is a relation between welding and a decline in brain functions and motor abilities. In this survey, various questionnaire and tests like neuropsychological tests were used [49].

Carcinogenic effects - There are some concerns regarding the presence of carcinogens in the welding fumes and gases. Sufficient evidences for carcinogenicity of nickel, cadmium, and chromium (VI) have been reported through experimental and epidemiological studies. These three metals have been categorized as carcinogen "Class 1" by the International Agency for Research on Cancer [51-52]. Ozone has been introduced as a suspect lung carcinogen in experimental animals, but there are very few documents about its long term effects on welders. The ultraviolet emissions resulting from welding arc can potentially cause skin tumors in animals and in overexposed individuals, however, there is no definitive evidence for this effect in welders [53].

Other health problems - Welding on surfaces covered with asbestos insulation may lead to risk of asbestosis, lung cancer, mesothelioma, and other asbestos-related diseases in exposed welders. The intense heat and sparks of welding can cause burns. Eye injuries are possible because of contact with hot slag, metal chips, and hot electrodes. Lifting or moving heavy objects, awkward postures, and repetitive motions result in strains, sprains and musculoskeletal disorders. High prevalence of musculoskeletal complaints (back injuries, shoulder pain, tendonitis, carpal tunnel syndrome, and white finger) is seen in welders [54].

5. EXPOSURE STANDARDS FOR WELDING EMISSIONS

Usually, exposure standards apply to long term exposure to a substance over an eight hour work per day for a normal working week, over an entire working life. Some organizations like American Conference of Governmental Industrial Hygienists (ACGIH), National Institute for Occupational Safety and Health (NIOSH), and Occupational Safety and Health Administration (OSHA) have published the exposure standards for various components in welding fumes and gases (table 2). According to Work Safe Australia exposure standards cannot be used as a fine dividing line between a healthy and unhealthy workplace. Adverse health effects below the exposure limits might be seen in some people because of individual susceptibilities and natural biological variation. ACGIH, however, recommends a TLV-TWA (Threshold Limit Value-Time Weighted Average) of 5 mg/m^3 for total welding fume, assuming that it contains no highly toxic components. Each metal or gas within the welding has its own exposure standard. As Table 2 indicates, biological media, Biological Exposure Index (BEI), and carcinogenicity class have been proposed for some welding emissions [55, 56].

6. WELDING MONITORING AND RISK ASSESSMENT

6.1. Monitoring Of Welding Emissions

Managing the risks of pollutants generated by welding process is carried out in some steps inculing identifying hazards, assessing the risks arising from these hazards, eliminating or minimising the risks via proper control ways, and checking the effectiveness of controls. Monitoring the welder's exposure is a main component of risk management process. Welding process leads to chemical exposures to fumes and toxic gases in enormous quantity. The hazard identification and risk assessment are necessary to work safely in a welding environment. Enough information, education, training and experience are required in this respect. In addition to the full-time welders, a large number of part-time welders who work in small shops and workers in the vicinity of the welding process may also be exposed. There is a greater potential for exposure due to welding in confined spaces with poor ventilation such as ship hulls, metal tanks and pipe, therefore, monitoring such welders should be seriously considered.

As it was stated previously, the level of welder's exposure to welding emission depends on some factors like the process type, process parameters, and consumables used. Materials and consumables used in welding determine the chemical composition of welding emissions. The specific toxicity of each element and the synergetic effect of generated constituents must be considered to evaluate the exposure status of welders. There are some other workplace specific factors, including the ventilation condition, welder position or posture, and the volume of welding room, that influence the exposure level. The emission rate and also its concentration in the breathing zone of the welder or in the work environment are directly related to the mentioned factors. When it is probable that the welders' exposure will be exceeded the prescribed limits, or when the workers' health and the environment are at risk, the monitoring of hazards and the risk assessment program are required. To evaluate the hazards caused by different welding emissions, collecting various information is recommended. Air monitoring and measuring related pollutants via personal and environmental sampling, biological monitoring, workplace assessment with regard to physical and chemical hazards, and occupational medical findings can be used to evalute the welder's exposure status compeletely [59-60].

Air Monitoring -Airborne pollutants generated by welding can threaten the worker's health and safety. Thus, during the health and safety program, air monitoring is used to identify and quantify welding emissions. To evaluate air contaminants, a sampling strategy is used for collection of exposure measurements. The choice of the best strategy is based on site-specific conditions. In a sampling strategy, some parameters like selection of workers for personal monitoring, sampling duration and required number of samples are important. The measurement of contaminants is carried out in the breathing zone of selected worker. The collected samples must be representative of the normal work activity and exposure of welder, because the sampling results are used to prevent overexposures. Air monitoring in welding processes includes the sampling and analysis of welding fumes and welding gases [61].

Table 2. Exposure limit of each individual constituent of welding components

Substance	OSHA PEL-TWA (mg/m3)	NIOSH REL-TWA (mg/m3)	ACGIH TLV-TWA (mg/m3)	ACGIH BEI	Carcinogenicity
Aluminum Fume	15 (Total) 5 (res)	5	5		
Arsenic	0.01	0.002 (Ceiling)	0.01	35 µg As/L	A1
Barium	0.5	0.5	0.5		
Beryllium	0.002	0.5 (Ceiling)	0.002		A1
Cadmium Fume	0.005	LFC (Ca)	0.01 (Total) 0.002 (Res)	5 µg Cd/g creatinine	A2
Cobalt	0.1	0.05	0.02	15 µg Co/L	A3
Chromium(VI)	--	0.001	0.05	25 µg Cr/L	A1
Chromium metal	1	0.5	0.5		A4
Copper Fume	0.1	0.1	0.2		
Iron Oxide	10 (as Fe)	5	5		A4
Lithium	--	--	--		
Manganese	5 (Ceiling)	1	0.2	range 0.5 to 9.8 mg/L; up to 50 mg/L for occupational exposure	
Molybdenum	5(Soluble) 15 (Insoluble)	--	5 (Soluble) 10 (Insoluble)		

Substance	OSHA PEL-TWA (mg/m3)	NIOSH REL-TWA (mg/m3)	ACGIH TLV-TWA (mg/m3)	ACGIH BEI	Carcinogenicity
Lead	0.05	0.1	0.05	30 µg /dL (whole blood)	A3
Nickel	1	0.015 (Ca)	1	10µmol/mol creatinine	Elemental (A5) Insoluble inorganic (A1)
Platinum	0.002 (Soluble)	1(Metal) 0.002 (Soluble)	1		
Selenium	0.2	0.2	0.2		
Silver	0.01	0.01	0.1		
Tellurium	0.1	0.1	0.1		
Thallium	0.1	0.1(Soluble)	0.1	50 µg Th/g creatinine	
Titanium Dioxide	15	LFC (Ca)	10		
Vanadium Pentoxide	0.1 (Ceiling)	0.05(Ceiling)	0.05	50 µg V/g creatinine	
Zinc Oxide	5	5	5		
Zirconium	5	5	5		
Total fumes	--	LFC (Ca)	5		
Carbon monoxide	50 ppm	35 ppm	25 ppm	3.5% of (Hemoglobin) 20 ppm (end-exhaled air)	
Nitrogen dioxide	5 ppm (ceiling)	5 ppm (ceiling) 1ppm (STEL)	3 ppm		
Ozone	0.1 ppm	0.1 ppm	0.08 ppm		

LFC=lowest feasible concentration; Res=Respirable; Ca=NIOSH potential occupational carcinogen [55, 57, 58]

Within recent years, standard practices have been developed to monitor exposures considering the occupational exposure limits for elements. Most measurements are made using personal monitoring systems with a pump at a proper flow rate connected to a cassette containing a membrane filter for a suitable period of time. To obtain the accurate result, filter cassette must be placed inside the welding helmet. Time-weighted average concentrations of total fumes is obtained by weighing the filter before and after exposure; the concentrations of elements are determined by chemical analysis methods provided by related organizations like American Welding Society and British Standards Institution [51], NIOSH Manual of Analytical Methods (NMAM) for metals in air and urine and OSHA Sampling and Analytical Methods are used to monitor the welding workplaces. In these methods, analysis of metals is performed by Inductively Coupled Argon Plasma-Atomic Emission Spectroscopy (ICP-AES) after sample preparation by acidic ashing [61, 62]. It is worth mentioning that the microwave digestion can be used instead of acidic ashing to prepare samples, leading to reduction in ashing time up to 90 percent, as well as cost saving and providing a healthier work environment for laboratory operators. Golbabaei et al. used the microwave digestion to prepare urine samples before urinary metal analysis by graphite furnace atomic absorption spectrometry [52].

As it was stated previously, there are different workplace conditions for workers who are welding in confined spaces compared to other welders. Limited access and little airflow or ventilation are the characteristics of a confined space. Hazardous concentrations of welding emissions can accumulate very quickly in such small spaces. Hazardous concentrations of welding emissions can accumulate very quickly in such small space. Thus, confined spaces should be monitored for toxic, flammable, or explosive emissions to evaluate welders' exposure. In some situations, continuous air monitoring may be necessary when workers are welding in a confined space with special conditions. Golbabaei et al. conducted an investigation to assess the risk related to welding pollutants for welders who work in confined spaces. Almost for all analyzed metals, there were significant differences between back welders and controls. Back welding is a task that workers perform welding inside the pipe as a confined space. Based on risk assessment, back welding was a high risk task [16]. These authors in another study assessed the welder's exposure to carcinogen metals (Cr, Cd, and Ni). The NIOSH methods were used for sampling and measurement of metals. Back welders group had maximum exposure to total fume and mentioned elements [52].

Determination of occupational exposures to gases must be based on workplace measurements, because the local ventilation and workplace design can affect the actual concentrations of toxic gases (ozone, carbon monoxide, nitrogen oxides) in the welders' breathing zone. Hariri et al. surveyed the appropriate personal sampling methods to measure the welding emissions in small and medium enterprises. They proposed NIOSH methods to evaluate the fumes and direct reading instruments for measurement of gases. Also, they offered some guidelines for correct assessment of welding workplaces [60]. Choonover et al. showed welders were exposed to higher concentrations of NO_2 and O_3 than controls. These gases were collected on pre-treated filters with proper solutions. Then, NO_2 and O_3 were analyzed by spectrophotometry and ion chromatography (IC), respectively [21]. Azari et al. conducted a study to evaluate exposure of mild steel welders to ozone and nitrogen oxides during TIG and MIG welding. OSHA ID214 and NIOSH 6014 methods were used to evaluate ozone and nitrogen oxides, respectively. High exposure of welders to these gases was reported in the study [64]. Golbabaei et al. also used OSHA and NIOSH methods as well as direct reading instruments for sampling and measurement of various gases [65].

Although there are various techniques for monitoring of welding emissions (both fumes and gases) in air samples, selecting the proper ones depends on some parameters. Availability of sampling media, sample storage time, and the simplicity, cost, time and sensitivity of analytical technique are essential to planning proper sampling strategies. It is necessary to consider those workers who probably have the highest exposures due to used materials and processes, the characteristics of their tasks, their postures during welding, the conditions of work environment, and other pollutants from processes in the vicinity of welding environment. It is known that high concentrations of some welding fumes and gases can also be explosive; therefore, the workplace should be tested to ensure a safe working environment [61, 66].

Biological Monitoring - Biological monitoring means the measurement of the concentration of a contaminant, its metabolites or other indicators in the tissues or body fluids of the worker. In some cases, biological monitoring may be a supplementary monitoring for the personal assessment [53]. Another advantage of the biological monitoring is the detection of biological effects of the chemical by monitoring reversible and irreversible biochemical changes. It can be used in the medical treatment to identify the real exposures of chemicals absorbed into the body of employees suspected of over-exposing to a chemical [58]. Airborne contaminants measurement and biological monitoring are complementary procedures used to prevent occupational disease, assess the risk to workers' health, and evaluate the effectiveness of control ways. Biological monitoring must be conducted based on a proper strategy. Careful considerations are required to select the best biological matrix for each component. To obtain valid results, timing sample collection, sample preparation and analytical method used to determine the concentration of components are critical. There are different methods for biological monitoring of some welding emissions. As it is indicated in Table 2, biological media and biological exposure indices (BEIs) have been recommended for some metals and gases emitted by welding processes. Totally, complete information can be provided by biological monitoring and air monitoring to assess the worker exposure to welding emissions.

Ellingsen et al. studied the concentration of manganese in whole blood and urine in welders. Concentration of Mn in whole blood (B-Mn) was about 25% higher in the welders compared to the controls. The increase in B-Mn and the dose-response relation between air-Mn and B-Mn in the welders are strong indicators of Mn. Long-term high exposure to welding fumes may lead to alterations of the urinary excretion of certain cations that are transported through the DMT1 transport system (divalent metal transporter 1 that is found on the surface of the lung epithelial cells) [67]. Kiilunen study showed the metal concentration in post shift urine samples were correlated with the personal air monitoring results. There were statistical significant correlations between urinary concentrations of chromium and nickel and the related total metal concentration in air in wire welding processes. Also, in MIG/MAG welding, chromium is accumulated in the body with a long half life. There is an association between the airborne concentration of nickel and its post shift urinary concentration. In welding, the nickel concentration in post shift urine samples can indicate the body burden [68]. In a study conducted by Hassani et al. the correlation between airborne Mn and urinary Mn was significant for all exposed subjects. The obtained result can introduce the urinary Mn as a biomarker for exposure to this element [69]. Azari et al. measured the serum level of malondialdehyde in welders. Serum MDA of welders was significantly higher than that of the control group. A significant correlation was detected between ozone exposure and level of serum MDA, but the correlation was not observed for nitrogen dioxide exposure [64]. Rossbach recommended the determination of Al in urine for biological monitoring because of the higher sensitivity and robustness of this marker compared to Al in plasma [70]. Golbabaei et al. analyzed the urinary metals among the different groups of welders. According to the results, exposure of welders to fume components leads to more accumulation of them at welders' bodies [52]. Based on different studies, the soluble metal compounds are accumulated in the body, affecting the critical organs. Urinary concentration of metal is used as a biomarker of metal exposure. Therefore, biomonitoring serves as an appropriate tool to monitor both the recent and past exposure and it can be related to the total chemical uptake through all exposure routes [69].

Health monitoring - In addition to the assessment of the airborne concentration of a particular contaminant and its comparison with standard limit, health monitoring may also be done for some hazardous chemicals to assess risks to exposed workers. Health monitoring means monitoring workers exposed to hazardous pollutants to identify changes in their health status and evaluate the effects of exposure. Health monitoring can provide effevtive information to implement proper ways for eliminating or minimizing the risk of exposure and improving control measures. Health monitoring considers all routes of exposure to contaminants [9, 66, 71]. Some tests including spirometry (lung function), audiometry (hearing), biochemical tests (e.g. kidney or liver function), cardiac function tests (heart function), nerve conduction velocity and electromyography tests (nerve and muscle function), and neurobehavioural tests (nerve and brain function) may be used in health monitoring. The type of test used will depend on the occupational hazards that the employee are exposed to [58]. Donaldson [72] and Antonini [73] surveyed lung functions in exposed welders and showed that exposure to welding fumes is associated with both pulmonary and systemic health endpoints, including decrease in pulmonary function, increased airway responsiveness, bronchitis, fibrosis, lung cancer and increased incidence of respiratory infection. In addition to these pulmonary effects, metal fume fever is frequently observed in welders. Exposure to metal fumes and irritating gases cause chronic obstructive pulmonary disease (COPD). Health monitoring of welders can help detect breathing problems and reduced lung functions in early stages, resulting in prevention of further damages. Spirometric tests are used by an occupational phisycian to assess lung functions [74]. Totally, health monitoring may include simple observation of the worker's skin to complicated tests in special cases. Health monitoring must be done by the experienced medical practitioner. An occupational physician can provide specialist services and testing such as spirometric tests, respiratory screening and chest X-rays. It is necessary to do the health monitoring before beginning work with a hazardous chemical to provide enough information for following changes in the worker's health during periods of exposure.

6.2. Risk Assessment Of Welding Emissions

Risk is defined as the possibility of occurance of an event leading to clear concequences. Evaluating risks to workers' safety and health is conducted in risk assessment process. It is performed in some steps including:

- Hazards identification and those at risk

- Evaluating the risks (qualitative or quantitative)

- Elimination or minimization of risks via implementing control measures and taking actions

- Monitoring and reviewing the effectiveness of adopted controls

The severity of hazard and the exopsure level determine the health risk and the type of chemical and nature of work are important factors in this regard. All workers in the vicinity of a special activity should be considered to assess the risk associated with chemical hazards, because they may potentially be at risk of chemicals emmitted by that activity.

In welding environments, employers are resposible to ensure the safety and health of welders and take proper measures for their protection. Although, preventing the occupational risks is the main purpose of risk assessment, it is not possible in all situations; therefore, risks should be reduced using control measures. There are different hazards related to welding process resulting in risks to welders. Chemical hazards, physical hazards, and those associated with ergonomics threaten the health of welders. Since this text deals with air pollution, the risk assessment of welding emissions i.e. fumes and gases is considered. Hazardous chemicals in the workplace result in different risks to workers.

There are different methods to do risk assessment of chemicals in which some principles should be considered. These principles include addressing all relevant hazards and risks and beginning the elimination of risks, if it is possible.

The ministry of manpower of Singapore has published a guideline intitled "semi-quantitave method to assess occupational exposure to harmful chemicals"[75]. This method may be useful to assess the risks resulting from welding emissions. Risk assessment is conducted for following purposes:

- Identifying the hazards related to each harmful chemical

- Evaluating the degree of exposure to chemical of interest

- Determining the likelihood of chemical adverse effects

A risk rating to different tasks can be designate using the mentioned method. After that, using risk rating matrix, hazards are ranked as negligible, low, medium, high and very high (legends 1 to 5) and required actions are prioritized to select appropriate controlling plans. This guideline deals with the health risk to workers exposed to chemicals via inhalation. There are eleven steps for hazard identification and rating, exposure evaluation, and assessing risk. The actual exposure level is required for determination of exposure rating and risk level. A step by step flow chart for assessing the risk, forms needed for completing some steps, and different tables and equations for evaluating the risk have been provided by guideline. All components to assess the risks are available in guideline and it can be used for risk assessment of welding emissions in a simple and fast way. Following, the process flow chart has been presented to understand the concept of risk assessment.

Golbabaie et al. used mentioned guideline to assess the health risks arising from metal fumes on back welders. Risk assessment was performed according to the steps previously explained. Cadmium concentration was ranked as "very high" group. Also, total fumes, total chromium, and nickel were ranked as "high" legend. Findings indicated back welding is a high risk task. High concentration of metals confirmed that working in confined spaces creates a great risk for welders. In some cases as in cadmium despite the rather low concentration of the pollutants, the risk is ranked as "very high" due to the carcinogenisity nature of this element. Therefore, it is not always possible to judge the health hazards of the pollutants based on their concentrations.

Following the risk assessment, employers can decide on required preventive measures, the working and production procedures, and also improving the level of welder protection. To complete risk assessment of welding chemicals, data related to air monitoring, biological monitoring, and health monitoring may be required for true judgement. Totally, risk assessment in workplace can result in some advantages. Workers do their tasks in a safe manner; employers provide appropriate programs to prevent high exposure and increase job satisfaction; regulators and related organizations can reliably present health and safety standards. The process of risk assessment is a basis for risk management to reduce welding hazards by choosing correct actions [76-77].

Air pollution control deals with the reduction of air pollutants emitted into the atmosphere using different technologies. Sometimes, managing the production process is used to control air pollutant emisstion, therefore, checking the production process can be useful for beginnig the air pollution control. Elimination of a hazard is the first aim to control related risk. In essence, keeping the pollutant emission at the minimum level during the process is the main purpose of controling the air pollution. Based on the risk assessment results, employers can decide for control of risk using proper ways. There are various ways to control the risk of chemicals like welding emissions. If the hazard elimination in not reasonably practicable, other approaches are used to minimize the risk. Substitution, isolation, engineering controls, work practices, and personal protective equipment (PPE) are used to reduce risks to the lowest practicable level in order of priority. Using personal protective equipment is the least recommended control way. To provide a layered safety net, a combination of several control ways may be adopted for preventing risks [66, 76, 78]. In the case of welding, if the elimination of fumes is not practicable, other controling measures should be applied. Modifying the welding process, improving working practices, ventilation, and using PPEs are considered in order to control of fumes.

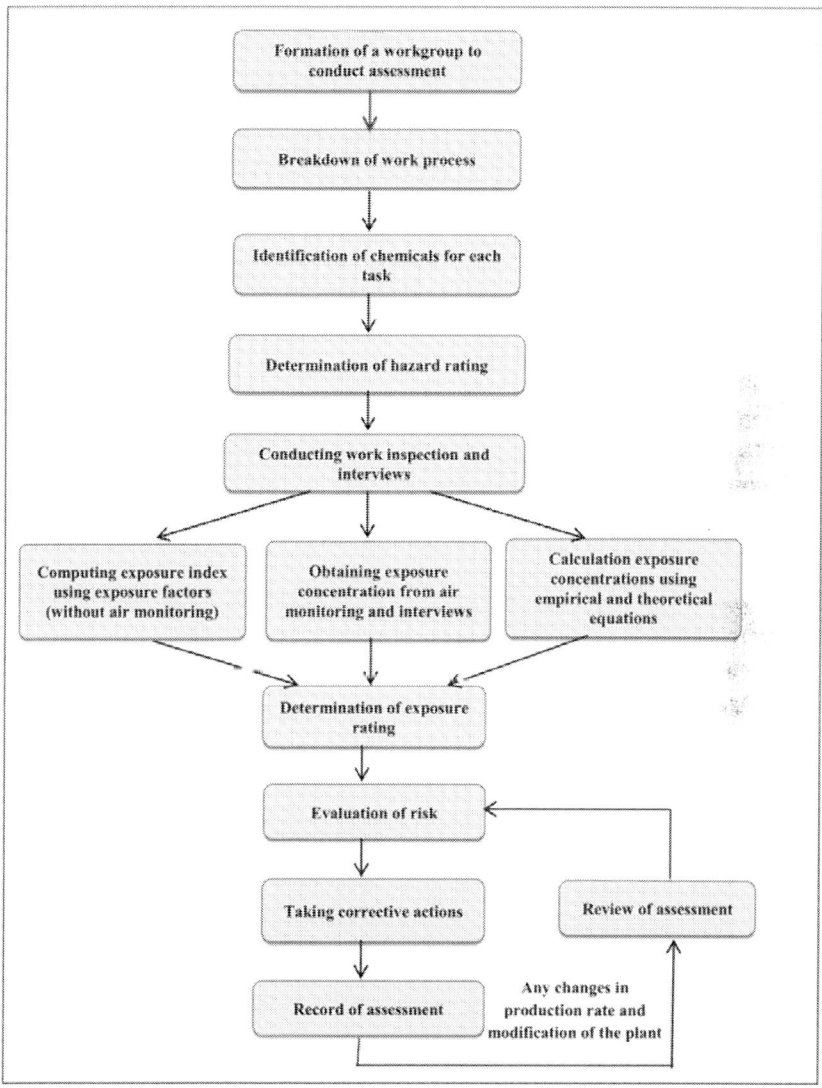

Figure 2. Process flow chart of semi quantitave method for chemicals risk assessment [75]

7. OCCUPATIONAL CONTROL

7.1. Choosing or Modifying The Welding Process

Employers can choose the welding type for production process based upon its efficiency, weld quality, available equipment, and economics. For instance, TIG welding generates less fume compared to MMA, MIG and FCAW processes, so, it can be a proper choice for welding operations. In order to modify the welding process, selecting consumables with minimum fume emissions and considering the welding parameters to minimize the emissions are recommended to employers. The generation of welding fumes is minimized using the lowest acceptable amperage. To optimize the process modification, paying attention to consumables, equipment, and control system is necessary. Selecting proper consumbles leads to minimizing the environmental impacts and controling risks to welders. Welding on non-painted or coated surfaces can also reduce the production of emissions. Process modification in welding results in decreasing needs for administrative controls and other expensive procedures, and also simplifying the process of risk assessment.

7.2. Improvement of Working Practices

Working practice, the way used to do work, can be improved for control of workers' exposure. Safe work practices are provided by company or organization to perform a task with minimum risk to workforce, environment, and process. Such practices control the manner of performing work and complete engineering measures. Placing the workpiece, as an improving measure, can keep the welders away from plume rising above the weld. Minimizing the welding in confined or enclosed spaces leads to reduction of exposure to pollutants. Proper training programs, housekeeping, maintenance, and doing task on time are the safe welding habits to reduce exposure. Consequently, welding based on safe practices and instructions results in healthier workplace and diminishing the risks of exposure to hazardous emmisions [79, 80].

7.3. Ventilation

Ventilation is the most effective way for removing welding emissions at source to reduce exposure to fumes and gases in welding operations. Designing the ventilation system in accordance with the types of hazardous emissions results in providing a safe atmosphere in the workplace. This control procedure is classified into dilution (general) ventilation and local exhaust ventilation (LEV).

The most efficient method to control welding emissions is the combination of LEV and dilution ventilation.

General or Dilution Ventilation -This type of ventilation uses the flow of air into and out of a working environment to dilute contaminants by fresh air. The required fresh air can be supplied by natural or mechanical ways. Dilution ventilation may not be sufficient to control exposure to welding emissions, because it cannot provide enough air movement to prevent the entry of fumes and gases into the welder's breathing zone before removing them from welding environment. In fact, the general ventilation is not suitable for controlling the toxic substances, specially when the worker is downstream of contaminant. To ensure the efficiency of the system, measuring airflow regularly and sampling contaminants to assess exposure are required. A well designed dilution system can be approprite for situations in which welding is done on clean, uncoated, mild steels. In dilution ventilation, draft fans or air-movers, wall fans, roof vents, open doors and windows may be used to move air through the work environment. Totally, if the generated contaminant is in low concentration and can be controlled to the standard exposure level, dilution systems will be effective enough as a control measure [66, 80-82].

Local Exhaust Ventilation - Local exhaust ventilation (LEV), as a primary engineering control, is used to remove contaminants before entering the breathing zone of workers. LEV can be used to control welding emissions close to the generation source. To be effective, LEV system should be well designed and installed, used correctly and properly maintained. Type of generated contaminants and characteristics of the process and work environment are crucial to design LEV [81]. To design a suitable system in welding process, some parameters should be considered, such as fume generation rate, arc- to-breathing zone distance, work practices and worker's exposure. Various parameters related to type of welding have important roles in the fume generation rate and fume composition. Therefore, considering these parameters is necessary to design LEV system [83-85].

For welding processes like stainless steel or plasma arc welding in which fumes containing heavy metals are generated, the LEV system can effectively be used to control worker's exposure. A local exhaust ventilation consists of a hood, fan, duct, and air cleaner. All parts of LEV system must be designed according to correct rules and requirements to remove air pollutants with appropriate efficiency. For instance, the ducting material and structure, air velocity through ducts, the number of branches, and the probability of the leakage and corrosion are important factors related to duct that can affect the LEV system. There are some considereations to select a suitable fan for the system. Some variables such as pressure, flow rate, power, noise, and rotation speed are the main characteristics influencing on the fan performance. Air cleaner is a device to capture welding emissions before it can escape into the ambient air. To select an appropriate air cleaner, some design considereations need to be addressed. Size and shape of welding space, pollutants generation rate, pollutant composition, cost of devices, process type, and the availability of equipment may be effective factors in this respect. In welding processes, source capture systems can be the ideal choise to control fume contaminants using the least air flow rate. In some situations, a source capture system cannot be used. For example situations in which worker has to work on mobile positions; there are a large number of small welding points producing hazardous emissions; welding must be done in confined spaces; and there are some obstructions like overhead cranes leading to problems with ducting installation. Dust collectors (filtration units) and electrostatic precipitators (ESP) can also be used as air cleaners to capture welding emissions before escaping into the environment. ESPs are ideal to collect submicron particles, especially in carbon steel welding. Although the efficiency of ESP is lower than filtration system, it needs very little maintenance and also there is no cost for filter replacement. ESPs are not recommended for stainless steel welding.

Some general considereations should be addressed to design a LEV system. Ducting system should be resistant to the captured emissions; the risks of contaminants accumulation and fire propagation in ducting system should be taken into account; exhausted air containing welding emissions should not be discharged where other workers or people are present; any draught from open doors or windows should be considered because of interference with hood performance. In addition, a maintenance program is required to ensure that control measures remain effective. For instance, regular inspections of LEVsystems should be carried out to check their effectiveness. As an other maintening plan, periodic air monitoring is done to ensure the system has proper performance. Therefore, as well as correct and completed design of LEV system, other elements like employee training, proper use, cleaning, and maintenance are required to achieve the effective protection.

Portable Systems - In some situations, portable systems may be used. These systems are used where welding is infrequently performed and the existing sysrem can be shared between working stations. Also, small mobile units may be used in confined spaces where installing the usual systems is not practical. In these cases, installing the hood close to the emissions point of origin, the hood placement and its distance from the source of welding emissions should be considered. Adequate ventiltion is essential in confined spaces, because the accumulation of hazardous emissions may lead to oxygen deficiency and also adverse effects related to generated fumes and gases. Commercially, there are different portable ventilation systems to use in confined spaces. Flexible air ducts and different kinds of portable fans are available for a variety of ventilation applications. In general, approximately 10 air exchanges per hour should be provided by ventilation in confined spaces. The volume of space and the flow rate of fan determine the time of each exchange. Before entry into the confined space for welding, that space should be ventilated for a minimum of five minutes. It is important to select a proper fan with enough capacity and position it in correct place. Some related organizations have provided procedures and instructions related to working in confined spaces, including ventilation equipment, confined spaces entry, emergency action plan, permit forms, and other requirements for working in these spaces [66, 81, 84, 86].

7.4. Respiratory Protection Equipments

Personal protective equipment (PPE) should not be used instead of other control measures, but sometimes they may be required along with engineering controls and safe work practices. Respiratory Protection Equipments (RPEs) are used to protect the workers against inhalation of hazardous emissions in the workplace, where exposures cannot adequately be controlled by other ways.

Using a respirator not selected appropriately leads to a false sense of protection for wearer and exposure to hazardous substances. It must be specific to the pollutant and fitted, cleaned, stored and maintained based on provided standards and guidelines for respirators. Each RPE has a protection factor (PF) that is determined as the ratio of the concentration of the pollutant outside the respirator to that inside the respirator. There is a wide range, from low to high, for protection factors. Some organizations like NIOSH have provided required equations and tables to calculate protection factors for respirators. There are different types of respirators and it is possible to select the most appropriate type for existing circumstances. In welding processes, respirators should be selected in accordance with generated emissions, welding type, welding task, and working conditions. For example, NIOSH recommends a self-contained breathing apparatus for welding in confined spaces because the oxygen concentration in the space may be reduced due to welding. Also, a combination of particulate/vapour respirator may be used because of the generation of both of fumes and gases during welding. A standard program is needed for using raspiratory protection devices. Some requirements are followed in this program including hazard assessment, selecting the appropriate respirators in respect of pollutants, respirator fitting test, worker training on how to use respirator correctly, inspection and maintenance of respirator, and recordkeeping. There are two types of RPE. The first type is respirators that clean workplace air before being inhaled and the second type is air-supplied respirators in which air supply is separate from workplace atmosphere. Totally, the suitable RPE for welding processes should be selected by an expert and based on fume concentration, presence of toxic gases, and the probability of oxygen deficiency. Selecting air-purifying respirators with correct filtration cartridge results in protection of welders from low levels of metal fumes and welding gases [87, 88].

8. CONCLUSION

Air pollution is contamination of the indoor or outdoor environment, leading to changes in the natural characteristics of the atmosphere. In all welding processes, various types of air pollutants are generated. Air pollutants created by welding include fumes and gases whose composition and emission level depend on some factors such as the welding method, welding parameters (current, voltage, shielding gas and shielding gas flow), base metal and other consumables. Exposure to excessive levels of fume and gases can cause different adverse health effects on workers. Since a large number of workers are exposed to welding emissions and also the generated pollutants have negative impacts on environment, a risk assessment program is required to protect workers and environment by suitable procedures. In an effective program, worker's safety and health is considered by management as a fundamental value.Taking different precautions can improve the welder's work situation. There are various techniques for evaluating and monitoring welding pollutants in air samples and biological matrices and also different procedures for their control. Selecting the proper engineering controls can lead to protection of workers and environment. During the risk assessment program and selection of control measures, it is necessary to consider nanoparticles emitted by welding operations. Particle sizes and size distributions of welding emission are critical to determine the efficient control devices. In some cases, breathing zone protection can be used. Health hazards can be reduced by choosing a correct welding helmet and by using the proper shielding gas and welding parameters. It is worth mentioning that proper information should be provided for workers about hazards of their tasks. The welder should be informed of operating techniques and all procedures that reduce welding fumes. The training programs should be included proper ways to perform tasks and proper work practices to reduce fumes. This program includes safety training, monitoring the good safety practices and good environmental practices. Also, the respirator and cartridge selection, fit-testing and respirator maintenance and storage are considered in a suitable training program. Furthermore, employers must be informed about industrial hygiene programs at workplaces and quantitative risk assessment for workers exposed to hazardous compounds. In recent years, different organizations have focused on climate change and environmental impacts of all industrial activities including welding. Various laws, instructions, and guidelines have been provided for protecting the air, environment, and water. Employers are responsible for the purchase of proper welding equipment to meet environmental requirements and choose more environmentally friendly processes.

REFERENCES

1. National Institute for Occupational Safety and Health (NIOSH). Nomination of Welding Fumes for Toxicity Studies. In: U.S. Department of Health and Human Services, Public Health Service: NIOSH; 2002.

2. El-Batanouny MM, Amin Abdou NM, Salem EY, El-Nahas HE. Effect of Exercise on Ventilatory Function in Welders. Egyptian Journal of Bronchology. 2009; 3(1): 103-11.

3. American Welding Society (AWS). Standard Welding Terms and Definitions. In: Committee TA, 12 ed. Miami: American Welding Society; 2009.

4. Stern RM. Process-Dependent Risk of Delayed Health Effects for Welders. Environmental Health Perspectives. 1981; 41: 235-53.

5. Husgafvel-Pursiainen I, Siemiatycki J. IARC Monograph on Welding fumes. 1990; 49.

6. Erdely A, Antonini JM, Salmen-Muniz R, Liston A, Hulderman T, Simeonova PP, et al. Type I interferon and pattern recognition receptor signaling following particulate matter inhalation. Particle and fibre toxicology. 2012; 9 (1).

7. Gonser M, Hogan, T., editor. Arc Welding Health Effects, Fume Formation Mechanisms, and Characterization Methods: InTech; 2011.

8. Carter GJ. Risk Assessment and Control of Exposure During Arc Welding of Steel. Intrernational Conference on Health and Safety in Welding and Allied Processes; 9-11 May; Kopenhagen, Denmark 2005.

9. Safe Work Australia. Guidance On The Interpretation Of Workplace Exposure Standards For Airborne Contaminants. 2012. p. 1-63.

10. Finneran A, O'Sullivan L. Force, posture and repetition induced discomfort as a mediator in self-paced cycle time. International Journal of Industrial Ergonomics. 2010; 40 (3): 257-66.

11. Canadian Centre for Occupational Health and Safety (CCOHS). Welding-Overview of Types and Hazards [Internet]; 2010. Available from: http://www.ccohs.ca/oshanswers/safety haz/

12. Turan E, Kocal T, unlugencoglu K. Welding technologies in shipbuilding industry. The Online Journal of Science and Technology. 2011; 1 (4).

13. Occupational Safety and Health Administration. Safety and Health Topic Criteria for a Recommended Standard: Working in Confined Spaces. Washington, DC Occupational Safety & Health Administration; 2007. Available from: www.osha.gov/SLTC/confinedspaces.

14. American Welding Society (AWS). Confined Spaces, Safety and health Fact Sheet No. 11. 1995.

15. NIOSH. Hazards to Outdoor Workers [Internet]. Centers for Disease Control and Prevention, Workplace Safety & Health Topics; 2013. Available from: http://www.cdc.gov/niosh/topics/outdoor/.

16. Golbabaei F, Khadem M, Hosseini M, et.al. Exposure to metal fumes among confined spaces welders. Ital J Occup Environ Hyg. 2012; 3 (4): 196-202.

17. Weman K. Welding processes handbook. bington Hall, Abington Cambridge, England: Woodhead Publishing Ltd and CRC Press LLC; 2003.

18. Kopeliovich D. Classification of welding processes: substech; 2012 [updated 2014/09/03]. Available from: http://www.substech.com/dokuwiki/doku.php

19. Flagan RC, Seinfeld JH. Fundamentals of Air Pollution Engineering. Englewood Cliffs, New Jersey Prentice Hall; 1988.

20. World Health Organization (WHO). Air Pollution; 2014. Available from: http://www.who.int/topics/air pollution/en/.

21. Schoonover T, Conroy L, Lacey S, Plavka J. Personal Exposure to Metal Fume, NO2, and O3 among Production Welders and Non-welders. Industrial Health. 2011; 49: 63-72.

22. Amza G, Dobrota D, Dragomir MG, Paise S, Apostolescu Z. Research on Environmental Impact Assessment of Flame Oxyacetylene Welding Processes. METALURGIJA 2013; 52 (4): 457-60.

23. Nakhla H, Shen JY, Bethea M. Environmental Impacts of Using Welding Gas. The Journal of Technology, Management, and Applied Engineering 2012; 28 (3): 1-11.

24. Kalliomaki PL, Hyvarinen HK, Aitio A, Lakoma EL, Kalliomaki K. Kinetics of the metal components of intratracheally instilled stainless steel welding

fume suspensions in rats. British journal of industrial medicine. 1986; 43 (2): 112-9.

25. AGA. FACTS ABOUT Fume and gases. AGA (a member of Linde Group), Report No. 110199 0912 – 1.3 HL.

26. Spear JE. Welding Fume and Gas Exposure. Magnolia, Texas J.E. Spear Consulting, LLC, 2004.

27. 27. Fuglsang K, Gram LK, Markussen JB, Kristensen JK. Measurement of ultrafine particles in emissions from welding processes. 16th International Conference on Joining of Materials; 10-13 May 2011; Elsinore, Denmark 2011.

28. Brand P, Klaus L, Uwe R, Kraus T. Number Size Distribution of Fine and Ultrafine Fume Particles From Various Welding Processes. Ann Occup Hyg. 2013; 57 (3): 305-13.

29. Keane M, Siert A, Stone S, Chen B, Slaven J, Cumpston A, et al. Selecting Processes to Minimize Hexavalent Chromium from Stainless Steel Welding. WELDING JOURNAL. 2012; 91: 241s-246s.

30. Gomes J. Albuquerque P, Miranda R, Vieira T. On the toxicological effects of airborne nanoparticles from welding processes. IIW European-South American School of Welding and Correlated Processes 18 –20th May 2011, Brazil: 2011.

31. Gomes J, Miranda R. Emission of airborne ultrafine particles during welding of steel plates. Ciencia & Tecnologia dos Materiais. 2014; 26 (1) :1-8.

32. Gomes JFP, Albuquerque PCS, Miranda RMM, Vieira MTF. Determination of airborne nanoparticles from welding operations. Journal of Toxicology and Environmental Health, Part A. 2012; 75 (13-15): 747-55.

33. A-Meo S, Al-Khlaiwi T. Health Hazards of Welding Fumes. Saudi medical journal. 2003; 24: 1-25.

34. De Flora S. Threshold Mechanisms and Site Specificity in Chromium (VI) Carcinogenesis. Carcinogenesis. 2000; 21 (4): 533-41.

35. Sowards JW, Ramirez AJ, Lippold JC, Dickinson DW. Characterization Procedure for the Analysis of Arc Welding Fume. Welding Research. 2008; 87: 76-83S.

36. Keegan GM, Learmonth ID, Case CP. Orthopaedic metals and their potential toxicity in the arthroplasty patient. The Journal of Bone and Joint Surgery. 2007; 89-B (5): 567-73.

37. pesch B, Haerting J, Ranft U, et.al. Occupational Risk Factor for renal Cell Carcinoma: Agent-Specific Results from a Case Control Study in Germany. International Journal of Epidemiology. 2000; 29: 1014-24.

38. Erhabor GE, Fatusi S, Obembe OB. Pulmonary functions in ARC-welders in Ile-Ife, Nigeria. Ast Afr Med J. 2001; 78 (9): 461-4.

39. Tenkate TD. Optical radiation hazards of welding arcs. Rev Environ Health. 1998; 13 (3): 131-46.

40. Mariutti G, Matzeu M. Measurement of ultraviolet radiation emitted from welding arcs. Health Phys. 1988; 54 (5).

41. Tenkate TD. Occupational exposure to ultraviolet radiation: a health risk assessment. Rev Environ Health. 1999; 14 (4): 187-209.

42. Ernst E, Bonde JP. Sex hormones and epididymal sperm parameters in rats following sub-chronic treatment with hexavalent chromium. Hum Exp Toxicol. 1992; 11 (4): 255-8.

43. Jensen TK, Bonde JP, Joffe M. The influence of occupational exposure on male reproductive function. Occupational Medicine. 2006; 56: 544-53.

44. Bonde JP. Subfertility in relation to welding. A case referent study among male welders. Dan Med Bull. 1990; 37 (1): 105-8.

45. Bonde JP, Hanse KS, Levine RJ. Fertility among Danish male welders. Scand J Work Environ Health. 1990; 16 (5): 315-22.

46. Mortensen JT. Risk for reduced sperm quality among metal workers, withspecial reference to welders. Scand J Work Environ Health. 1988; 14 (1): 27-30.

47. Sheiner EK, Sheiner E, Hammel RD, Potashnik G, Carel R. Effect of Occupational Exposures on Male Fertility: Literature Review. Industrial Health. 2003; 41: 55-62.

48. Iregren A. Manganese neurotoxicity in industrial exposures: proof of effects, critical exposure level, and sensitive tests. Neurotoxicology. 1999; 20 (2-3): 315-23.

49. Bowler RM, Gysens S, Diamond E, Booty A, Hartney C, Roels HA. Neuropsychological Sequelae of Exposre to Welding Fumes in a Group of Occupationally Exposed Men. Int J Hyg Environ Health. 2003; 206: 517- 29.

50. [50]Wang X, Yang Y, Wang Xi, Xu S. The Effect of Occupational Exposure to Metals on the Nervous System Function in Welders. J Occup Health. 2006; 48: 100-6.

51. International Agency for Research on Cancer (IARC). welding fumes and gases. lARC Monograhs 49; 1987.

52. Golbabaei F, Seyedsomea M, Ghahri A, Shirkhanloo H, Khadem M, Hassani H, Sadeghi N, Dinari B. Assessment of Welders Exposure to Carcinogen Metals from Manual Metal Arc Welding in Gas Transmission Pipelines, Iran. Iranian J Publ Health. 2012; 41 (8): 61-70.

53. National Occupational Health and Safety Commission. Welding: Fumes and Gases. In: Commonwealth of Australia, editor: Ambassador Press Pty Ltd; 1990.

54. Division of Workers' Compensation. Welding Hazards Safety Program. In: Texas Department of Insurance, editor; 2012.

55. American Conference of Governmental Industrial Hygienists (ACGIH). Threshold Limit Values for chemical substances and physical agents & biological exposure indices.: ACGIH, Cincinnati, OH; 2010.

56. Safe Environment-Managing Property Risk. Welding Fume [Internet]. Safe Environment, Sydney; 2012. Available from: http://www.safeenvironments.com.au/welding-fume/.

57. Ashby HS. Welding Fume in the Workplace, Preventing Potential Health Problems through Proactive Controls. Professional Safety. 2002; Apr: 55-60.

58. Department of Consumer and Employment Protection. Risk-based health surveillance andbiological monitoring. Resources Safety, Department of Consumer and Employment Protection, Western Australia; 2008.

59. Pires I, Quintino L, Miranda R, Gomes J. Fume emissions during gas metal arc welding. Toxicological and Environ Chemistry. 2006; 88 (3): 385-94.

60. Spiegel-Ciobanu V. Occupational health and safety regulations with regard to welding and assessment of the exposure to welding fumes and of their effect. Welding and Cutting. 2012; 11 (1): 61.

61. Occupational Safety and Health Administration (OSHA). OSHA Technical Manual, Personal Sampling for Air Contaminants In: U.S. Department of Labor, Washington, DC OSHA; 1999.

62. Education and Information Division. National Institute for Occupational Safety and Health. Occupational Exposure Sampling Strategy Manual. In: Department of Health Education and Welfare, Cincinnati: NIOSH; 1977.

63. Hariri A, Yusof MZM, Leman AM. Sampling Method for Welding Fumes and Toxic Gases in Malaysian Small and Medium Enterprises (SMEs). Energy and Environment Research. 2012; 2 (2): 13-20.

64. Azari MR, Esmaeilzadeh M, Mehrabi Y, Salehpour S. Monitoring of Occupational Exposure of Mild Steel Welders to Ozone and Nitrogen Oxides. Tanaffos. 2011; 10 (4): 54-9.

65. Golbabaei F, Hassani H, Ghahri A, Arefian S, Khadem M, Hosseini M, Dinari B. Risk Assessment of Exposure to Gases Released by Welding Processes in Iranian Natural Gas Transmission Pipelines Industry. International Journal of Occupational Hygiene (IJOH). 2012; 4 (1): 6-9.

66. Safe Work Australia. Managing Risks of Hazardous Chemicals in the Workplace. Australia; 2012.

67. Ellingsen DG, Dubeikovskay L, Dahl K, et al. Air exposure assessment and biological monitoring of manganese and other major welding fume components in weldersw. J Environ Monit. 2006; 8: 1078-86.

68. Kiilunen M. Use of biological monitoring for exposure assessment in welding Health and safety in welding and allied processes Brondby, Denmark; 2005.

69. Hassani H, Golbabaei F, Ghahri A, et al. Occupational Exposure to Manganese-containing Welding Fumes and Pulmonary Function Indices among Natural Gas Transmission Pipeline Welders. J Occup Health. 2012; 54: 316-22.

70. Rossbach B, Buchta M, Csanady GA, et al. Biological monitoring of welders exposed to aluminium. Toxicology Letters. 2006; 162: 239-45.

71. Safe Work Australia. Health Monitoring for Exposure to Hazardous Chemicals, Guide for Persons Conducting -A Business or Undertaking. Australia; 2013.

72. Donaldson K, Tran L, Jimenez LA, Duffin R, et al. Combustion-derived nanoparticles: A review of their toxicology following inhalation exposure. Particle and Fibre Toxicology. 2005; 2 (10): 1-14.

73. Antonini JM, Lewis AB, Roberts JR, Whaley DA. Pulmonary Effects of Welding Fumes: Review of Worker And Experimental Animal Studies. American Journal of Industrial Medicine. 2003; 43: 350-60.

74. Health and safety in welding and allied processes - assessing and controlling the exposure risk and complying with the COSHH regulations for welding fume [Internet]. Technical Knowledge 2014. Available from: http://www.twi-global.com/technical-knowledge/faqs/health-and-safety-faqs

75. Ministry of manpower. A semi-quantitative method to assess occupational exposure to harmful chemical. In: Occupational Health and safety Division, Singapore; 2005.

76. Europian Commision. Guidance on Risk assessment at Work. In: Directorate General Employment and Social Affairs, Luxembourg: Office for Official Publication; 1996.

77. National Academy of Sciences. Occupational Health and Safety in the Care and Use of Nonhuman Primates. Washington DC: The National Academies Press; 2003.

78. Overseas Environmental Cooperation Center. Air Pollution Control Technology Manual. Japan: Environmental Agency, Government of Japan; 1998.

79. Hewitt PJ. Strategies for risk assessment and control in welding: challenges for developing countries. Annals of Occupational Hygiene. 2001; 45 (4): 295-8.

80. The welding Institute (TWI). Control of welding fume, Health, safety and accident prevention. [Internet]. The Welding Institute: UK; 2014. Available from: http://www.twi-global.com/technical-knowledge/job-knowledge.

81. Government of Alberta-Employment and Immigration. Welder's Guide to the Hazards of Welding Gases and Fumes Canada: Workplace Health and Safety Bulletin; 2009.

82. Health and Safety Executive. General Ventilation in the Workplace, Guidance for Employers. Sheffield: Health and Safety Executive; 2000.

83. American Welding Society (AWS). Fumes and gases in the welding environment. Miami, FL: AWS; 1979.

84. Chevron MCBU. Confined Space Entry. USA: HSE, 2012 Contract No. CPL-HES 201

85. Flynn MR, Susi P. Local exhaust ventilation for the control of welding fumes in the construction industry-a literature review. Annals of occupational hygiene. 2012: 1-13.

86. OSHA. OSHA and Welding Exhaust In: General Industry Standards and Interpretations, Marine Chemist Service, Inc; 1979.

87. Health and Safety Authority. A Guide to Respiratory Protective Equipment. Dublin; 2010.

88. Employment and Social Development Canada. A Guide to Health Hazards and Hazard Control Measures with Respect to Welding and Allied Processes. Canada, Quebec; 2007.

CHAPTER 3

Discontinuous and Continuous Indoor Air Quality Monitoring in Homes with Fireplaces or Wood Stoves as Heating System

Gianluigi de Gennaro [1,2,*], Paolo Rosario Dambruoso [2], Alessia Di Gilio [2], Valerio Di Palma [1], Annalisa Marzocca [2] and Maria Tutino [2]

[1] Department of Biology, University of Bari Aldo Moro—Via Orabona 4, Bari 70126, Italy
[2] Apulia Regional Agency for Environmental Prevention and Protection—Corso Trieste 27, Bari 70126, Italy

ABSTRACT

Around 50% of the world's population, particularly in developing countries, uses biomass as one of the most common fuels. Biomass combustion releases a considerable amount of various incomplete combustion products, including particulate matter (PM) and polycyclic aromatic hydrocarbons (PAHs). The paper presents the results of Indoor Air Quality (IAQ) measurements in six houses equipped with wood burning stoves or fireplaces as heating systems. The houses were monitored for 48-h periods in order to collect PM_{10} samples and measure PAH concentrations. The average, the maximum and the lowest values of the 12-h PM_{10} concentration were 68.6 μg/m^3, 350.7 μg/m^3 and 16.8 μg/m^3 respectively. The average benzo[a]pyrene 12-h concentration was 9.4 ng/m^3, while the maximum and the minimum values were 24.0 ng/m^3 and 1.5 ng/m^3, respectively. Continuous monitoring of PM_{10}, PAHs, Ultra Fine Particle (UFP) and Total Volatile Organic Compounds (TVOC) was performed in order to study the progress of pollution phenomena due to biomass burning, their trends and contributions to IAQ. The results show a great heterogeneity of impacts on IAQ in terms of magnitude and behavior of the considered pollutants' concentrations. This variability is determined by not only different combustion technologies or biomass quality, but overall by different ignition mode, feeding and flame management, which can also be different for the same house. Moreover, room dimensions and ventilation were significant factors for pollution dispersion. The increase of PM_{10}, UFP and PAH concentrations, during lighting, was always detected and relevant. Continuous monitoring allowed singling out contributions of other domestic sources of considered pollutants such as cooking and cigarettes. Cooking contribution produced an impact on IAQ in same cases higher than that of the biomass heating system.

KEYWORDS

fireplace; stove; wood combustion; indoor air quality; ultrafine particles

1. INTRODUCTION

Indoor Air Quality requires attention as it relates to the health and comfort of people that spend most of their time indoors [1,2,3]. Heating, cooking, smoking, cleaning as well as furnishings or building materials are important indoor sources of gaseous pollutants and particles [4,5,6,7,8]. The impact of these sources is linked to the amount and hazard of the emitted pollutants [9,10]. Moreover, several factors such as occupant's behavior, microclimatic and ventilation condition and outdoor intrusion can influence indoor pollution levels [11]. Great interest is paid to particulate matter (PM) in relation to its concentration, chemical composition and the duration of exposure [8]. A large number of indoor particle sources were identified and investigated by many studies. Among these, resuspension of particles by human activities and pet movements, dusting, vacuuming and showering contributes to the coarse mode of indoor particles [12,13,14]. Tobacco smoking, cooking, kerosene heating [15], gas burners [15,16], burning of candles [17], incense sticks [18] and biomass in open fireplaces [8,19,20,21] are the main indoor sources of fine and ultrafine particles [22,23,24,25]. In recent studies, a considerable attention was paid to indoor biomass combustion [26] that releases a considerable amount of pollutants, including carbon monoxide (CO), nitric oxides (NO_x), sulfur dioxide (SO_2), formaldehyde (HCHO), volatile organic compounds (VOC), particulate matter (PM) and polycyclic aromatic hydrocarbons (PAHs) [27,28,29]. Ventilation systems and other heat sources can influence indoor air quality determining higher pollutant levels than outdoors [30,31,32,33]. Moreover it was found that combustion processes contribute poorly to 24-h mean PM_{10} levels

[8,34]. However, number concentrations of emitted fine particles ($PM_{2.5}$) and UFP (particles with diameter less than 100 nm) are relevant and thus, may be a more appropriate predictor of health effects [35,36]. Therefore, several authors have paid particular attention to the formation of PM and UFP in indoor air when operating wood-burning fireplace ovens and stoves [36,37,38,39,40,41]. These studies proved that these heating systems were potential sources of particles [42]. In detail, the number concentration and chemical composition of the particles emitted by open fireplace and stoves are key elements for indoor exposure assessment and for developing appropriate mitigation strategies. Therefore, this work aimed to evaluate the impact of wood fireplaces and stoves on indoor pollutant concentrations and to study the dynamics of pollution phenomena. In particular, 12-h PM_{10} samples were collected in six residential houses located in the hinterland of Bari (Southern Italy) in order to measure their mass and PAH concentrations for evaluating the residents' exposure. Simultaneously, real time monitoring of PM_{10}, PAHs, UFP and Total Volatile Organic Compounds (TVOC) was carried out in order to study the progress of pollution phenomena due to biomass burning.

2. EXPERIMENTAL SECTION

Indoor air quality was measured in six houses in the Apulia Region characterized by a heating system based on wood burning stoves or fireplaces. Houses 1 and 3 have cast iron wood stoves while the other houses have open fireplaces as heating system, respectively. The house and heating system characteristics and information regarding type and weight of wood are reported in Table 1.

Table 1. Type of heating systems used, amount and type of wood burned and room volume for each house.

Monitored Houses	Heating System	Amount of Wood Burned (kg)	Type of Wood Burned	Room Volume (m^3)
House 1	Wood stove	16	Olive tree wood	98.0
House 2	Fireplace	15	Olive and almond tree wood	72.0
House 3	Wood stove	20	Olive and pine wood	42.0
House 4	Fireplace	18	Olive and almond tree wood	40.0
House 5	Fireplace	12	Olive tree wood	103.0
House 6	Fireplace	18	Olive tree wood	56.0

Monitoring campaign of 48-h periods was performed in each house for assessing indoor PM_{10} concentrations and particle size distribution. For the duration of the monitored periods, two 12-h PM_{10} samples were collected in each house during biomass burning and two 12-h PM_{10} samples during no-burning periods. In particular, PM_{10} was collected by a Sequential Air Sampler (SILENT

Sequential Air Sampler—FAI Instruments S.r.l., Roma, Italy) for 12 h on polycarbonate fiber filters (47 mm diameter Whatman, Buckinghamshire, UK) equipped with sampling heads operating at a flow rate of 10 L/min with a relative uncertainty of 5% of the measured value. A total amount of 24 PM_{10} samples were collected and stored in a freezer at -4 °C. PM_{10} samples were analyzed for determining PAH concentrations. The extraction of PAHs was conducted with a mixture of acetone/hexane through a microwave assisted solvent extraction by a Milestone, model Ethos D device (Milestone s.r.l., Sorisole (BG) Italy), which allowed the simultaneous extraction of up to 10 samples under the same conditions. The extracted samples were analyzed using an Agilent 6890 PLUS gas chromatograph (Agilent Technologies, Inc., Santa Clara, CA, USA) equipped with a programmable temperature vaporization injection system (PTV) and interfaced with a quadrupole mass spectrometer, operating in electron impact ionization (Agilent MS-5973 N, Inc., Santa Clara, CA, USA. The identification of each PAH (benzo[a]anthracene (BaA), benzo[b+j]fluorene (Bb+jF), benzo[k]fluoranthene (BkF), benzo[a]pyrene (BaP), benzo[g]perylene (BgP), indenopyrene (IP) and dibenzoanthracene (DBA)) was performed using perylene D_{12} (PrD, 264) as the internal standard (IS). The analytical performance of the whole procedure (extraction recovery, extraction linearity, analytical repeatability, Limit Of Detection-LOD) was verified in our previous study [43].

Moreover, the high time resolved concentrations of PM_{10} and ultrafine particles were measured. The real time PM_{10}concentration was provided by an Optical Particle Counter (OPC Multichannel Monitor—FAI Instruments, Roma, Italy). Number concentration and size distribution of ultrafine particles ranged from 5.6 nm to 560 nm were measured by using a Fast Mobility Particle Sizer (FMPS 3091—TSI, Buckinghamshire, UK. Simultaneously the real time concentrations of total VOC and total PAHs were monitored by a PID PhoCheck TIGER (Ion Science Inc., Cambridge, UK) and by an Ecochem PAS 2000 (SARAS S.p.A., S.p.A., Milno, Italy), respectively. The instruments were installed in front of the heating system, at a height of 1.5 ± 0.1 m from the ground and 1.0–1.5 m away from any door or vent. All sampler inlets were placed at least 1 m from any other combustion source (Figure 1).

A questionnaire was administered to habitants in order to obtained information about stove or fireplace use, the amount of wood used during the sampling period, ventilation conditions and the occupants' normal behavior (cooking and heating appliances, hygiene and personal care, use of sanitation and cleaning products, etc.).

(a)   **(b)**

Figure 1. Instrument positioning in (**a**) House 1 and (**b**) House 5.

3. RESULTS AND DISCUSSION

The average PM_{10} and PAH concentrations obtained during the biomass burning periods (two 12-h PM_{10} filters) and no-burning periods (two 12-h PM_{10} filters) are showed in Table 2.

Table 2. PM_{10} and PAH concentrations obtained during biomass burning (lighting) and no-burning (no lighting) periods in house operating wood stove (Houses 1 and 3) and fireplace (Houses 2, 4–6).

Monitored Houses	Activities	B(a)A	B(b+j)F	B(k)F	B(a)P	DBA	IP	BgP	$\sum PAH$ (ng/m³)	PM_{10} (ng/m³)
House 1	Lighting	73.8	31.1	9.7	19.7	20.9	19.5	38.0	212.7	66.3
	No lighting	29.5	26.6	8.7	16.6	15.1	15.8	27.6	199.9	54.2
House 2	Lighting	97.0	22.5	9.1	12.9	11.0	10.6	22.8	185.9	74.8
	No lighting	34.3	12.1	3.9	6.5	8.7	7.4	16.1	89.0	54.7
House 3	Lighting	46.6	12.0	4.0	4.1	7.3	5.9	13.0	92.9	212.3
	No lighting	4.7	5.1	2.6	3.2	3.3	3.1	5.5	27.5	52.9
House 4	Lighting	99.5	17.7	5.4	16.6	12.8	9.3	21.5	182.8	80.7
	No lighting	24.6	14.7	5.1	9.3	9.9	8.6	20.0	92.3	53.8
House 5	Lighting	52.1	19.7	14.1	11.9	14.2	9.4	24.3	145.7	38.2
	No lighting	30.3	11.1	4.0	5.0	14.7	7.5	17.3	89.9	22.8
House 6	Lighting	45.5	11.6	3.2	5.6	8.6	6.1	12.5	93.0	67.1
	No lighting	1.7	2.6	1.3	1.5	2.0	1.7	2.9	13.7	45.9

The results obtained during the monitored periods showed that the highest concentrations of PM_{10} were detected when operating the wood burning fireplace or stove, highlighting the impact of biomass burning on indoor air quality. PM_{10} concentrations were higher than the limit value (50 µg/m³) established by the Directive 2008/50 European Commission for outdoor environments, except for House 5. The highest concentrations were determined in House 3 (212.3 µg/m³ and 52.9 µg/m³ with the fireplace on and off, respectively) where a cast iron closed wood stove was placed in an environment with lower ceilings. On the contrary, the lowest concentrations of House 5 (38.2

$\mu g/m^3$ and 22.8 $\mu g/m^3$ during biomass burning and no-burning, respectively) could be due to the larger and better ventilated monitored environment. The same findings were observed for PAHs. The sum of PAH concentrations ranged from 92.9 to 212.7 and from 13.7 to 139.9 ng/m^3 during the two monitored periods. The highest values were determined in House 1 where a cast iron wood stove was used for domestic heating. BaP concentrations were always higher than the target yearly value of 1 ng/m^3 established by Directive 2008/50 European Commission for outdoor air, reaching a maximum value of 19.7 ng/m^3 in House 1 during lighting periods. Moreover, high concentrations of PAHs were detected in the House 5, characterized by the lowest levels of PM_{10}. Therefore, this finding suggested that PM_{10} concentration is not a good indicator of indoor air quality.

The comparison between subsequent lightings performed in the same indoor environment showed that many factors could influence the emission process. In detail, it was found that indoor concentrations of PM_{10} and BaP during the two burning periods were different although operating under the same conditions: same heating system, wood burned and indoor environment (see Figure 2).

This finding highlighted that the efficiency of the burning process (combustion duration and temperature, and smoldering or flaming combustion), greatly influence the emissions from fireplaces or wood stoves [11,38].

PAH diagnostic ratios have recently been used as tools for identifying and assessing pollution emission sources [44]. Huang *et al.* found BaP/BgP and IP/[IP+BgP] ratios equal to 1.4–2.0 and 0.64, respectively, during rice straw burning [44,45]. Hays *et al.* showed that BaP/BgP and BaP/IP ratios are strongly linked to wheat residue burning in an experimental chamber [46]. In our previous study of sources conducted in olive tree fields, the BaP/BgP, IP/[IP+BgP], BaP/IP and IP/BgP ratios were found [47]. These average ratio values for olive wood burning were compared with those of this study during biomass burning and no-burning periods (Table 3). Good agreement between the diagnostic ratios obtained from olive wood combustion and those in indoor environments was not found, suggesting that several sources in indoor environments contribute to PAH concentrations influencing the diagnostic ratio.

Moreover, no significant differences were observed between the diagnostic ratios associated with biomass burning and no-burning periods, both for fireplaces and wood stoves. This result suggested that the impact of biomass burning source on IAQ was also relevant when the fireplace and stove were not operated.

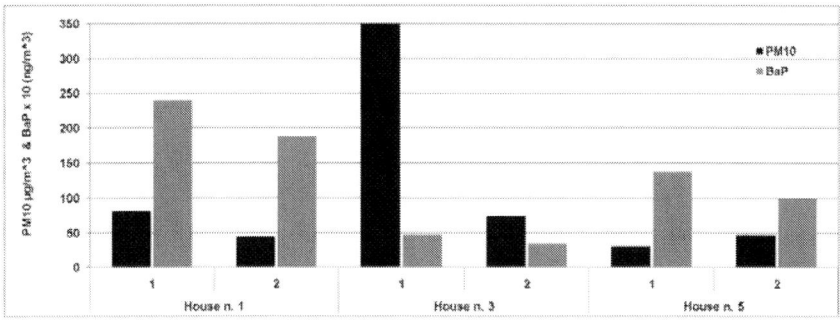

Figure 2. PM_{10} and BaP concentrations for each lighting in three houses.

Table 3. The average of diagnostic ratios comparison between this study (lighting and no lighting) and biomass burning in olive trees field.

Diagnostic Ratios	Biomass Burning in Olive Tree Fields [47]	This Study	
		Lighting	No Lighting
IP/BgP	1.12	0.5	0.5
IP/(IP+BgP)	0.53	0.3	0.3
BgP/BgP	1.55	0.5	0.5
BgP/IP	1.38	1.1	0.9

Therefore, the real time concentrations of PM_{10}, UFP, total PAHs and VOCs were monitored in each house. The VOC concentration results are not included in the paper because no relevant information concerning biomass burning as an indoor VOC source were obtained. On the contrary, the high time resolved data of PM_{10}, UFP and total PAH concentrations, highlighted the impact of several indoor sources such as lighting, cigarettes and cooking. As an example, Figure 3 shows the temporal trend of these pollutants in House 1. The two biomass burning periods showed different trends of investigated pollutants in terms of intensity and duration, confirming the above. Real time monitoring of pollutants also showed that the habitants' behavior may have a great influence on indoor pollutant emissions and thus on indoor air quality. In fact, it was found that closing the windows during the night after the biomass burning determined a pollutant increase due to their stagnation in the indoor environment. This finding suggests that ventilation frequency and duration, fireplace characteristics, design and location are the key elements for improving IAQ. In detail, it was found that PAH, PM_{10} and UFP concentrations were lower in large and well-ventilated environments. In particular, in House 5 the fireplace was characterized by a strong chimney draft and it was positioned in a big room, an open space where there was access to three other rooms and a stairway leading upstairs.

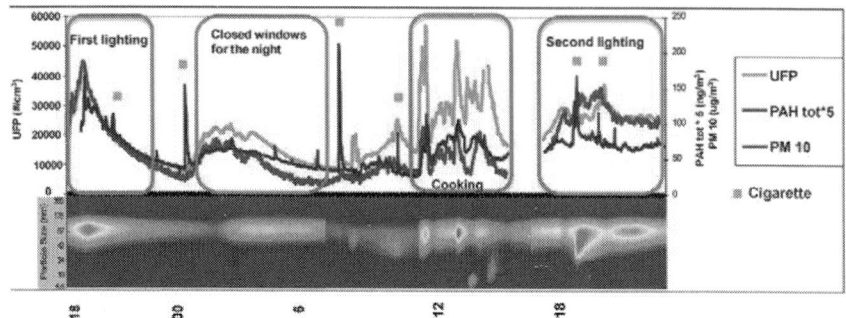

Figure 3. Temporal trend of PM₁₀, UFP and total PAH concentrations in House 1. The FMPS spectrum is reported in the bottom picture.

In addition, UFP concentrations determined in all houses provided the same results concerning BaP: the highest concentrations were determined in House 1 (daily mean: 2.2×10^4 particles/cm³), while the lowest one in House 3 (daily mean: 1.4×10^4 particles/cm³). Moreover, the UFP concentrations during biomass burning in fireplaces ranged from 1.8×10^4 to 4.5×10^4 particles/cm³ and the highest concentrations were detected during food cooking, while sharp peaks were observed during cigarette smoking. In particular, the mean UFP concentration during food cooking was 4.8×10^4 particles/cm³ and it reached maximum values of 5.8×10^4 particles/cm³. The number of UFP emitted during cigarette smoking was variable and reached a maximum value of 5.2×10^4 particles/cm³. The FMPS spectrum also revealed the nucleation and accumulation period of UFP. In particular, the first lighting resulted in midday "blobs" in the spectrum suggesting nucleation bursts of particles around 80 nm. On the contrary, the second lighting was characterized by a nucleation event with "banana-like" growth characteristics, indicating that coagulation occurred when emitted particles reached high concentrations.

In fact, as reported in previous works, coagulation occurs when Brownian motion determines collisions with surrounding gas molecules producing a shift in UFP size [48,49,50]. The differences between the two lightings could be due to the better dispersion conditions during the first event that do not allow particle agglomeration and growth. The other two nucleation events with the "banana-like" growth characteristics were determined in the hours when coffee and lunch were cooked.

The size distribution of UFP emitted by biomass burning in fireplaces was unimodal with a primary mode from 70 to 90 nm. The same trend was determined for the UFP distributions during cooking (in Figure 4), when the UFP concentrations reached 5.8×10^4 particles/cm^3. However, even if the particles in accumulation mode were comparable between biomass burning and cooking sources, higher concentrations of UFP with diameters ranging from 10 to 20 nm were registered for cooking sources. This result is in agreement with previous studies where frying produced peak number concentrations of UFPs at about 70 nm, with a secondary peak at 10 nm [51,52,53]. These results confirmed the high impact of cooking on indoor air quality.

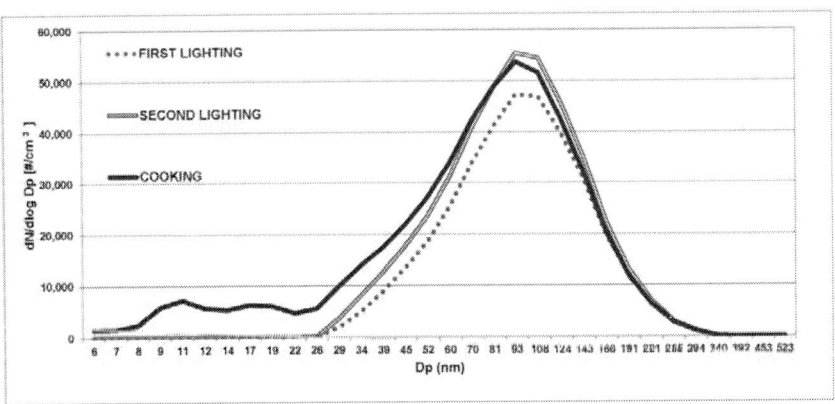

Figure 4 Size distribution of UFP emitted during the two lightings and cooking for House 1 (for example).

4. CONCLUSIONS

The present study enables us to highlight the impact of biomass burning on indoor air quality. In fact, an increase of PM_{10} and BaP concentrations were detected when operating open wood burning fireplaces and stove, reaching values higher than the limit and target values set by Directive 2008/50 European Commission. The 12 h PM_{10} sampling and PAHs diagnostic ratio highlighted the relevant impact of biomass burning source on IAQ, even if fireplaces and stoves were not operated. The real time monitoring was a useful tool: (a) to study the dynamics of pollution phenomena due to biomass burning; (b) to identify the different indoor sources and (c) to evaluate the factors influencing the indoor pollutant concentrations. The study showed that ventilation frequency and duration, fireplace characteristics, design and location could be the key elements for improving the indoor air quality and preserving human health.

AUTHOR CONTRIBUTIONS

All authors contributed equally to this work.

REFERENCES

1. Zhang, J.J.; Smith, K.R. Indoor air pollution: A global health concern. *Br. Med. Bull.* **2003**, *68*, 209–225.

2. Ashmore, M.R.; Dimitroulopoulou, C. Personal exposure of children to air pollution. *Atmos. Environ.* **2009**, *43*, 128–141.

3. Weschler, C.J. Changes in indoor pollutants since the 1950s. *Atmos. Environ.* **2009**, *43*, 153–169.

4. Park, H.S.; Ji, C.; Hong, T. Methodology for assessing human health impacts due to pollutants emitted from building materials. *Build. Environ.* **2016**, *95*, 133–144.

5. Isaxon, C.; Gudmundsson, A.; Nordin, E.Z.; Lönnblad, L.; Dahl, A.; Wieslander, G.; Bohgard, M.; Wierzbicka, A. Contribution of indoor-generated particles to residential exposure. *Atmos. Environ.* **2015**, *106*, 458–466.

6. Bekö, G.; Weschler, C.J.; Wierzbicka, A.; Karottki, D.G.; Toftum, J.; Loft, S.; Clausen, G. Ultrafine particles: Exposure and source apportionment in 56 Danish homes. *Environ. Sci. Technol.* **2013**, *47*, 10240–10248.

7. Airborne Particles: Exposure in the Home and Health Effects. Available online: http://www.einsten.net/pdf/0100250502.pdf (accessed on 23 October 2015).

8. He, C.; Morawska, L.; Hitchins, J.; Gilbert, D. Contribution from indoor sources to particle number and mass concentrations in residential houses. *Atmos. Environ.* **2004**, *38*, 3405–3415.

9. Arku, R.E.; Adamkiewicz, G.; Vallarino, J.; Spengler, J.D.; Levy, D.E. Seasonal variability in environmental tobacco smoke exposure in public housing developments. *Indoor Air* **2015**, *25*, 13–20.

10. Elliott, T.G.; Ellison, M.C.; Matt Earnest, C.; Brent, S. Indoor air pollution in developing countries: Research and implementation needs for

improvements in global public health. *Am. J. Public Health* **2013**, *103*, e67–e72.

11. De Gennaro, G.; Dambruoso, P.R.; Loiotile, A.D.; di Gilio, A.; Giungato, P.; Tutino, M.; Marzocca, A.; Mazzone, A.; Palmisani, J.; Porcelli, F. Indoor air quality in schools. *Environ. Chem. Lett.* **2014**, *12*, 467–482.

12. Monn, C.; Fuchs, A.; Kogelschatz, D.; Wanner, H. Comparison of indoor and outdoor concentrations of PM_{10} and $PM_{2.5}$. *J. Aerosol Sci.* **1995**, *26*, S515–S516.

13. Tucker, W.G. An overview of $PM_{2.5}$ sources and control strategies. *Fuel Process. Technol.* **2000**, *65–66*, 379–392.

14. Verriele, M.; Schoemaecker, C.; Hanoune, B.; Leclerc, N.; Germain, S.; Gaudion, V.; Locoge, N. The MERMAID study: Indoor and outdoor average pollutant concentrations in ten low energy school buildings in France. *Indoor Air* **2015**,*11*.

15. Carteret, M.; Pauwels, J.F.; Hanoune, B. Emission factors of gaseous pollutants from recent kerosene space heaters and fuels available in France in 2010. *Indoor Air* **2012**, *22*, 299–308.

16. Wallace, L.; Wang, F.; Howard-Reed, C.; Persily, A. Contribution of gas and electric stoves to residential ultrafine particle concentrations between 2 and 64 nm: Size distributions and emission and coagulation rates. *Environ. Sci. Technol.* **2008**, *42*, 8641–8647.

17. Wallace, L.; Ott, W. Personal exposure to ultrafine particles. *J. Expo. Sci. Environ. Epidemiol.* **2011**, *21*, 20–30.

18. Glytsos, T.; Ondracek, J.; Dzumbova, L.; Kopanakisa, I.; Lazaridis, M. Characterization of particulate matter concentrations during controlled indoor activities. *Atmos. Environ.* **2010**, *44*, 1539–1549.

19. Wang, B.; Lee, S.C.; Ho, K.F.; Kang, Y.M. Characteristics of emissions of air pollutants from burning of incense in Temples, Hong Kong. *Sci. Total Environ.* **2007**, *377*, 52–60.

20. Lahiri, T.; Ray, M.R. Effects of indoor air pollution from biomass fuel use on women's health in India. In *Air Pollution*; Gujar, B.R., Molina, L.T., Ojha, C.S.P., Eds.; CRC Press: Boca Raton, FL, USA, 2010; pp. 135–163.

21. Tuckett, C.J.; Holmes, P.; Harrison, P.T.C. Airborne particles in the home. *J. Aerosol Sci.* **1998**, *29*, S293–S294.

22. Long, C.M.; Suh, H.H.; Koutrakis, P. Characterization of indoor particle sources using continuous mass and size monitors. *J. Air Waste Manag. Assoc.* **2000**, *50*, 1236–1250.

23. Morawska, L.; Zhang, J. Combustion sources of particles: Health relevance and source signatures. *Chemosphere* **2002**,*49*, 1045–1058.

24. Abt, E.; Suh, H.H.; Catalano, P.; Koutrakis, P. Relative contribution of outdoor and indoor particle sources to indoor concentrations. *Environ. Sci. Technol.* **2000**, *34*, 3579–3587.

25. Afshari, A.; Matson, U.; Ekberg, L.E. Characterization of indoor sources of fine and ultrafine particles: A study conducted in a full-scale chamber. *Indoor Air* **2005**, *15*, 141–150.

26. Brauer, M.; Hirtle, R.; Lang, B.; Ott, W. Assessment of indoor fine aerosol contributions from environmental tobacco smoke and cooking with reportable nephelometer. *J. Expo. Anal. Environ. Epidemiol.* **2000**, *10*, 136–144.

27. Bruce, N.; Perez-Padilla, R.; Albalak, R. Indoor air pollution in developing countries: A major environmental and public health challenge. *Bull. World Health Organ.* **2000**, *78*.

28. Smith, K.R.; Samet, J.M.; Romieu, I.; Bruce, N. Indoor air pollution in developing countries and acute lower respiratory infections in children. *Thorax* **2000**, *55*, 518–532.

29. Siddiqui, A.R.; Lee, K.; Bennett, D.; Yang, X.; Brown, K.H.; Bhutta, Z.A.; Gold, E.B. Indoor carbon monoxide and $PM_{2.5}$ concentrations by cooking fuels in Pakistan. *Indoor Air* **2009**, *19*, 75–82.

30. Zuk, M.; Rojas, L.; Blanco, S.; Serrano, P.; Cruz, J.; Angeles, F.; Tzintzun, G.; Armendariz, C.; Edwards, R.D.; Johnson, M.; *et al.* The impact of improved wood-burning stoves on fine particulate matter concentrations in rural Mexican homes. *J. Expo. Sci. Environ. Epidemiol.* **2007**, *17*, 224–232.

31. Albalak, R.; Bruce, N.; McCracken, J.P.; Smith, K.R.; de Gallardo, T. Indoor respirable particulate matter concentrations from an open fire, improved cookstove, and LPG/openfire combination in a rural Guatemalan community. *J. Environ. Sci. Technol.* **2001**, *35*, 2650–2655.

32. Albalak, R.; Keeler, G.J.; Frisancho, A.R.; Haber, M. Assessment of PM_{10} concentrations from domestic biomass fuel combustion in two rural Bolivian highland villages. *Environ. Sci. Technol.* **1999**, *33*, 2505–2509.

33. Bruce, N.; McCracken, J.; Albalak, R.; Schei, M.; Smith, K.R.; Lopez, V.; West, C. Impact of improved stoves, house construction and child location on levels of indoor air pollution exposure in young Guatemalan children. *J. Expo. Anal. Environ. Epidemiol.* **2004**, *14*, S26–S33.

34. Ardkapan, S.R.; Nielsen, P.V.; Afshari, A. Studying passive ultrafine particle dispersion in a room with a heat source.*Build. Environ.* **2014**, *71*, 1–6.

35. Logue, J.M.; Price, P.N.; Sherman, M.H.; Singer, B.C. A method to estimate the chronic health impact of air pollutants in U.S. residences. *Environ. Health Perspect.* **2012**, *120*, 216–222.

36. Penttinen, P.; Timonen, K.L.; Tiittanen, P.; Mirme, A.; Ruuskanen, J.; Pekkanen, J. Ultrafine particles in urban air and respiratory health among adult asthmatics. *Eur. Respir. J.* **2001**, *17*, 428–435.

37. Stephenson, D.; Seshadri, G.; Veranth, J.M. Workplace exposure to submicron particle mass and number concentrations from manual arc welding of carbon steel. *AIHA J.* **2003**, *64*, 516–521.

38. Ward, T.; Noonan, C. Results of residential $PM_{2.5}$ sampling program before and after woodstove changeout. *Indoor Air* **2008**, *18*, 408–415.

39. Noonan, C.W.; Navidi, W.; Sheppard, L.; Palmer, C.P.; Bergauff, M.; Hooper, K.; Ward, J. Residential indoor $PM_{2.5}$ in wood stove homes: Follow-up of the Libby changeout program. *Indoor Air* **2012**, *22*, 492–500.

40. McNamara, M.; Thornburg, J.; Semmens, E.; Ward, T.; Noonan, C. Coarse particulate matter and airborne endotoxin within wood stove homes. *Indoor Air* **2013**, *23*, 498–505.

41. Carvalho, L.; Wopienka, E.; Pointnera, C.; Lundgren, J.; Verm, V.K.; Haslinger, W.; Schmidl, C. Performance of a pellet boiler fired with agricultural fuels. *Appl. Energy* **2013**, *104*, 286–296.

42. Jalava, P.I.; Salonen, R.O.; Nuutinen, K.; Pennanen, A.S.; Happo, M.S.; Tissari, J.; Frey, A.; Hillamo, R.; Jokiniemi, J.; Hirvonen, M.R. Effect of combustion condition on cytotoxic and inflammatory activity of residential wood combustion particles. *Atmos. Environ.* **2010**, *44*, 1691–1698.

43. Bruno, P.; Caselli, M.; de Gennaro, G.; Tutino, M. Determination of polycyclic aromatic hydrocarbons (PAHs) in particulate matter collected with low volume samplers. *Talanta* **2007**, *72*, 1357–1361.

44. Viana, M.; López, J.M.; Querol, X.; Alastucy, A.; García-Gacio, D.; Blanco-Heras, G.; López-Mahía, P.; Piñeiro-Iglesias, M.; Saenz, M.J.; Sanz, F.; *et al.*

Tracers and impact of open burning of rice straw residues on PM in Eastern Spain.*Atmos. Environ.* **2008**, *42*, 1941–1957.

45. Huang, L.; Wang, K.; Yuan, C.S.; Wang, G. Study on the seasonal variation and source apportionment of PM_{10} in Harbin, China. *Aerosol Air Qual. Res.* **2010**, *10*, 86–93.

46. Hays, M.D.; Fine, P.M.; Geron, C.D.; Kleeman, M.J.; Gullet, B.K. Open burning of agricultural biomass: Physical and chemical properties of particle-phase emissions. *Atmos. Environ.* **2005**, *39*, 6747–6764.

47. Dambruoso, P.R.; de Gennaro, G.; di Gilio, A.; Tutino, M. The impact on infield biomass burning on PM levels and its chemical composition. *Environ. Sci. Pollut. Res.* **2014**, *21*, 13175–13185.

48. Nazaroff, W.W. Indoor particle dynamics. *Indoor Air* **2004**, *14*, 175–183.

49. Spilak, M.P.; Boor, B.E.; Novosela, A.; Corsi, R.L. Impact of bedding arrangements, pillows, and blankets on particle resuspension in the sleep microenvironment. *Build. Environ.* **2014**, *81*, 60–68.

50. Spilak, M.P.; Frederiksen, M.; Kolarik, B.; Gunnarsen, L. Exposure to ultrafine particles in relation to indoor events and dwelling characteristics. *Build. Environ.* **2014**, *74*, 65–74.]

51. Wallace, L.A.; Emmerich, S.J.; Howard-Reed, C. Source strengths of ultrafine and fine particles due to cooking with a gas stove. *Environ. Sci. Technol.* **2004**, *38*, 2304–2311.

52. Buonanno, G.; Morawska, L.; Stabile, L. Particle emission factors during cooking activities. *Atmos. Environ.* **2009**, *43*, 3235–3242.

53. Zhang, Q.; Gangupomu, R.H.; Ramirez, D.; Zhu, Y. Measurement of ultrafine particles and other air pollutants emitted by cooking activities. *Int. J. Environ. Res. Public Health* **2010**, *7*, 1744–1759.

CHAPTER 4

Low-Cost Sensors Calibration for Monitoring Air Quality in the Federal District—Brazil

Erick Frederico Kill Aguiar, Henrique Llacer Roig, Luís Henrique Mancini, Eduardo Neiva Caetano Botelho de Carvalho

Institute of Geosciences (IG), University of Brasilia, Brasilia, Brazil

ABSTRACT

Critical situations that cannot be solved by conventional approaches (traditional air quality monitoring networks), have the possibility of being managed quickly by a wide network of portable systems with sensors. The purpose of this research was to calibrate and validate low-cost sensors. Pilot indoor and outdoor areas, in the central area of Brasilia (Brazil's capital city) were chosen for corporative performance evaluation of the sensors. The CO at 99.999% volumetric injection method has been used in a gas test box, among two MiCS-5521 (CO/VOC) sensors, one being new and the other one with a short useful life. The number of injections adopted to each volume (from 1 ml to 6 ml) was 10, rising each sensor's confidence interval mean. A increase of the injected volume (ml) of CO resulted in significant decrease in a resistance (Ohms), as shown by a good inverse relationship on the interaction of these two variables (r = 0.88), with good measurement accuracy, when compared to the manufacturer's reference datasheet. Finally, a geospatial management system was built for the pollution data measured by the low-cost sensors.

KEYWORDS

Low-Cost Sensors, Monitoring Air Quality, Calibration of Micro-Sensors MiCS-5521 (CO/VOC), WebGIS

1. INTRODUCTION

The atmospheric urban pollution is a major concern in modern cities, especially in developing countries [1] [2] , where pollutants affect directly human health and cause various respiratory and cardiovascular diseases, when there is long term exposure to pollution [3] -[5] . The World Health Organization (WHO) reported that high concentrations of gases and particulates were the cause of 223,000 lung cancer deaths around the world in 2010 [6] . In the last decades, many studies demonstrated positive associations between air pollution and mortality [7] -[9] .

In this context, information about pollutant emissions released in urban areas is very important to public health policies for human health and environmental protection [2] [4] . Currently, monitoring is done by static measuring stations (subsequently called base stations) that are operated by official authorities, such as governmental environment organs. This monitoring has high reliability in terms of data generation and it is capable of measuring, with precision, a wide variety of atmospheric pollutants using traditional analytic instruments, like mass spectrophotometers and gas chromatograph. However, the disadvantages of such measuring methodology are its complexity and high maintenance cost [10] [11] , for this reason it is not available in many urban centers [8] .

Two basic limitations exist in the approach used to control and publish air quality data: First, the spatial resolution sampling is low, making it necessary to use mathematical models to estimate the concentrations of pollutants in not monitored metropolitan areas. Second, pollutants concentration observations do not reflect actual exposure suffered by people, due to spatial heterogeneity in pollutant concentrations and the individual mobility patterns [12] .

Measuring protocols and monitoring sensors are extremely new and much research is still needed in order to integrate these technologies and improve environmental information systems. An important point to improve air quality monitoring is sharing of environmental data gathered from different sources (public and private companies), into a real-time system, in order to merge data from different sensor networks [10] [12] [13] .

In Sydney, Australia [12], the Project "HazeWatch" involved citizens participation in the management of the pollution they are exposed to, utilizing customized tools, such as micro systems controlled by low-cost sensors, to generate real-time information. The research achieved satisfactory results regarding the understanding of urban air quality, as information about exposure to determined type of pollutant during one's quotidian is presented to the user of the system, characterizing it as a mean of increasing environmental awareness.

In Brazil there are a total of 5570 cities, but only 1.7% of them have an air pollution monitoring network. Nationally, there are 252 monitoring stations, but not every station monitors all important pollutants [14] . The city which is best monitored in Brazil is São Paulo [15] [16] . This Metropolitan area has a traditional monitoring network (stationary and mobile stations), creating reports on regional situations, which can be accessed on the web, through the institutions portals. The advantage of this type of system is the quality level of the information that can be correlated with several variables, allowing more precise environmental modeling.

On the other hand, regions such as the Federal District (Brazilian Capital) have an inefficient network regarding data generation and monitoring of the local air quality [17] -[19] , which opens the opportunity of investments on alternative mobile portable monitoring networks, diminishing the costs invested by the government [20] .

Thus the research fits the perspective of generating distributed monitoring systems, if possible with the participation of collaborative networks.

2. MATERIALS AND METHODS

In order to acquire the response to the resistance signals of the micro controlled plate on the experiment, the software used was from MiCS-EK1 kit, manufacturer E2V [21] (Figure 1(a)), that shows the reading of two metal oxide sensor slots, with results in resistance units (KOhms or Ohms). Other two fields make the working temperature of the plate and the humidity reading (factory calibrated sensors). With that, a data logging is generated, which can be defined in the time scale desired by the user. The final data are saved in a CSV (.csv) file.

At the sensors calibration stage, an area from the Geochronology laboratory was used, as well as a CO gas cylinder, the pressure regulator, syringes for the volumetric injections, needles and digital thermometer, detailed below.

The Carbon Monoxide cylinder with volume capacity of 8.5 m^3 in the laboratory, possesses concentration of 99.999% - 5.0 analytical. The removal of CO from the cylinder was made through the outlet of the pressure regulator, where the duct outlet had a silicone cap, which would open the cap to release the gas and then use the needle on the syringe to remove the volume to be used in each measurement.

For the CO injection, there were used glass hypodermic syringes with 3 ml, 5 ml and 10 ml volumes and needles measuring 16 G 1.5 (1.60 mm × 40 mm).

The SR # 3 (Figure 1(b)) is a gas testing box, made by the Japanese industry Figaro [22] , with the specifications Power Source: 110X A.C 50/60 Hz; Consumption: 10 W; Dimensions: 235 mm × 180 mm × 210 mm; Effective volume: 5.4 L; and Material: Acrylic.

The sensor used was MiCS-5521 from E2V [21] , which is indicated for the detection of gases such as carbon monoxide (CO), hydrocarbons (HC) e volatile organic compounds (VOC). Sensor 1 was never used and sensor 2 was used for a year and a half. According to the producer, the sensor has a two year useful life-time. An expe- rimental micro controlled system (Figure 2(a), Figure 2(b)), was used, developed by Geosignals company, with the intent of collecting on field data, with information pairing via bluetooth to mobile platforms (smartphones) and sample collected dispatch through mobile or wireless networks. The prototype is ready, however it needs ca- libration tests with the sensors and adjustment of the algorithm with the data resulting from this research analysis. It utilizes reading system similar to the one used by the plate on MiCS-EK1 kit, performing the sensor's data reception on resistance (Ohms) form and , with that, being able to apply mathematical modelings in order to obtain ppm form data. The prototype also has the function of reading other metal oxide sensors for environment va- riables monitoring (humidity, temperature and atmospheric pressure) and gasses (CO_2, SO_2, NO_2, O_3) [23] [24] .

The calibration procedure was performed at Universidade de Brasília's Geocronology Lab, linked to the Geosciences Institute—IG/UNB on the days (DD/MM/YYYY) 16/05/2014, 20/05/2014, 23/05/2014 and 10/06/ 2014, divided in five stages, described as:

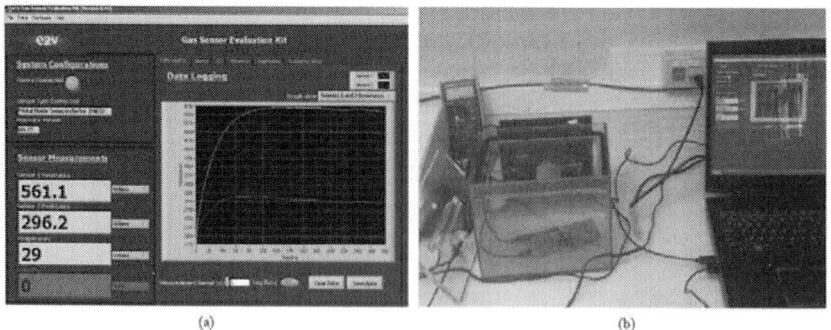

(a)   (b)

Figure 1. Kit MiCS-EK1 used to calibration the sensor. (a) Software Interface with readings of the slots; (b) Box of gas test (SR # 3) with the kit MiCS-EK1 connected to evaluation software.

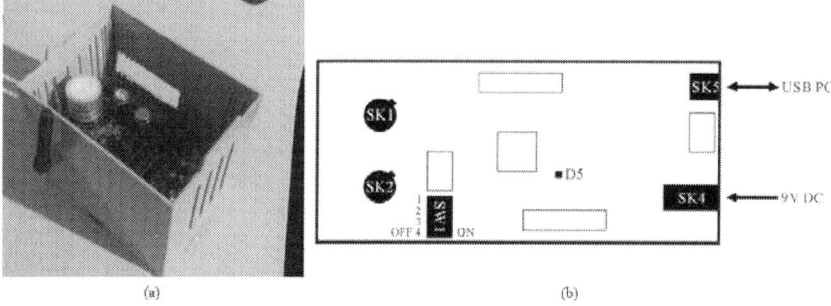

(a)   (b)

Figure 2. Prototype micro-controlled system. (a) For analysis of air quality; (b) Basic inteface kit MiCS-EK1.

- Baseline;
- Controlled Volume Measuring;
- CO (Carbon Monoxide) Concentrations Conversion;
- Calibration Curve Generation;
- Simple Linear Regression Analysis;
- Monitoring Test

To achieve the response time, both MiCS-5521 have been exposed to ambient air in the laboratory, performing readings during the period of continuous hours, with readings every second. The result was a data logging showing the stabilization time of the sensor's stabilization. For the use of the SR # 3 box, the method specified by the manufacturer was used, keeping the stall opened in a clean environment and turning on the mixer for 2 - 3 minutes to guarantee that all contaminants had been removed from the box. After that, the box was closed with its lid. Subsequently, the syringe was filled with the volume of CO extracted from an output shaft of the gas tank valve. Thus, CO was injected in the box through a silicon septum. After that, the mixer (fan) was turned on for 30 seconds, 30 seconds of wait before the output reading observation. The box had its lid removed so it could return to the 2 - 3 minutes cleaning cycle. The CO volumes utilized on the calibration were 1 ml, 2 ml, 3 ml, 4 ml, 5 ml and 6 ml. To each volume (ml), the calibration was repeated ten times, in order to obtain a mean value for each volumetry.

Volume fraction or molar fraction unities are frequently utilized for gas concentration. In the analysis, the box's volume was added to each ml injected. The most utilized fraction of value is the ppm (parts per million in volume), defined by Equation (1) [23] :

$$ppm_v = \frac{v_i}{v_{total}} \times 10^6$$

(1)

where v_i is the gas volume and v_{total} air volume. The conversion from ppm to mg/m^3 [23] , is described by Equation (2):

$$mg/m^3 = \frac{ppm_v \times M}{24.45}$$

(2)

where ppm_v is the mol value of the solution, M the molecular mass of the air pollutant and the value 24.25 is a conversion factor that represents a mol of gas' volume.

The conversion equations depend on the temperature the conversion is desired (usually around 20°C to 25°C). At a 1 atm atmospheric pressure (101,325 KPa or 1.01325 bar), and the general Equation (3) [24] :

$$ppm_v = \left(mg/m^3\right) \times \frac{\left(273.15 + °C\right)}{\left(12.187\right) \times \left(M\right)}$$

(3)

where mg/m^3 is the amount of milligrams of the pollutant per cubic meter of the ambient, ppm_v the air pollutant concentration, as in parts per million per volume (the volume of pollutant gas per 10^6 ambient air volumes), °C the ambient air temperature in Celsius degrees, 12.187 the value of universal gas constant, 273.15 is the T_0 in Kelvin and M the pollutant's molecular mass.

2.1. Analysis of Linear Regression

The regression and the correlation are procedures utilized to estimate relations between variables that may exist in a certain population. The analysis of correlation and regression is done by studying sample data in order to understand if and how two or more variables are related one to the other in population. A regression model establishes a cause and effect relationship between two or more variables [24] . To estimate the expected value, a model is used to determinate the relationship between both variables by Equation (4):

$$Y_i = \beta_0 + \beta_1 X_i + \varepsilon_i$$

(4)

Y_i is an explanatory variable (dependent); is the value you want to achieve, β_0 is a constant, that represents a straight line intersection with the vertical axis, β_1 is another constant, that represents the straight line slope, X_i the explanatory variable (independent), represents the explanatory factor in the equation, ε_i the term that includes all residual factors more the possible measurement errors.

2.2. Monitoring Test

Day 06/06/2014, Brasilia's Bus Station was used as an outdoor reference point for measuring. Through it, the plugin Hawths Tools was generated, in the Arc GIS Info 10.2.2 ambient [25] , a regular grid (grid) 20 m × 20 m (Sampling Tools, in "creator vector grid" algorithm) (Figure 3(b)) and random samples ("generate random points"), with the intent of distributing peripheral points for the collection of air quality data. The entry to the random sampling was eight measurement points. The outdoor samples were performed with 5 minutes time for sample collection in each point.

Later, on 13/06/2014, there were made outdoor collections in the vincinity of Darcy Riberiro Campus, from Brasilia's University, on the main peripheral pathways of "minhocão" (Figure 3(c)). Another part of the outdoor collections were made during the World Cup (Switzerland × Equator), with the raise of 4 points on the event's surrounding area, which happened on 15/06/2014 at 1 pm, at Brasilia's National Stadium (Figure 3(d)). In order to collect indoor data, the University of Brasilia (UnB) was selected, located on Darcy Ribeiro Campus. There were divided into two measuring environments: ground floor and garage. The measurements made in ICC South Wing, ICC Central Wing and ICC North Wing. The indoor samples were performed on 13/06/2014, with 5 minutes time for sample collection in each point (Figure 4).

2.3. Data Availability

Once the air quality data gathering was over, the analysis spatialization into a dynamic digital map (WebGIS). Initially, a geodatabase was created using the indoor and outdoor analysis, in a vector format (point), with the fields described below (Table 1).

The publishing of the maps service on the tool ArcGIS Viewer of Flex [25] , makes it easier to configure the final layout for the web application (Figure 5). The layers of base maps were inserted, comparative graphics tool, interactive design, identifier, subtitles, layer list, searcher, data printing and changing color themes. This set of interactive tools allows graphic analysis and dynamic queries, with the possibility of new entries of vector variables for future correlations with the data analyzed. Queries can be exported from the system in various formats (.xls, .doc, .pdf, etc.).

3. RESULTS AND DISCUSSION

3.1. Baseline

The baseline shows similar behavior from both sensors still, their resistance values were different. That variation may be linked to a wear over time on the material used on the resistance, which may also can undergo interferences due to impurities on it, decreasing its reading capability. The Geocronology laboratory' ambient showed no major variation on its CO concentration, for it is a closed room. There is the temperature control (25°C) and a minimum variation on the humidity, that remained in the range of 50% (recommended humidity for the datasheet calculation of a R_S/R_0) [26] .

An important fact on the process of the baseline construction is that after 9 hours collecting data (Figure 6) (Table 2) in the laboratory's ambient, both sensors had their readings stabilized and when a sudden change occurred, and then a new stabilization at a higher level. This event may have been caused by a crossed sensibility that happened after the laboratory's glassware was cleaned with alcohol, in a closed ambient. This interference had been observed previously in other studies. Alcohol and humidity sensors are necessary when a micro controlled system that uses those kinds of machinery to calculate the interferences of both varieties. As a consequence, it is indicated that a controlled humidity environment is created for the system's module [21] [22] .

Table 1. Description of fields defined in the table of attributes of the analyses.

Type	Data	Fields	Description	Environment	Date of acquisition
Point	Location	Date (Datetime); Hour (Datetime); Temperature (String); Humidity (String); Sensor_1_R (String); PPM_Senso (Double); Sensor_2_R (String); PPM_Sen_1 (Double); Latitude (Double); Longitude (Double)	Local description	Outdoor or indoor	Day/Month/Year

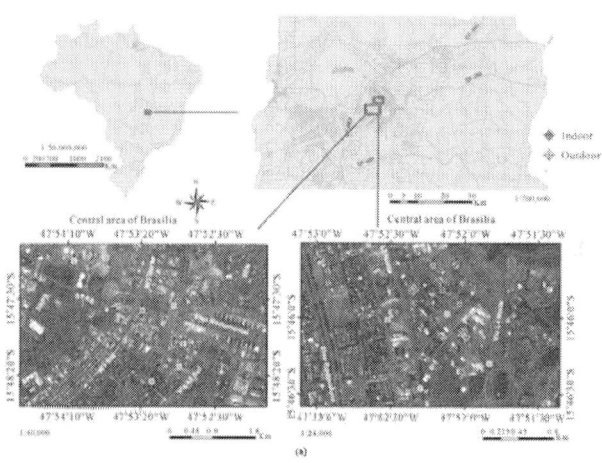

(a)

(b)

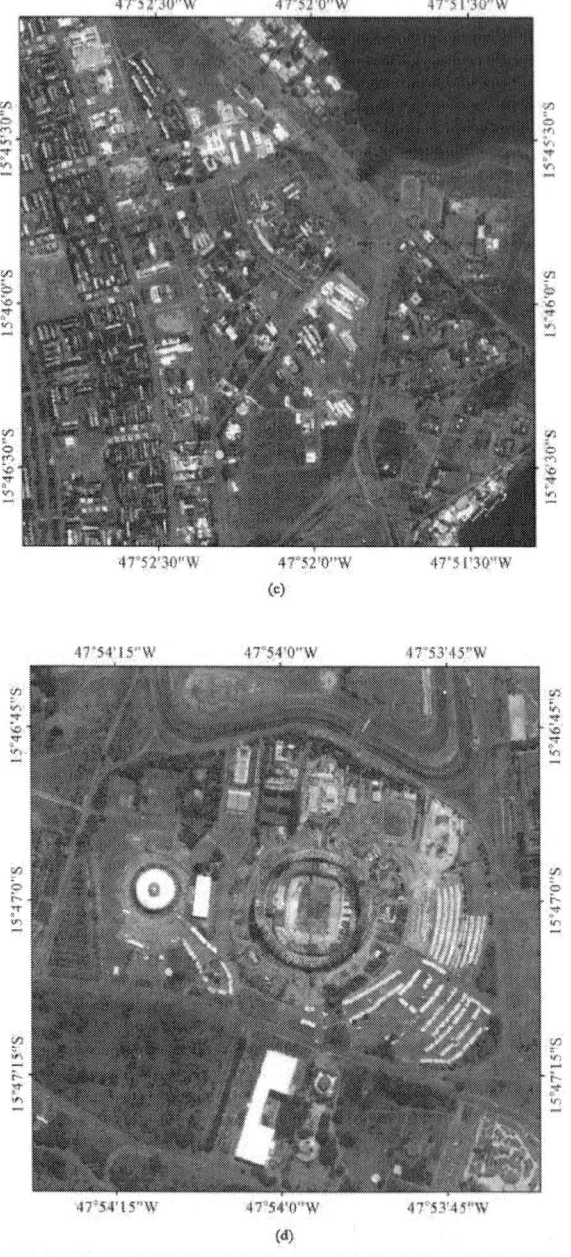

Figure 3. Map with points of outdoor and indoor collections in: (a) Map with the spot selected to available the system; (b) The hot spots in the central Brasilia; (c) Campus Darcy Ribeiro, the University of Brasilia; (d) Brasilia National Stadium.

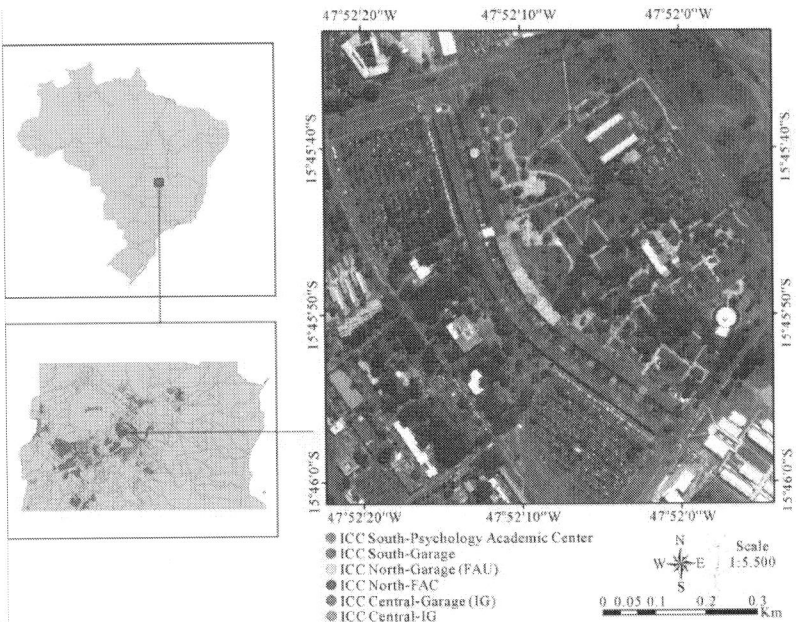

Figure 4. Map with points of indoor collections in University of Brasilia.

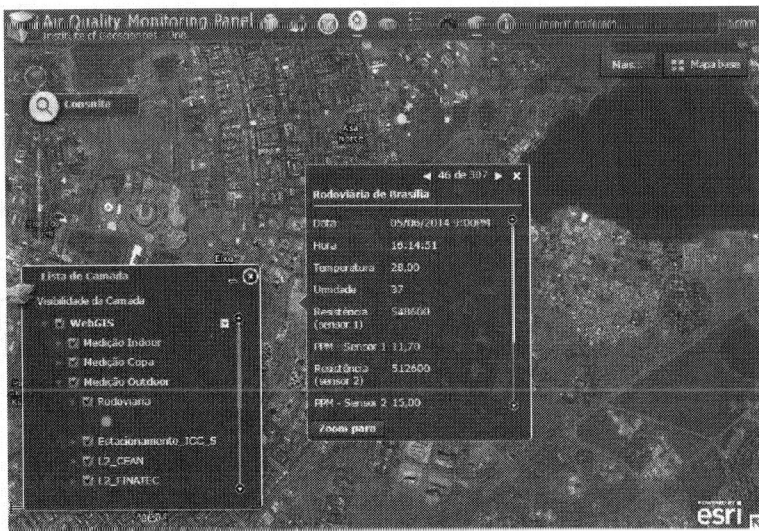

Figure 5. WebGIS system screen with the information of the outdoor and indoor measurements.

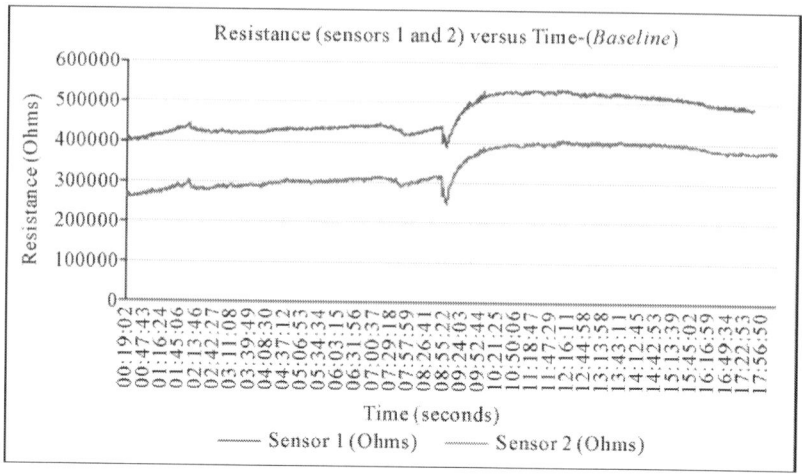

Figure 6. Baseline resistance (Ohms) obtained over approximately 18 hours.

Table 2. The mean baseline for sensors 1 and 2 (R_0).

Sensor 1 (Ohms)	Sensor 2 (Ohms)
468,958	341,631

3.2. Responses to CO (ml) Injections

This procedure was carried out using the test box Figaro (Figure 1) acrylic material with a total volume of 5.4 liters being confined for the CO (ml) injections. Two sensors were used, one being unused (sensor 1) and the other with about a year and a half of use (sensor 2). The relation between enclosed volume the (5.4 liters) and the concentration ppm$_v$ of CO box, calculated according to (Figure 7). Metal oxide sensors demonstrate a response to the presence of CO gas from concentrations above 10 ppm. According to the sensors manucturer's datasheet [21] , the detection limit is 1000 ppm. Using such information and the calculation of the volumetric data (Figure 8(a),Figure 8(b)) the box's volume reaches this concentration at 6 ml.

It can be seen that upon injection of 6 ml (Figure 8(a), Figure 8(b)), the resistance decreases to the point wherein the metal oxide material becomes unable to detect the target gas, having the resistance in its lowest reading.

The higher the temperature reached by the resistance (700°C to 900°C), the higher the sensitivity and selectivity of the sensor at low concentrations of CO. A greater amount of energy must be provided to the controlled micro system, so it can reach that temperature [21]. It can be seen that the sensor behaves in an inverse relationship. The larger the target gas volume, the smaller are the resistance values (Figure 9). From these data we carried out a basic statistical analysis in order to verify data quality. The mean (10 injections) was calculated, the standard deviation and the variation coefficient of resistance (Ohms) of the sensors for each volume (ml) injected by Equation (5).

$$CV = \frac{\sigma(x)}{\overline{x}} \times 100$$

(5)

where $\sigma(x)$ is the standard deviation and $\overline{x}$ is mean.

The coefficient of variation values are low for all injections (ml), indicating that for both sensors, the data dispersion from the average is small, i.e., the dispersion is relative low and the results can be considered good (Table 3, Table 4).

The result of the correlation coefficient (r) appeared around 0.96 among the resistance achieved in readings between both sensors (Table 5). The sum of the products, covariance of the population and sample covariance did not indicate significant variations, demonstrating that the resistance kept close variations, of the total of injections (ml). However, the sensor 2 presented significant differences of values of resistance (Ohms) read, when compared to the sensor 1 and may be related to the interference mentioned previously. This fact must be taken into consideration and the calibration curve should be reviewed, as there is signal degradation and consequently a smaller concentration reading.

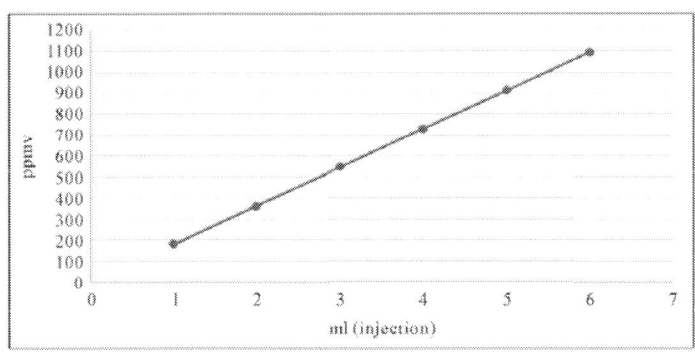

Figure 7. Relationship between the ppm$_v$ on a volume of the box by volumetry of the injections.

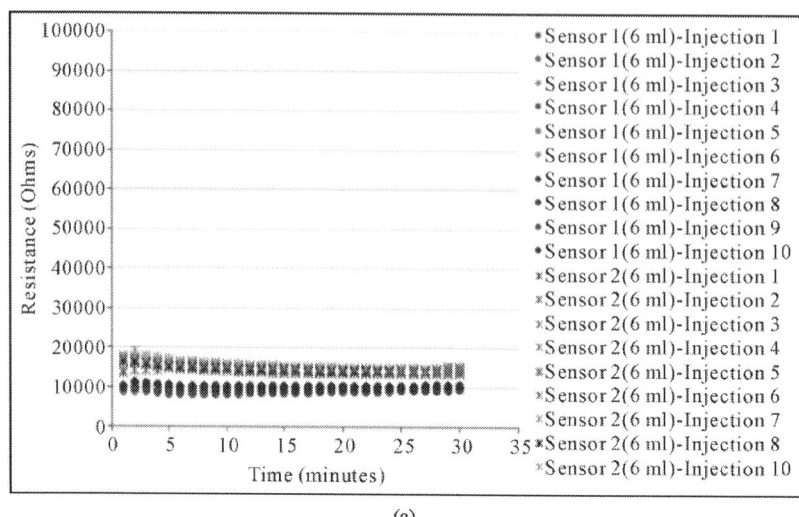

(a)

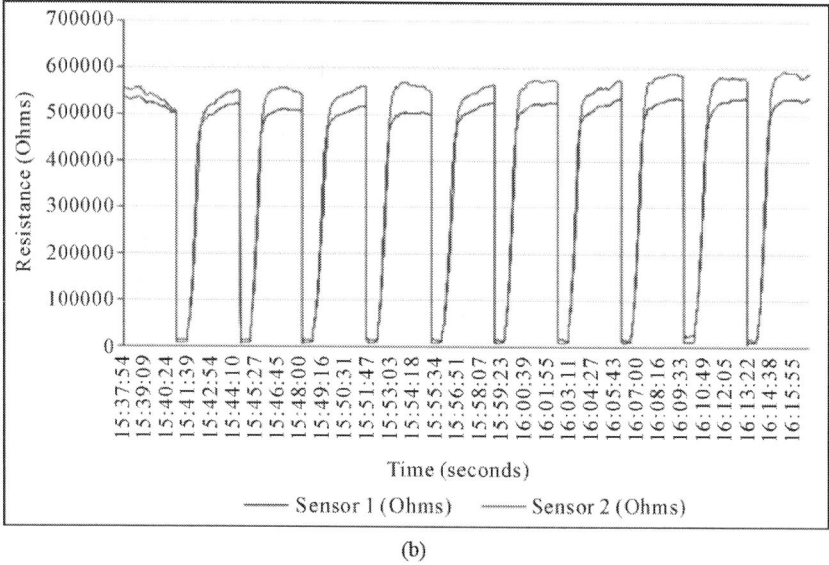

(b)

Figure 8. (a) Response of the resistance (Ohms) of the sensors 1 and 2 to 6 ml injection (10 repetitions) of CO (30 seconds); (b) Response of the resistance R_s (Ohms) the 10 repetitions of injections (6 ml CO).

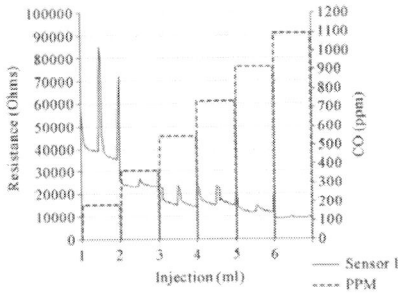 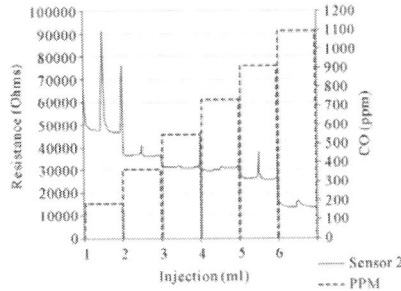

Figure 9. Response of resistance R_s (Ohms) in 2 moments of each injection (ml), demonstrating the inverse relation with the concentration. Sensor 1—left graphic; Sensor 2—right graphic.

Table 3. Mean, standard deviation and coefficient of variation (%) of resistance (Ohms) injected volume (ml)—sensor 1.

Sensor 1 (Ohms)	Mean	Standard deviation	Coefficient of variation (%)
1 ml	41,833	1341	3.20
2 ml	25,910	622	2.40
3 ml	23,894	481	2.01
4 ml	15,978	130	0.81
5 ml	13,366	410	3.07
6 ml	9350	139	1.49

Table 4. Mean, standard deviation and coefficient of variation (%) of resistance (Ohms) injected volume (ml)—sensor 2.

Sensor 2 (Ohms)	Mean	Standard deviation	Coefficient of variation (%)
1 ml	51,068	605	1.19
2 ml	35,879	752	2.10
3 ml	31,424	434	1.38
4 ml	30,766	602	1.96
5 ml	17,553	415	2.37
6 ml	14,271	162	1.14

Table 5. Comparison between the means of the resistance (Ohms) of the injections (1 ml to 6 ml) of the sensor 1 and sensor 2, their covariances and the correlation coefficient.

	Sensor 1	Sensor 2
Mean (Resistance)	21,722	30,160
Sum of products	−103,981	−119,810
Covariance of the population	−17,330	−19,968
Sample covariance	−20,796	−23,962
Sample covariance	21,722	30,160
Correlation (Resistance (Ohms))	$r = 0.96$	

3.3. Calibration by the Linear Regression Method

For this study, two variables were correlated, the resistance (Ohms) and the volume of injections (ml). There is a cause-effect relationship observed in the procedure, when there is injection (ml) in the box, the resistance (Ohms) tends to fall. Good correlation between the resistance reading data per injected volume to the sensor 1 ($r = 0.88$) also found for the sensor 2 ($r = 0.89$) (Figure 10).

Around 88% (sensor 1) and 89% (sensor 2) the variability of the resistance (Ohms) can be explained by variability in CO concentration (ppm). The remainder (12% to 11%) can not be explained by other factors present, such as cross-sensitivity of interference with other gases, lifetime reduction, foreign particles in the layer of the material that makes up the resistance, which are factors causing significant changes in the readings. The straight line's behavior for both sensors indicate that the higher the concentration (CO), the lower the resistance (Ohms). It was observed that after the box volume (5.4 liters) was mixed with 6 ml, equivalent to 1096 ppm$_v$, exceeding the limit of detection for this sensor, the response sensitivity of the resistance is reduced with particles dispersing in the microenvironment. By analyzing the CO sensor data with the MiCS-5521 from another study, it was observed, comparing the metal oxide sensors, that some have a small influence of temperature, while others tend to have a great influence and may have a positive or negative correlation with this variable. Furthermore, the influence of the sensor's temperature may change over time with use and seasonal changes. Also mentions that the results of measurements of the sensors MiCS-5521 sensors next to the station a traditional station, differ significantly from the results obtained in the laboratory. For the same sensor model, the author found very strong correlation ($r = 0.99$) when comparing the resistance reading by volume of gas (CO) in the laboratory [27] - [29] .

3.4. Results of Indoor and Outdoor Analysis

The outdoor analyzes sought to demonstrate the local situations of interactions with CO readings. The objective was to assess behaviors adverse to those obtained in the laboratory, in environments that have heterogeneous characteristics. Such a proposal is given in order to take readings with similar changes for the two sensors. The monitoring of air quality in outdoor environment involves preparation for the micro-controlled system, in order to avoid interference in reading. The intention of the field analysis for this study is to understand the interference factors in the resistance material from MiCS-5521 sensor. With that, auxiliate in future work with this sensor model [24] . One of the points assessed was the square of the SQS 202 (Figure 11(a)), in which the sensors demonstrated very close responses in resistance readings. Being a region with residential character, with reduced car traffic in the analyzed time, the resistance has not oscillated considerably.

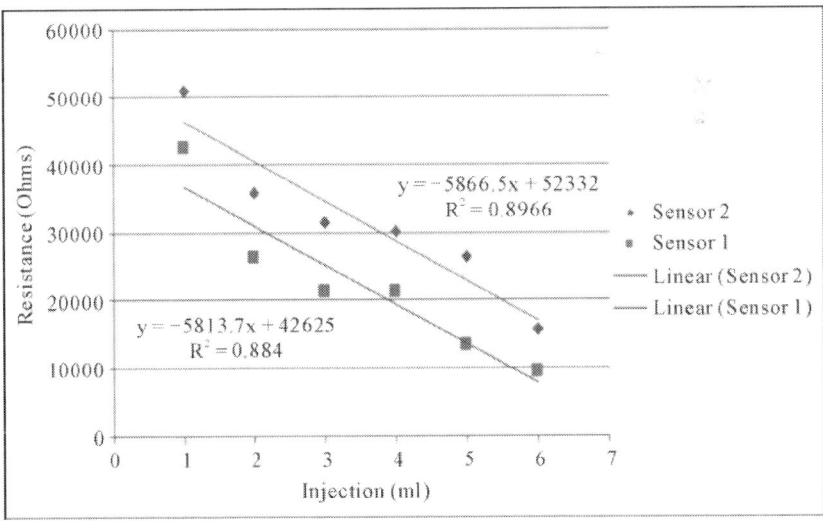

Figure 10. Coefficient of determination of resistance (Ohms) in relation to injection (ml) of CO.

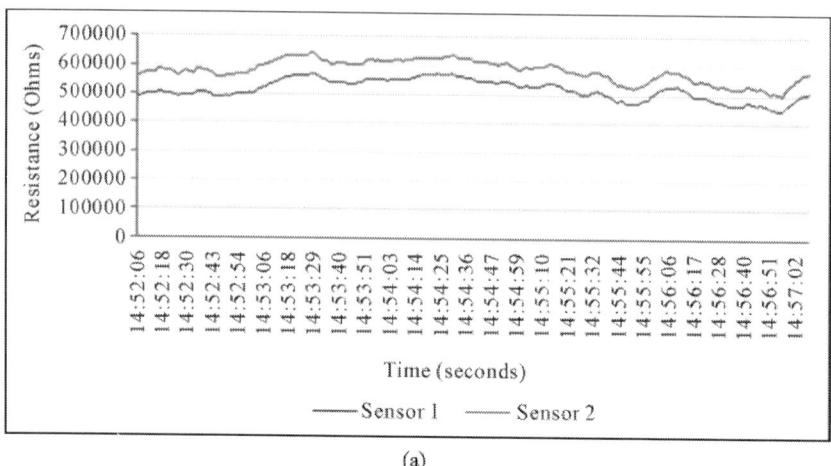

(a)

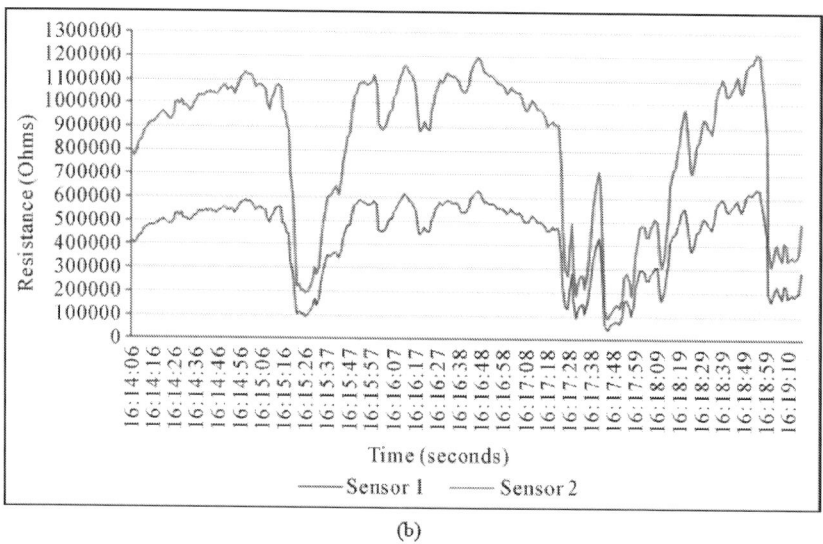

(b)

Figure 11. Outdoor sampling: (a) In the square of the SQS 202; (b) Bus central station of Brasilia.

Another rated point was the main bus station of Brasilia, which is a critical point of emissions in central Brasilia. At times, the values of fluctuations in the resistance of the sensors were almost four times higher than that presented in other regions analyzed. The fleet circulating in this region is diverse, with large cars (public bus), and be located in the road axis (Figure 11(b)).

The variations found in the ICC South Wing, near the Psychology Academic Center and garage, did not show large fluctuations in the resistance of the sensors, analyzing it in the garage, the reading has remained largely stable and in the vicinity of the Academic Center, showing decay in reading. However, when there is movement of vehicles on the internal via of the garage that can be subject to alteration (Figure 12).

The calibration of the sensors (1 and 2) indicated a strong correlation between the resistance (Ohms) and the injection of volumetric (ml), with values r exceeding 0.8. The use of these sensors in micro-controlled systems for monitoring air quality, the use of the generated calibration equations is needed. Its application is given directly to the reading software, where the data processing algorithms (readings) are stored. However, the results of the present study were similar to the results of other studies, which had the low-cost sensors (MiCS-5521) calibration as its purpose [27] -[29] .

3.5. WebGIS Panel Results

The results analyzed in indoor and outdoor surveys were georeferenced in order to make the data available in an online panel format (WebGIS). In this sense, tools (widgets) have been configured for interaction with the available information. A consultation widget was implemented to allow end users to query information by executing a predefined query. At the end-user level, query execution is simple and is performed with a single button click, working in a single layer for consultation (Figure 13(a)). For graphical analysis of the variations of the resistance (Ohms) and time, was implemented graphic widget, so that comparisons have been collected at each point (Figure 13(b)).

4. CONCLUSIONS

The calibration of the sensors (1 and 2) showed strong correlation between the resistance (Ohms) and the injection (ml) volumetric, the r values exceeding 0.8. The use of these sensors in micro-controlled systems for monitoring air quality, the use of the generated calibration equations is needed. Its application occurs directly in the reading software, which stored input data processing algorithms (reads). For the sensor 1 have the following Equation (5) with r = 0.88.

$$y = -3634.9x + 29491$$

(5)

For the sensor 2 (r = 0.89), Equation (6).

$$y = -5866.5x + 52333$$

(6)

However, the results of the present study were similar to the results of other studies, which had the goal of calibration low-cost sensors (MiCS-5521) [21] . Sensors MiCS-5525 are for obtaining CO reading data. To do this, we conducted a linear regression analysis of sensed data generated by sensor MiCS-5521. Based on the results of linear regression, a calibration equation was created, used to correct the readings of sensor MiCS-5525 from the sensor MiCS-5521, which showed a strong correlation (r = 0.85) [22] .

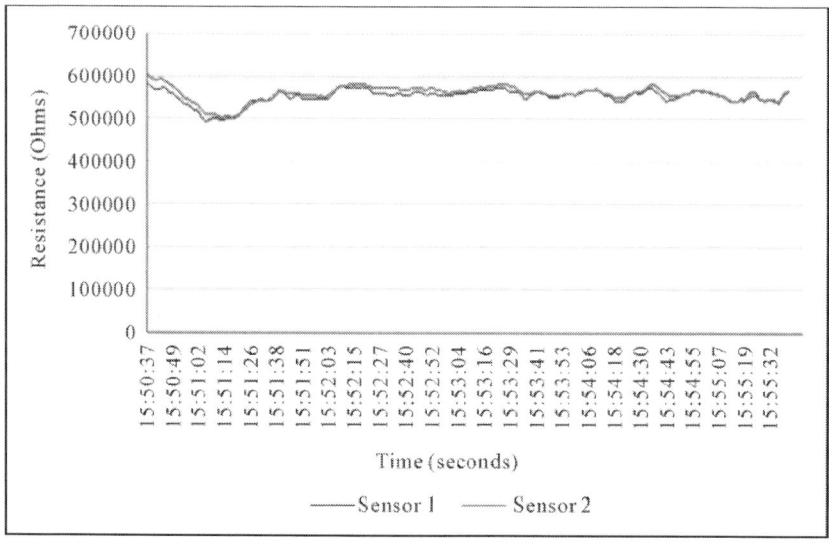

Figure 12. Indoor sampling in Academic Center of Psychology (CA).

Thus, with some restrictions, the presented hypothesis is confirmed—there is a strong correlation (r) in the volumetric samples taken for the sensors. However, the restriction to confirm the hypothesis made on the calibration test, which proved to be the sensor 1 is within the range (datasheet) as established by the manufacturer of the oscillating resistance in CO (ml) concentration injected. The sensor 1 is new, use of wear-free. Therefore, it is emphasized that compared the responses analyzed, the hypothesis was confirmed, and the 6 ml volumes larger than the sensor 1 responds with the decay close to the zero resistance (Ohms), indicating that it has more CO concentration readings in these ranges. Thus, it can be said that both sensors have the potential to be used in emission measurements generated in urban traffic, as alternative equipment monitoring air quality, adjusted sensor 2 depending on the sensor 1 accuracy. However, they are not suitable for use as an air quality autonomous sensor due to cross-sensitivity problems. When combined with other sensors in a multisensor system can eliminate the interference. Another detail that makes it feasible for a monitoring system is its low power consumption (less than 100 mW).

The results of the analyses pointed outdoor critical areas, with high variations in resistance of the sensors, such as the Central Bus station of Brasilia and the Commercial Sector South.

It is emphasized that the possibility of comparison with traditional monitoring stations of air quality that have CO sensors (reference), the data can be validated with the precision found in these detectors. It is emphasized that continuous tests in indoor analysis check, over time, the sources that may come to contribute to CO in the ICC environment. Analyses did not detect any source that contributed significantly to a considerable variation in the concentration of CO.

The WebGIS system (Panel) was presented as a suitable platform for the provision of data collected in the environments mentioned above. It demonstrated the dynamic configuration capabilities in question tools (widgets) customization. It showed that WebGIS has eased in upgrading with the addition of new data in an automatic manner, and can connect from a preconfigured database with the application, or from the update a project built on ArcGIS Desktop environment, with connection to ArcGIS Server. Numerous configuration possibilities for displaying the same platform indicate that the created panel can suit any setting that may be redesigned, from new needs.

The research presented few limitations, such as not monitoring environment variables, which can cause cross- sensitivity, contained in lab space. It takes a field calibration with the reference stations, for comparison with the procedure performed in a controlled environment. If calibration is important to a larger number of sensor units MiCS-5521, reaching a result with a larger universe sample and then data with greater confidence. Finally, assessments in different settings of indoor and outdoor environments have the CO behavior panorama, in the diversity of these situations.

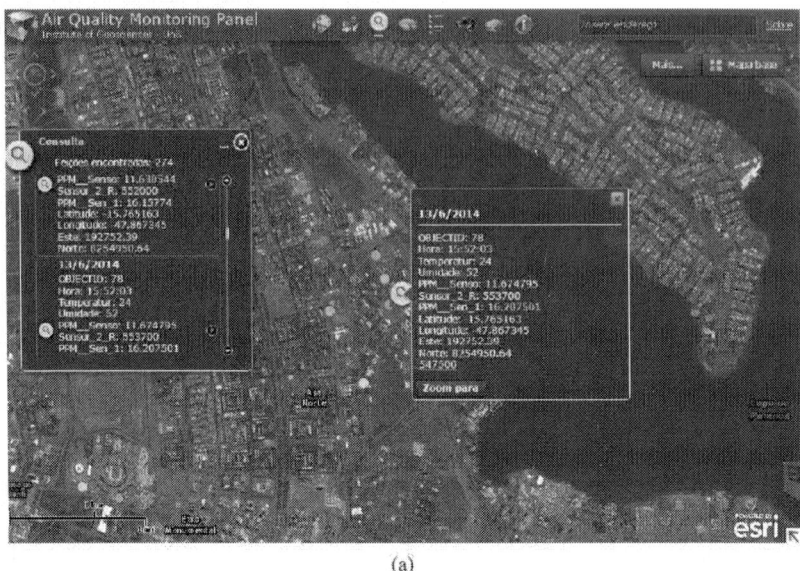

(a)

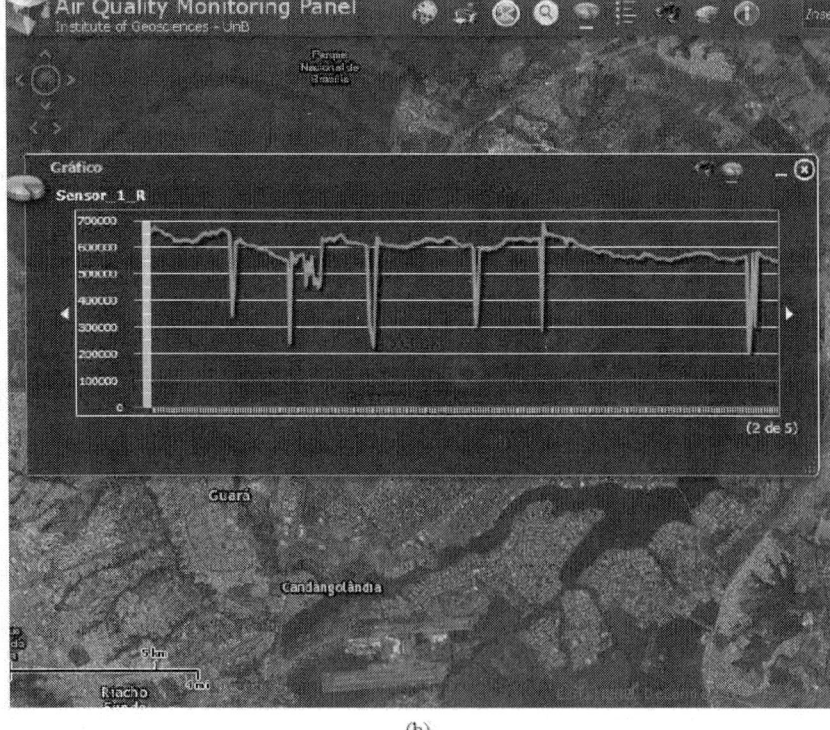

(b)

Figure 13. (a) WebGIS interface panel screen with the query tool; (b) Panel screen to the graphic tool show the concentration CO variation (line type).

ACKNOWLEDGEMENTS

The CAPES for the scholarship to the first author during the development his master. The Laboratory of Remote Sensing and Spatial Analysis (LSRAE) and the Geocronology for the technical support to this research. ESRI for providing the package of tools that make up the through ArcGIS 10 family of the contract number 2011 MLK 8733 and IMAGEM for the support and feasibility of establishing the terms of use and between the IG and the support of ESRI software.

REFERENCES

1. Cohen, A.J., Anderson, H.R., Ostra, B., Pandey, K.D., Krzyzanowski, M., Kunzli, N., Guschmidt, K., Pope, A., Romieu, I., Samet, J.M. and Smith, K. (2005) The Global Burden of Disease Due to Outdoor Air Pollution. Journal of Toxicology and Environmental Health, Part A, 68, 1-7.

2. Thepanondh, S. and Toruska, W. (2011) Proximity Analysis of Air Pollution Exposure and Its Potential Risk. Journal of Environmental Monitoring, 13, 1264-1270.

3. Gorai, A.K., Francis Tuluri, F. and Tchounwou, P.B. (2014) A GIS Based Approach for Assessing the Association between Air Pollution and Asthma in New York State, USA. International Journal of Environmental Research and Public Health, 11, 4845-4869

4. ATSDR (2014) Agency for Toxic Substances and Disease Registry. Air Pollution. Atlanta, United States of America.

5. Nandasena, S., Wickremasinghe, A.R. and Sathiakumar, N. (2012) Respiratory Health Status of Children from Two Different Air Pollution Exposure Settings of Sri Lanka: A Cross-Sectional Study. American Journal of Industrial Medicine.

6. WHO (2013) Outdoor Air Pollution a Leading Environmental Cause of Cancer Deaths.http://www.iarc.fr/en/media-centre/iarcnews/pdf/pr221_E.pdf

7. Levy, J.I., Hammitt, J.K. and Spengler, J.D. (2000) Estimating the Mortality Impacts of Particulate Matter: What Can Be Learned from between Study Variability? Environmental Health Perspectives, 108, 109-117.

8. Schwartz, J. (2004) The Effects of Particulate Air Pollution on Daily Deaths: A Multi-City Case-Crossover Analysis. Occupational and Environmental Medicine, 61, 956-961.

9. Pope III, C.A., Burnett, R.T., Thurston, G.D., Thun, M.J., Calle, E.E., Krewski, D. and Godleski, J.J. Cardiovascular Mortality and Long-Term Exposure to Particulate Air Pollution: Epidemiological Evidence of General Pathophysiological Pathways of Disease. Circulation, 109, 71-77.

10. Hasenfratz, D., Saukh, O., Sturzenegger, S. and Thiele, L. (2012) Participatory Air Pollution Monitoring Using Smartphones. Zurich.

11. Kanaroglou, P.S., Jerrett, M., Morrison, J., et al. (2005) Establishing an Air Pollution Monitoring Network for Intra- Urban Population Exposure Assessment: A Location-Allocation Approach. Atmospheric Environment, 39, 2399-2409.

12. Sivaraman, V., Carrapetta, J., Hu, K. and Luxan, B. (2010) HazeWatch: A Participatory Sensor System for Monitoring Air Pollution in Sydney. 38th Annual IEEE Conference on Local Computer Networks-Workshops, Australia, 56-64.

13. Hasenfratz, D., Saukh, O. and Thiele, L. (2012) On-the-Fly Calibration of Low-Cost Gas Sensors. Computer Engineering and Networks Laboratory, Zurich.

14. Réquia, W.J.R., Koutrakis, P. and Roig, H.L. (2015) Spatial Distribution of Vehicle Emission Inventories in the Federal District, Brazil. Atmospheric Environment, in press.

15. CETESB. Environmental Sanitation Technology Company (2013) Air Quality. São Paulo.

16. CAESB. Environmental Sanitation Company of the FD (2011) Base Data: Location and Classification of Hydrometers Are Federal District. Brasilia.

17. Réquia Jr., W.J., Roig, H.L. and Koutrakis, P. (2015) A Spatial Multicriteria Model for Determining Air Pollution at Sample Locations. Journal of the Air & Waste Management Association, 65, 232-243.

18. Réquia Jr., W.J. and Roig, H.L. (2014) Multi-Criteria Model in Geographic Information Systems for Determination of Sampling Points of Air Pollution. Urban Environmental Pollution. Climate Change and Urban Environment—Con- ference, 45-46.

19. Aguiar, E.F.K. and Roig. H.L. (2014) Air Pollution Dispersion Mapping by Remote Sensing: Case Study from the Federal District. GEO Processing 2014: The Sixth International Conference on Advanced Geographic Information Systems, Applications, and Services, Barcelona, 15-23.

20. Rada, E., Ragazzi, M., Brini, M., Marmo, L., Zambelli, P., Chelodi, M. and Ciolli, M. (2012) Perspectives of Low-Cost Sensors Adoption for Air Quality Monitoring. UPB Scientific Bulletin, Series D, 74, 241-250.

21. E2V (2010) e2v Metal Oxide Semiconductor (MOS) Gas Sensor Evaluation Kit. User Guide, Vol. 44, United States of America, 1-29.

22. Figaro Engineering Inc. (2014) Gas Test Box SR#3. Japan.

23. NIST. National Institute of Standards and Technology (2011) Chemistry WebBook. U.S. Secretary of Commerce on Behalf of the United States of America, Gaithersburg.

24. Beychok, M. (2005) Fundamentals of Stack Gas Dispersion. 4th Edition, California.

25. ESRI (Environmental Systems Resource Institute) (2014) ArcInfo 10.2.2. ESRI, Redlands.

26. Yamauti, M. (2013) Simple Linear Regression in Statistical Textbooks for Courses of Administration: A Didactic Study. PUCSP, São Paulo.

27. Loreto, V. (2012) Report on: Sensor Selection, Calibration and Testing. Enhance Environmental Awareness through Social Information Technologies, Italy, 1-73.

28. Li, X., Ramasamy, R. and Dutta, P. (2009) Study of the Resistance Behavior of Anatase and Rutile Thick Films towards Carbon Monoxide and Oxygen at High Temperatures and Possibilities for Sensing Applications. Sensors and Actuators B: Chemical, 143, 308-315.

29. Yoo, K.S. (2011) Gas Sensors for Monitoring Air Pollution. Department of Materials Science and Engineering, University of Seoul, Seoul.

CHAPTER 5

Towards the Development of a Low Cost Airborne Sensing System to Monitor Dust Particles after Blasting at Open-Pit Mine Sites

Miguel Alvarado [1,*], Felipe Gonzalez [2], Andrew Fletcher [3] and Ashray Doshi [4]

[1] Centre for Mined Land Rehabilitation, Sustainable Mineral Institute, The University of Queensland, Brisbane 4072, Australia
[2] Science and Engineering Faculty, Queensland University of Technology (QUT), Brisbane 4000, Australia
[3] Centre for Mined Land Rehabilitation, Sustainable Mineral Institute, The University of Queensland, Brisbane 4072, Australia
[4] Faculty of Engineering, Architecture and Information Technology, School of Information Technology and Electrical Engineering, The University of Queensland, St. Lucia 4072, Australia

ABSTRACT

Blasting is an integral part of large-scale open cut mining that often occurs in close proximity to population centers and often results in the emission of particulate material and gases potentially hazardous to health. Current air quality monitoring methods rely on limited numbers of fixed sampling locations to validate a complex fluid environment and collect sufficient data to confirm model effectiveness. This paper describes the development of a methodology to address the need of a more precise approach that is capable of characterizing blasting plumes in near-real time. The integration of the system required the modification and integration of an opto-electrical dust sensor, SHARP GP2Y10, into a small fixed-wing and multi-rotor copter, resulting in the collection of data streamed during flight. The paper also describes the calibration of the optical sensor with an industry grade dust-monitoring device, Dusttrak 8520, demonstrating a high correlation between them, with correlation coefficients (R^2) greater than 0.9. The laboratory and field tests demonstrate the feasibility of coupling the sensor with the UAVs. However, further work must be done in the areas of sensor selection and calibration as well as flight planning.

KEYWORDS

PM10; monitoring; blasting; fixed-wing UAV; quadcopter; optical sensor

1. INTRODUCTION

The mining and coal seam gas industries in Australia and around the world are important economic activities. Coal exports from Queensland from March 2013 to March 2014 totaled more than $24.5b [1]. These activities generate particles and gases such as methane (CH_4), carbon dioxide (CO_2), nitrogen oxides (NO_x), and sulfur oxides (SO_x) that have potentially dangerous environmental and health impacts.

Blasting in particular includes effects such as airblast, ground vibration, flyrock, toxic gases and particulate matter [2,3]. Particulate matter, aerosols, ammonia, carbon dioxide (CO_2), nitrogen, nitrogen oxides (NO_x) and sulfur oxides (SO_x) are the primary residues produced by blasting events at mining sites. In an ideal situation, the exothermic reaction produces CO_2, water vapor and molecular nitrogen (N_2); however, due to environmental and technical factors, other noxious gases are often produced in a range of concentrations [4].

In this paper, we propose the use of small unmanned aerial vehicles (UAV) carrying air quality sensors to allow precise characterization of blasting plumes in near-real time. This approach may lead to actionable data for harm avoidance or minimization. Most pollution dispersion models use predefined estimates of pollution sources and atmospheric conditions; near-real time information from within the plume has been practically impossible to collect. Flight instrument data transmitted as telemetry from the UAV provides high resolution instantaneous micrometeorological data that can assist interpretation of concentrations detected by on-board air quality sensors. In addition, this information including location, micrometeorological data and air quality, can be delivered in real time to analytical software. The data stream may therefore be used to feed flight path-planning algorithms or atmospheric dispersion models in near-real time.

In order to assess this approach, fixed-wing and multi-rotor UAVs were used. These UAVs were developed at The University of Queensland for ecological investigations. The platforms were capable of autonomous predetermined flight path planning or semi-autonomous direction. These platforms have weight restrictions and require sensors with high temporal sampling resolution (<1 s that can be digitally sampled but allow air quality sensors to be integrated and tested. In this paper we tested light-emitting diode (LED)-based optical sensors due to the combination of essential characteristics including rapid response, light weight and ease of data digitization. To date, two dust sensors have been tested with the UAV (SHARP GP2Y10 and Samyoung DSM501A) [5,6].

Characterizing blasting plumes and predicting dispersion using this approach requires integration of a number of factors:

- Development or modification of micro UAV platforms that can be safely operated near active mine blasts.

- Identification of sensors with necessary sampling rates (<1 s), weight (<500 g), data output format and sufficient sensitivity (1 mg/m^3 PM10).

- System endurance sufficient to capture plume evolution and dynamics (>20 min).

- Integration and formatting of data streams necessary for mathematical predictive models via live telemetry.

This project aims to develop tools that inform, cross calibrate and validate plume models for particulate and gaseous pollutants associated with blasting activities.

This paper is organized as follows: Section 2 reviews current methods to monitor blasting plumes, dust and gases after blasting at open-cast mine sites, and the use of UAVs for environmental monitoring and modeling approaches; Section 3describes the current sensing system that has been developed; Section 4 describes progress in the integration of the dust sensor system with UAVs and flight testing; and, Section 5 outlines current conclusions and further work.

2. BLAST-ASSOCIATED AIR SAMPLING

2.1. Methods to Monitor Blasting Plumes

Blast-associated dust is a significant potential hazard, and novel monitoring methods are continuously explored. Roy et al.developed a multi-platform system using ground-based dust samplers in combination with balloon-carried samplers near open pit mines. The data collected informed multiple regression and neural network models how to monitor and predict the drifting of blast plumes [7,8].

As samplers were static during blasting, this approach required detailed site-blasting plans and favorable weather conditions to determine their interconnectivity. Under this configuration, neural network models performed better than multiple regression models in predicting outcomes [9].

Furthermore, fugitive NO_2 and PM10 emissions of coal mining in the Hunter Valley, Australia have been examined using gravimetric and LIDAR methods. LIDAR provided long-path laser-integrated concentration signal with very low limit of detection, but required a fixed location [10]. Attalla *et al.* used a different approach by implementing NDIR (non-dispersion infrared) and mini-DOAS (differential optical absorption spectroscopy) for prediction of NOx and other pollutant gases. This method also required a fixed-location ground-based sensing apparatus. Modelling in AFTOX (Air Force toxics model) resulted in overestimation of plume concentrations at a distance [4].

Richardson (2013) assessed particulate fractions using a scintillation probe dust sensor (Environmental Beta Attenuation Monitors—EBAMs) and a real-time laser photometer (Dusttrak) in Hunter Valley and Central Queensland (Goonyella Riverside) and confirmed that PM2.5 is a small fraction of the overall suspended blast-associated particles, while PM10 is dominant [11].

2.2. Dust Sampling Sensors

The method and type of sensors used to measure contaminant gas or dust emissions will vary according to the type of emission, concentration range of concern, and required response time. Sensors are commonly based on ultrasound, optical, and electrochemical sensing elements [12,13,14]. These sensors can either be handheld, installed in vehicles, or form ground-based network systems. Table 1 shows different examples of sensors and their characteristics classified by the way they are implemented. Network systems are very useful when specific receptors or areas are to be monitored [15,16]. However, effective monitoring diameter, costs of installation, operation and maintenance are important considerations that may limit their use and procurement.

A complex criteria matrix must be considered when selecting an airborne sensor to monitor blast plumes. Factors include dimensions (weight and size); tolerance of vibration and movement given mounting on a UAV platform (up to 15 m/s); concentration range of sensor as well as the accuracy and limitations of the sensor (e.g., response time, mean square deviation, calibration, interference of other gases, humidity and temperature).

Table 1. Example of sensing technology used for monitoring gases in the mining, oil and gas industries.

Instrument	Description	Gases/Particles	Characteristics
	Handheld		
Dräger X-am 5600 [17]	Compact instrument for the measurement of up to 6 gases; complies with standard IP67; IR sensor for CO_2 and electrochemical for other gases.	O_2, Cl_2, CO, CO_2, H_2, H_2S, HCN, NH_3, NO, NO_2, PH_3, SO_2, O_3, Amine, Odorant, $COCl_2$ and organic vapors.	Dimensions: 4.7 × 13.0 × 4.4 cm Weight: 250 g
	Installed in ground vehicles		
Picarro Surveyor [18,19]	Cavity ring-down spectroscopy (CRDS) technology, sensitivity down to parts-per-billion (ppb); survey gas at traffic speeds and map results in real time; real-time analysis to distinguish natural gas and other biogenic sources.	CO_2, CO, CH_4, and water vapor	Dimensions: Analyzer 43.2 × 17.8 × 44.6 cm, external pump 19 × 10.2 × 28.0 cm Weight: 24 kg + vehicle Power: 100–240 VAC
	Stationary		
Tapered Element Oscillating Microbalance (TEOM) [20,21].	Continuous particle monitoring. The tapered element consists of a filter cartridge installed on the tip of a hollow glass tube. Additional weight from particles that collect on the filter changes the frequency at which the tube oscillates.	Total suspended particles (TSP), PM10, PM2.5	Dimensions: 43.2 × 48.3 × 127.0 cm) Weight: 34 kg Power: 100–240 VAC
	Networks		
AQMesh [22]	Wireless monitor; high sensitivity (levels to ppb); designed to work through a network of arrayed monitors.	NO, NO_2, O_3, CO, SO_2, humidity and atmospheric pressure.	Dimensions: 17.0 × 18.0 × 14.0 cm Weight: <2 kg Power: LiPo batteries
	Airborne		
Yellow scan [23]	LIDAR technology with a total weight of 2.2 kg; 80,000 shots/s; resolution of 4 cm; class 1 laser at 905 nm.	Dust and aerosols.	Dimensions: 17.2 × 20.6 × 4.7 cm Weight: 2.2 kg Power: 20 W

Optical LED particulate sensors are potentially suitable devices that could be used to explore the proposed system. LED sensors have the advantages of low power consumption, high durability, compact size, and easy handling and have been tested as a reliable source light for DOAS [24]. They also have demonstrated the ability to reduce internal stray light and can be used as a light source for applications requiring numerous kilometers of total light path [25]. Several other authors have also highlighted their advantages over other types of sensors [26,27].

2.3. Use of Unmanned Aerial Vehicles (UAVs) for Environmental Monitoring

Researchers are identifying advantages of UAVs to undertake investigations in difficult terrain/landscape areas, where health and safety risks exist, or where there is a lack of resources (human and/or economic). Gas sensing with small-micro UAVs using electrochemical and optical sensors is still not well established due to rapid development of sensors and UAVs. Sensing of CO_2,

CH_4 and water vapor [28], NO_2 and NH_3 [29], ethanol and CH_4 [30,31], have been conducted using rotary-winged platforms. Watai et al. [32], used a kite plane to monitor CO_2 using a NDIR gas analyzer which had a response time of 20 s. A spectrum-specific video camera has been developed to visualize SO_2 emissions from volcanoes by Brown et al.[33]. Lega et al. integrated a multi-rotor-sensing platform that monitors air pollutants in real time and provides 3D visualization [34]. Several variations of this platform, StillFly and BiLIFT, detect gases like CO, C_6H_6, NO_2, O_3, SO_2, NO_x and PM10, as well as thermal IR images to detect sewage discharges along the coastline of Italy [34,35]. Fixed-wing systems capable of achieving real time monitoring and providing indexed-linked samples are also currently possible [34]. Target sampling locations and source scales will be important to platform and sensor selection due to the fundamental differences between fixed-wing and multi-rotor UAVs, such as hovering capacity, endurance, and flight envelope.

Other approaches have been taken to characterize and track fugitive emission contamination plumes and register their concentrations [4,7,29,36,37,38,39]. However, to the authors' knowledge, UAVs have never been used to understand dust or gas emission associated with mine blasting. Small UAV platforms and real-time air quality sampling impose a number of novel and complex sampling requirements that still need to be addressed:

- Rapid sensor response time is important in mobile sensing platforms that move relative to both air and ground.

- Limited power requires flight efficiency and sensors with low power consumption.

- Sensors that require extended duration equilibration times (~20 s) require sampling chambers, and delayed response times result in difficult or impossible flight performance required to return to the estimated plume location.

- Multi-rotor platforms operate via GPS, and thus approximate a ground-based sampler independent of the fluid it is sampling. Fixed-wing platforms move through a defined volume of air in a given time regardless of ground location.

- Moving platforms require flow control for sample chambers to ensure calibrated values can be reported.

- The use of mathematical models is an essential element when monitoring air quality and atmospheric contamination. Defining appropriate models given the type of data collected is important. There are several approaches commonly used for air (emission factors, Gaussian, Lagrangian,

Eulerian, *etc.*), each of which have limitations to their performance [40,41,42].

- Improve data visualization since current attempts to map pollutants in the atmosphere are presented as snapshots, not as a dynamic environment with concentration measurements that change before and after the moment a reading is produced.

3. DESIGN OF SENSING SYSTEM

The sensor system consists of a gas-sensing node, the UAV, and a data integration and visualization interface.

3.1. System Architecture

The system architecture for the fixed-wing UAV with integrated dust sensor (Figure 1) and a multi-rotor carrying a telemetered dust sensor (Figure 2) are necessarily different due to the use of different autopilots. Micro meteorological data deduced from UAV platform flight control is a novel and detailed source of data for interpretation of air quality measurements. However, it requires integration with gas sensor data to allow a meaningful application.

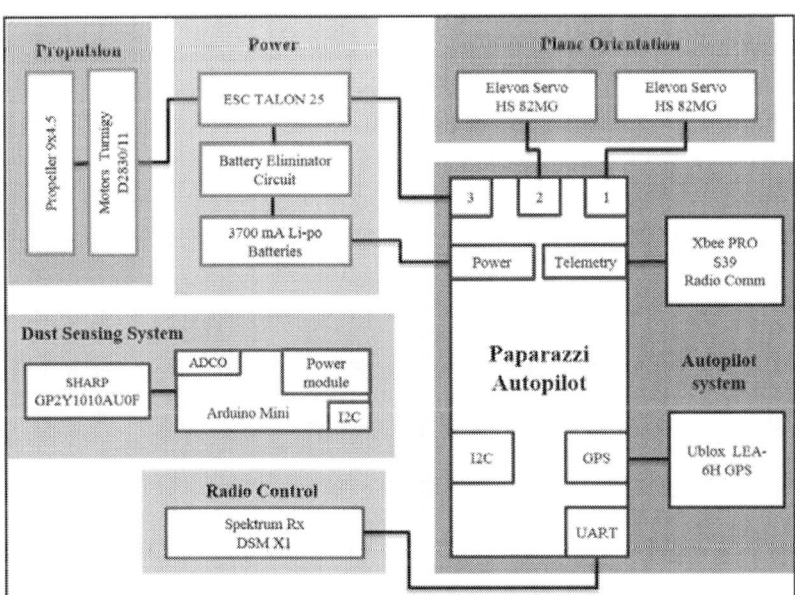

Figure 1. System architecture for the fixed-wing UAV with dust sensor.

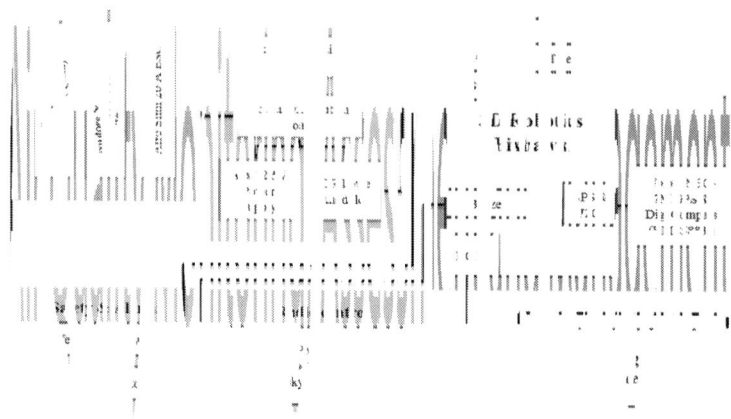

Figure 2. System architecture for quadcopter UAV with independent gas-sensing system.

3.2. Gas-Sensing Node

Two sensors were experimentally assessed to date: GP2Y10 (SHARP) and DSM501A (Samyoung) for PM10. The SHARP and Samyoung sensors tested were connected through an Arduino microcontroller to integrate sensor-telemetry data streams. The sensor and associated electronics are low weight and constrained size to allow simple installation in other multi-rotor or fixed-wing platforms. The dust-sensing module was placed on the top side of the quadrotor platform to minimize high velocity air flow that is fundamental to similar quadcopters [43,44]. Figure 3 shows the system architecture for the modular dust sensor in detail.

The system is constructed around an Arduino MEGA 2560, powered by a 7.4 V lithium polymer battery, data telemetry is via XBee Pro S1 (2.4 GHz) radio transmitter while a GP-635T provides a timestamp for serial port data. Sensors include a SEN51035P temperature and humidity sensor and GP2Y10 SHARP dust sensor (Figure 3). All data was transmitted and logged on a ground station which displays received raw values and PM10 concentration readings in real time.

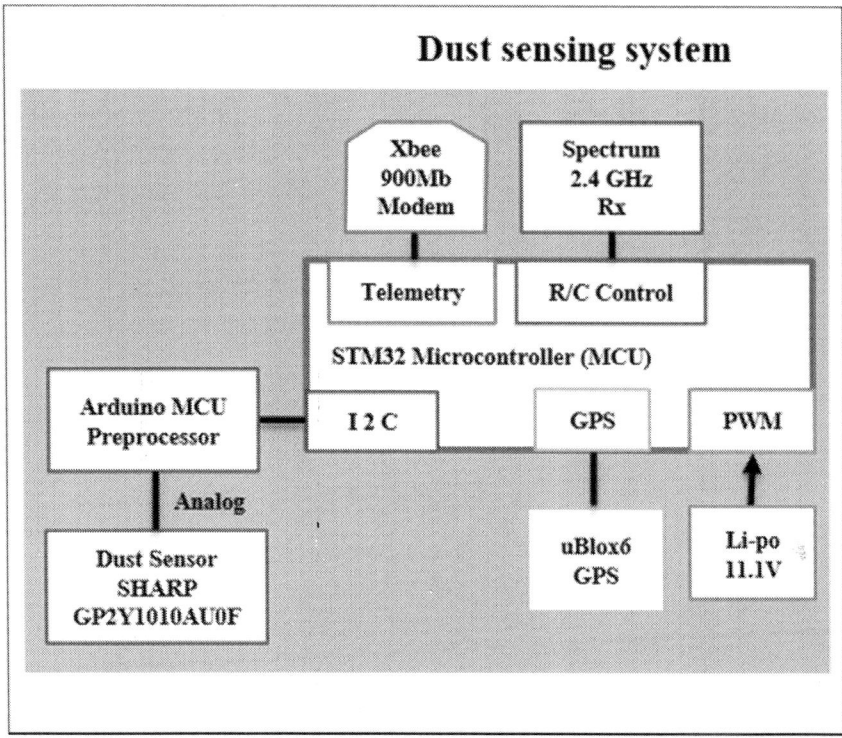

Figure 3. System architecture for the modular dust sensor.

3.3. UAV Platforms

Both a fixed-wing and a multi-rotor UAV were selected to develop the sensing system. These aerial vehicles operate in fundamentally different ways with fixed-wing UAVs traversing a set volume of air in a given time while the hovering ability of rotary-winged UAVs allow collection of data at specific locations in space and time; however, they experience wind.

The suitability of three fixed-wing platforms was considered for integration with the optical and/or electrochemical sensors. All UAVs are constructed of expanded polypropylene (EPP) and composite materials that have been demonstrated as safe and robust platforms in the mining industry environment. Specifications for the models considered during this investigation are provided in Table 2 and Figure 4.

All UAVs listed in Table 2 have low kinetic energy (<50 joules) and low air speed (<60 km/h). Low kinetic energy and speed improve safety and simplifies data acquisition performance but require reasonably calm conditions to operate. All have a pusher-propeller design providing access to clean airflow for sensors. The Paparazzi autopilot used on Teklite and GoSurv records altitude, platform coordinates, speed and direction. They also estimate wind speed and direction by response difference. The autopilot of the Swampfox platform records airspeed using a pitot tube, as well as speed and direction with the GPS. Air speed and geolocation data are drawn from the autopilot telemetry that is integral to all small UAV operations.

Table 2. Characteristics of UAVs identified as feasible platforms for this investigation.

Model	Wingspan (mm)	Length (mm)	Flying Weight (g)	Endurance (min)	Approx. Payload [3] (g)
Teklite [1]	900	575	900–950	45	200
GoSurv [2]	850	350	900–1200	50	>300
Swamp Fox [45] [1]	1800	1000	4500	40	1000

[1] Commercially available platform; [2] Fixed-wing platform designed at UQ SMI-CMLR; [3] Determined through experimental procedures.

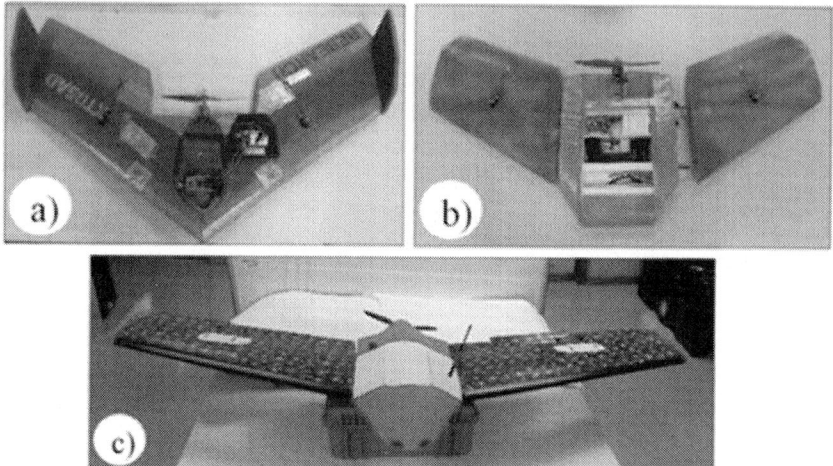

Figure 4. Fixed-wing UAV platforms, (**a**) Teklite; (**b**) GoSurv; and (**c**) Swamp Fox.

The Teklite was selected as the best platform for the type of test to be conducted due to its portability, ease of integration of sensors, successful flight testing, light weight and low (<100 ft) target flight altitude. The UAV is controlled using a ground control station. The flight plan is preloaded from the ground station that displays the flight parameters of the UAV, flight route and atmospheric pollution readings in real time. The flight plan can also be modified manually using a handheld radio transmitter and/or by altering the parameters through the ground station.

The UAV can be flown from as far as 1.5 km from the ground control station. Weather conditions (wind speed and direction, temperature, *etc.*) are used to pre-plan the UAV flight path to follow and characterize the blasting plume. If required, a flight path can be modified and uploaded into the ground station based on post-blasting observations. The UAV is restricted to fly more than 35 m above ground level as a safety factor to avoid collision with trees or infrastructure.

A multi-rotor platform was used to record readings below 35 m above ground level. The system was designed for agricultural and air-monitoring surveys. Figure 5 shows the quadcopter integrated with the modular dust sensor. The multi-rotor platform has an average flight time of 20 min and a total weight of 2.5 kg (with batteries). The modular dust sensor had a total weight of 150 g and was placed inside a plywood case.

Figure 5. Quadcopter and modular gas-sensor system integrated.

4. BENCH TESTING OF OPTICAL SENSORS

A gas chamber (see Figure 6) based on the work of Budde *et al.* [46] was constructed in order to expose the sensor node to different concentrations of particles and compare the readings against a calibrated dust-monitoring device—Dusttrak 8520. The Dusttrak has a response time of 1 s, a resolution of

0.001 mg/m³ and is capable of monitoring PM10 and PM2.5. Smoke from standard incense sticks was used as an airborne particulate source. The Samyoung sensor produced a low correlation coefficient (R^2) of 0.5 and was therefore deemed an unsuitable option for the integration of the dust-monitoring module and UAVs.

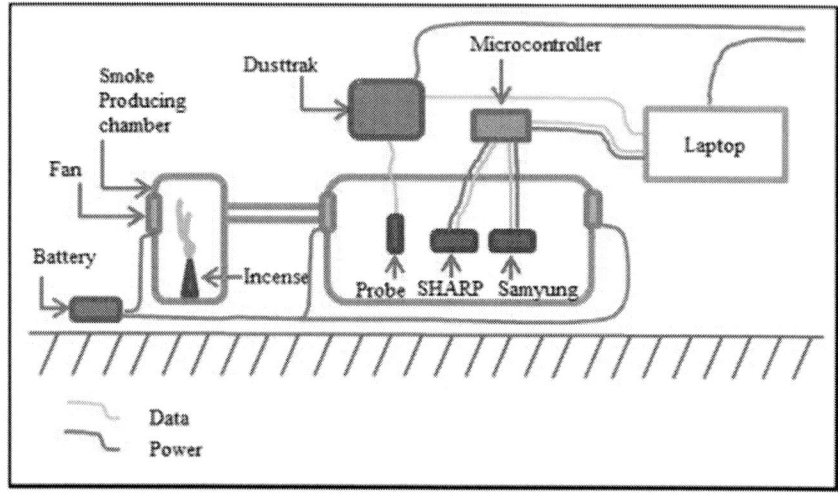

Figure 6. Gas chamber for sensor testing and calibration.

Results for SHARP (GP2Y10)

Tests for PM10 and PM2.5 where undertaken for the SHARP dust sensor. An initial data collection test was used to correlate the raw values obtained from the sensor, which is the voltage modified by the light absorption of the receiver, with the values registered by the Dusttrak (see Figure 7). A linear and second-degree calibration equation, with correlation coefficients greater than 0.9, were obtained and applied to the sensor data.

The original algorithm of M. Chardon and Trefois [47] developed to use the SHARP sensor with an Arduino board was modified to take readings every second. The objective of this test was to check that the data collected by the SHARP sensor was comparable to the Dusttrak readings, results are shown in Figure 8. The offset observed in the initial test was reduced having a satisfactory match between sensors. Percentage errors were calculated obtaining 38.0% and 13.6% for PM10 linear and quadratic fits respectively. PM2.5 errors were 11.9% and 9.96% for linear and quadratic fits respectively.

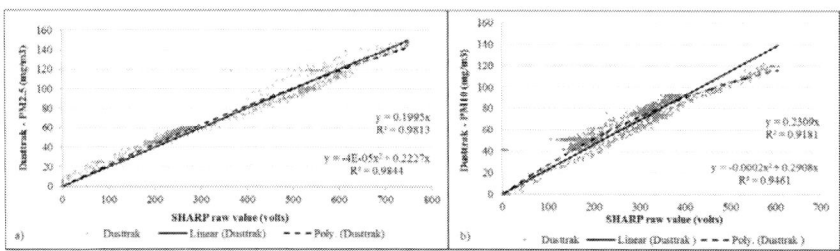

Figure 7. Correlation of raw values obtained with SHARP sensor for (**a**) PM2.5 and (**b**) PM10 *vs.* readings collected with Dusttrak (mg/m³).

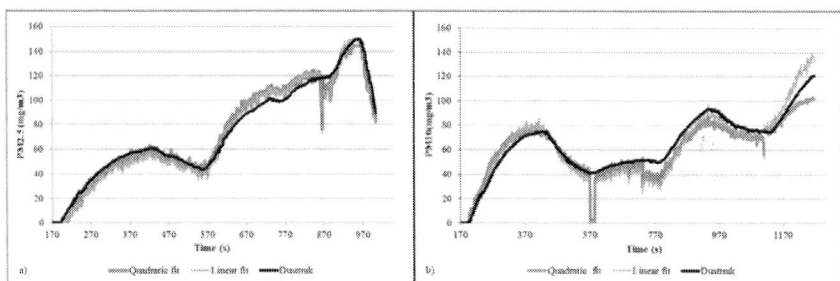

Figure 8. Linear and quadratic linear fit for raw SHARP values of (**a**) PM2.5 and (**b**) PM10 particle concentrations.

A third test was done using two SHARP sensors and the Dusttrak (Figure 9). An offset between the SHARP sensors was also observed; however this error was reduced after correlating data with the linear equation fitted previously. For this test the linear fit produced a lower percentage error for SHARP A and B, of 19.3% and 12.5% respectively, however the second degree fit produced very similar results with errors of SHARP A:21.5% and SHARP B:14.9%.

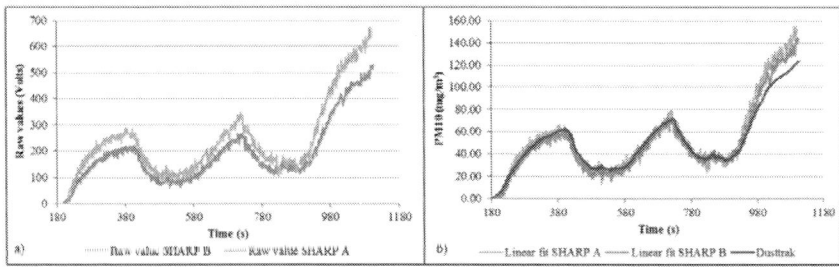

Figure 9. Dual SHARP and Dusttrak test showing (**a**) raw values data and (**b**) corrected particle measurements against Dusttrak readings.

Sensor variability due to temperature changes [5,30], was not considered for the experiment, however it will be undertaken in further tests.

5. FLIGHT TEST

5.1. SHARP Sensor Integrated to Fixed-Wing UAV

The air intake and discharge were modified to produce a continuous flow inside the SHARP sensor chamber (see Figure 10). Air sampling intake was through a carbon fiber scooped cowl on the top surface of the wing directly over the sensor inlet. Sample exhaust was through a 4 mm tube attached to the sensor outlet and extended through the lower surface of the platform.

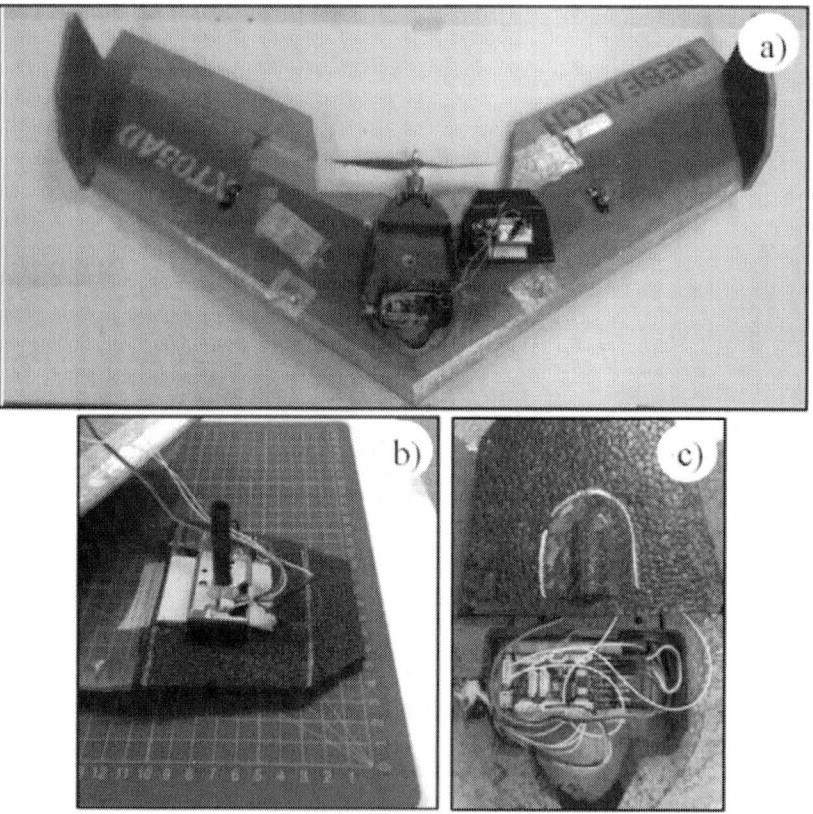

Figure 10. Modifications made to Teklite and SHARP sensor for flight, (**a**) Teklite UAV and SHARP sensor; (**b**) Air outlet for SHARP sensor; (**c**) Air intake for SHARP sensor.

5.2. Test 1: Sensor Integration

Several flights were made to test the feasibility of integrating the SHARP sensor with the Teklite platform. The first test was conducted on 6 June 2014 in order to evaluate the integration of the system. The test used a fire in an open area as an airborne particulate source. The UAV was programmed to fly around the fire for approximately 30 min. Data collected from the UAV and the air quality sensor are shown in Figure 11 and Figure 12. The data did not report variations in particulate matter concentration in the atmosphere, as it is observed in Figure 12. Analysis of the data indicated that electrical noise caused by motor and onboard electronics was interfering with the output.

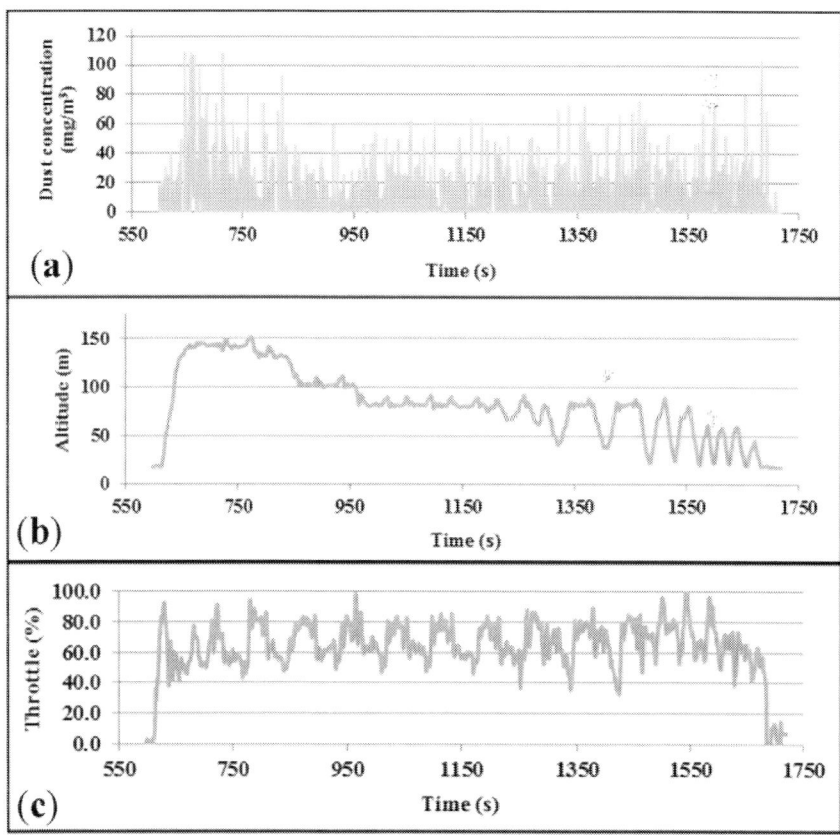

Figure 11. Data collected from Teklite flight with SHARP sensor attached. (**a**) Dust concentration; (**b**) Altitude; (**c**) Throttle.

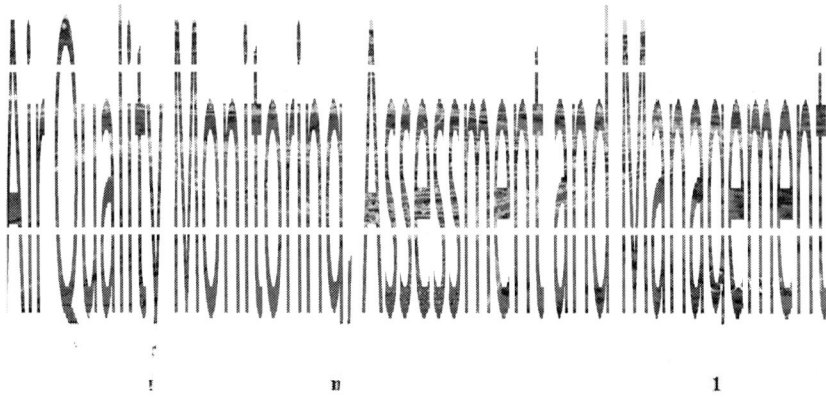

Figure 12. 3D visualization of the Test 1 data collected with Teklite- SHARP sensor.

High frequency noise consistent with electrical switching of motors and servos was filtered by installing a 50 V (0.1 μF) capacitor to the power source.

5.3. Test 2: PM10 Monitoring

A second field test was conducted on 13 October 2014 using "talcum powder" lifted into the atmosphere using a petrol-powered leaf blower (STIHL BG 56 Blower—max of 730 m³/h). Talcum powder was used due to its safe handling and availability. Talcum powder is composed of 0.2–0.3 mass fraction with a particle diameter no greater than 10 μm [48,49].

In order to determine PM10 concentrations measured during the flight, the data was processed using the particle correlation (Figure 7) obtained from laboratory testing of the SHARP sensors. Figure 13 shows the distribution of PM10 concentrations in the atmosphere registered by the optical sensor by using top and side 3D visualization of the particulate plume. The wind direction was towards the west–southwest and concentrations ranged from 15 mg/m³ to 66 mg/m³, describing the shape of the plume when dispersed by the wind.

Tests 1 and 2 demonstrated the functionality and feasible integration of the system; however, the need for systematic characterization of a particulate plume of known composition and size remained. This is required to demonstrate the ability to calculate particulate emission rates, as most parameters can be independently measured using a constant powder emission, constant emission rate, known atmospheric conditions and particle size distribution of the source.

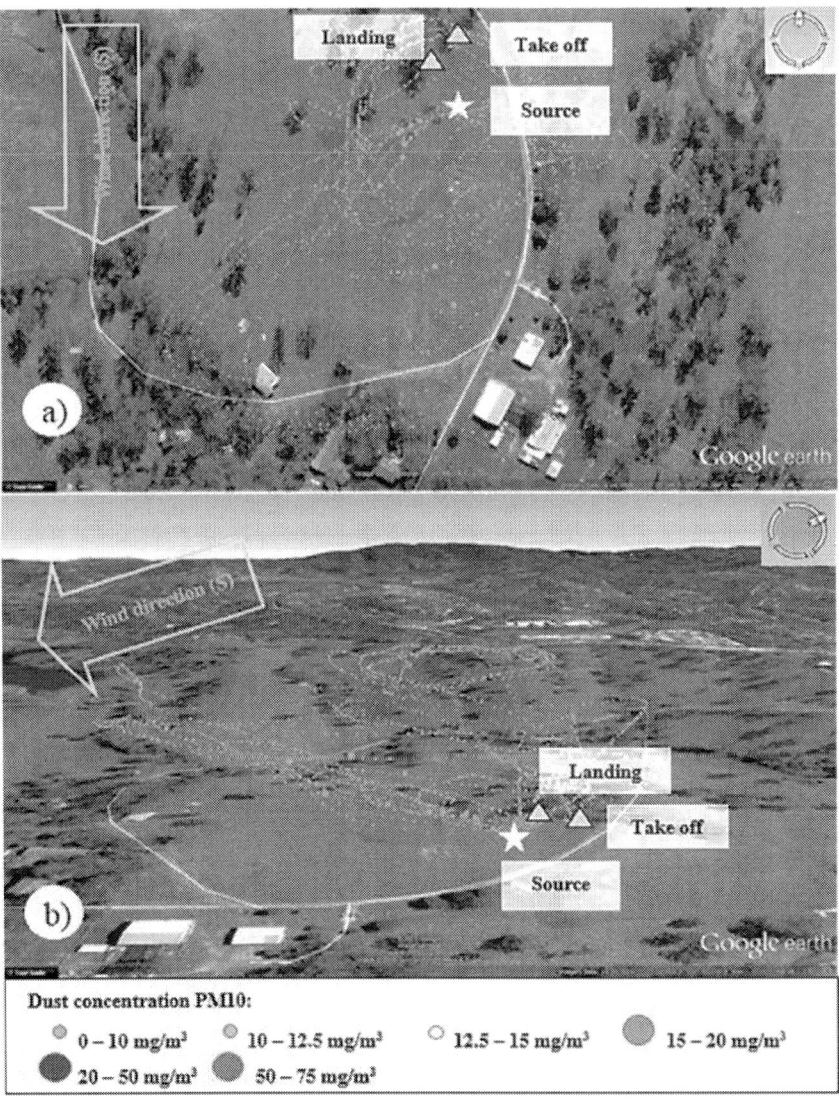

Figure 13. 3D visualization of Test 2 data collected with Teklite-SHARP sensor for PM10 (**a**) Overview and (**b**) Side view.

To achieve a systematic plume characterization, it was necessary that the UAV reproduced a fixed experimental flight pattern to aid spatial calculations and also exclude biased measurements that could easily be made when flying manually into the visible plume produced by the powder ejected. A flight path consisting of concentric circles at different heights and radius was planned for Test 3. The flight path ensures the UAV covers the designated area around the source. This ensures that the sensor intersects the plume and tests the ability of the data to describe the behavior of the plume in the air space surrounding it.

5.4. Test 3: Mixed Fixed and Rotary Wings

Test 3 was undertaken on 3 March 2015. The setup for the field experiment was based on Test 2, incorporating modifications to satisfy UAV flight and rigorous plume modelling requirements. The fixed-wing and multi-rotor UAVs were able to fly following the patterns programmed for the tests. Table 3 shows the radius and heights used for the test. These parameters were defined according to the capabilities of each UAV and to collect complementary datasets at two spatial scales.

Table 3. Programmed flight parameters and UAV capabilities.

Parameters	Quadcopter	Fixed-Wing
Max. Height *	120 m	120 m
Max. Radius	100 m	200 m
Programmed Heights (MAGL)	7, 14, 21	35, 45, 55
Programmed Radius	5, 15, 35	45, 55, 65, 75, 85

* Determined by UAV height flight restrictions [50].

The talcum powder plume was generated using a petrol-powered fan connected to a 5.5 m long and 0.05 m diameter PVC stack. The powder was loaded into the airstream through an intersection custom made for the powder containers at an approximate rate of 300 g/min (Figure 14).

For Test 3, the SHARP sensors where recalibrated due to the different characteristics including color and particle diameter that smoke and talcum powder have. The calibration procedure previously used for smoke particles was repeated for the talcum powder. A correlation equation was calculated using a polynomial fit by processing the data obtained with the Dusttrak 8520 and with the SHARP sensor (Figure 15). Integrated datasets from each platform were post-processed to visualize the concentrations measured by the fixed-wing and quadcopter during experimental flights (Figure 16).

Figure 14. Powder ejection system setup.

PM10 concentration values ranged from 0.5 mg/m³ to 19 mg/m³ and their distribution described the path followed by the powder plume to the west, downwind from the source (Figure 16a,b). Mid-range concentrations to the east (downwind) and north of the source are likely the result of petrol motor exhaust particles and potentially spilled talcum powder. Experimental equipment modification using battery-powered fans and venture effect powder loading are being developed.

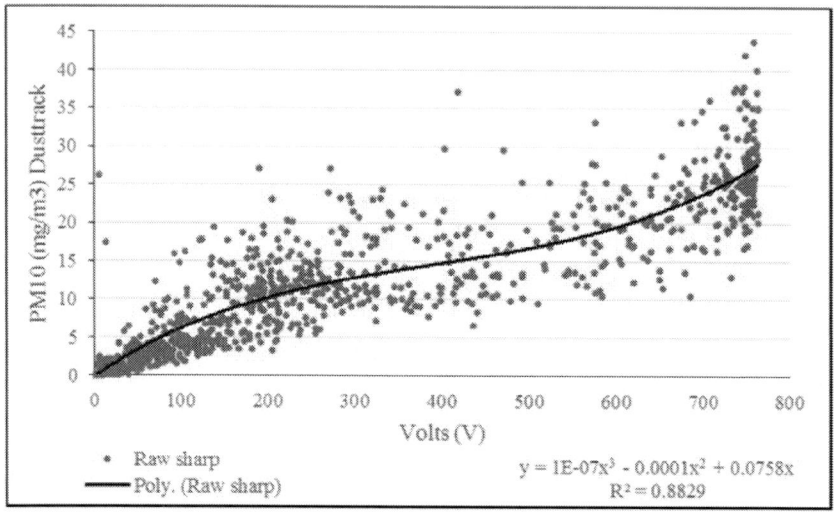

Figure 15. Correlation between talcum powder particles and raw value readings from the SHARP sensor and Dusttrak.

Future tests will also include measurements of background levels during flight monitoring periods to determine their influence in the UAV readings.

For safety reasons and the complexity involved in flying two UAVs simultaneously, the quadcopter and the fixed-wing UAVs were not flown simultaneously. The fixed-wing UAV was flown after the quadcopter and recorded maximum concentrations of 2.0 mg/m³ without an observable pattern (Figure 16c). Weather conditions with wind speed ranging from 7 to 9 m/s prevented the powder plume rising to the minimum programmed height of 35 m; therefore, it is unlikely that detectable particulates associated with the plume were present.

Figure 17 shows the contour plots of the powder distribution at a height of 18 m above ground level and 30 m to the west of the source. The contour plots together with the volume rendering produced with the software Voxler aid in the interpretation of the data. They produce a model of the plume which can be challenging to interpret when plotting all readings independently, due to the high density of information. Higher concentrations of PM10 particles are shown in red color which are located in the western side of the source located at the center of the plot.

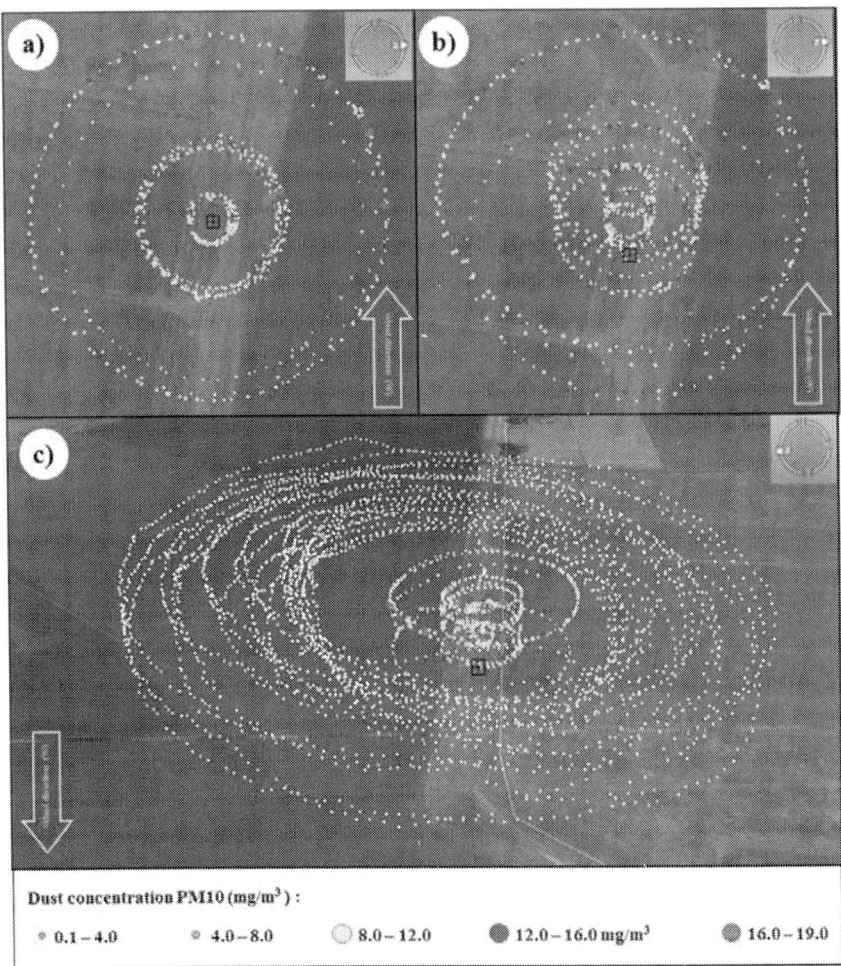

Dust concentration PM10 (mg/m³) :

  ⊛ 0.1 – 4.0   ⊛ 4.0 – 8.0   ○ 8.0 – 12.0   ● 12.0 – 16.0 mg/m³   ● 16.0 – 19.0

Figure 16. Flight path and PM10 concentrations monitored with the UAV quadcopter (**a**) top view and (**b**) side view; and (**c**) fixed-wing and quadcopter (overlapped flights).

6. CONCLUSIONS AND FURTHER WORK

The sensor systems developed to date are technically capable of delivering data comparable to industrial quality dust-monitoring devices but require individual calibration equations for each sensor used to characterize dust plumes. The use of talcum powder is primarily a detection exercise at this stage as most particulate matter in this product has a diameter greater than 10 μm. System testing at PM2.5 will require a chemical source such as a smoke generator.

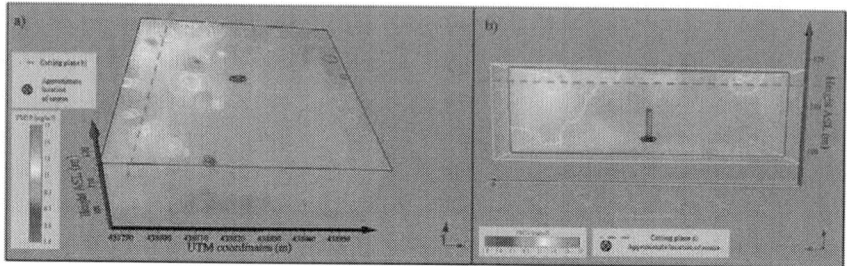

Figure 17. Volume rendering and contour plots created with quadcopter dataset (**a**) top view 18 m above ground level (from the East) and (**b**) side view 30 m away from the source (from the west).

The tests described in this paper only measured concentrations with a precision of 1 mg/m³; more precise readings of smaller concentrations will require the use of a different optical sensors and reference calibration with more precise equipment. Cross-contamination sources will require to be controlled in further experiments, and background levels will need to be measured to determine their content percentage in the final concentrations. These measurements will allow the programming of different flight patterns which could be focused in the intersection of the plume and will provide additional flight time.

Current experimental work indicates that integration of air quality sensor and autopilot data is feasible and will characterize airborne particulates in time and space.

Further work will be focused on the analysis of near-real time data to feed atmospheric modeling software and for flight path-planning algorithms.

ACKNOWLEDGMENTS

The first author would like to acknowledge the support of the Centre for Mine Land Rehabilitation (CMLR) at UQ SMI, the Science and Engineering Faculty at QUT, and from the UQ eGatton program to Pfsr. Kim Bryceson, Armando Navas Borrero and Sotiris Ioannou.

AUTHOR CONTRIBUTIONS

Miguel Angel Alvarado Molina was responsible for conducting the literature review, design and coordination of experimental procedures. He had the main role in the writing of this paper.

Felipe Gonzalez provided advice on experimental design, sensor integration and data interpretation. He also provided guidance on the structure and editing of the research article.

Andrew Fletcher provided overall project guidance and advice on experimental design, sensor integration and data interpretation. He also provided guidance regarding the formatting and editing of the research article.

Ashray Doshi assisted in the experimental design and sensor integration. He also provided advice in codes development for operation of the modular dust-sensing system and coordinated the integration of the dust sensor to UAV platforms.

REFERENCES

1. Department of Natural Resources and Mines (DNRM). *Queensland Monthly Coal Report*; Queensland Government, The State of Queensland: Brisbane, Australia, 8 July 2014.

2. Raj, R. Sustainable mining systems and technologies. In *Sustainable Mining Practices*; Taylor & Francis: Oak Brook, USA, 2005; pp. 91–178.

3. NSWEPA. EPA Investigating Reports of Blasting Fumes from Wambo Coa. Available online: http://www.epa.nsw.gov.au/epamedia/EPAMedia 14051501.htm (accessed on 29 July 2014).

4. Attalla, M.; Day, S.; Lange, T.; Lilley, W.; Morgan, S. *NOx Emissions from Blasting in Open Cut Coal Mining in the Hunter Valley*; Australian Coal Industry's Research Program, ACARP: Newcastle, Australia, 2007.

5. Sharp. Opto-Electronic Devices Division Electronic Components Group. Available online: http://www.dema.net/pdf/sharp/PC847XJ0000F.pdf (accessed on 10 August 2015).

6. SYhitech. DSM501A Dust Sensor Module. Available online: http://i.publiclab.org/system/images/photos/000/003/726/original/t mp_DSM501A_Dust_Sensor630081629.pdf (accessed on 8 August 2014).

7. Roy, S.; Adhikari, G.; Renaldy, T.; Singh, T. Assessment of atmospheric and meteorological parameters for control of blasting dust at an Indian large surface coal mine. *Res. J. Environ. Earth Sci.* **2011**, *3*, 234–248.

8. Roy, S.; Adhikari, G.R.; Singh, T.N. Development of Emission Factors for Quantification of Blasting Dust at Surface Coal Mines. *J. Environ. Protect.* **2010**, *1*, 346–361.

9. Roy, S.; Adhikari, G.; Renaldy, T.; Jha, A. Development of multiple regression and neural network models for assessment of blasting dust at a large surface coal mine. *J. Environ. Sci. Technol.* **2011**, *4*, 284–299.

10. Bridgman, H.; Carras, J.N. *Contribution of Mining Emissions to NO_2 and PM10 in the Upper Hunter Region*; ACARP: NSW, Australia, 2005.

11. Richardson, C. *PM2.5 Particulate Emission Rates From Mining Operations*; Australian Coal Industry's Research Program, ACARP: Castle Hill, Australia; March; 2013.

12. Koronowski, R. FAA Approves Use of Drones by ConocoPhillips to Monitor Oil Drilling Activities in Alaska. Available online: http://thinkprogress.org/climate/2013/08/26/2524731/drones-conocophillips-alaska/ (accessed on 22 January 2013).

13. Fernandez, R. Methane Emissions from the U.S. *Natural Gas Industry and Leak Detection and Measurement Equipment*; Available online: http://arpa-e.energy.gov/sites/default/files/documents/files/Fernandez_Presentation_A RPA-E_20120329.pdf (accessed on 10 August 2010).

14. Nicolich, K. High Performance VCSEL-Based Sensors for Use with UAVs. Available online: http://www.princeton.edu/pccmeducation/undergrad/reu/2012/Nicolich.pdf (accessed on 4 May 2014).

15. DECCW. Upper Hunter Air Quality Monitoring Network. Available online: www.environment.nsw.gov.au/aqms/upperhunter.htm (accessed on 10 October 2013).

16. DEHP. Air Quality. Available online: http://www.ehp.qld.gov.au/air/ (accessed on 10 August 2014).

17. Dräger X-am 5600. Drägerwerk AG & Co. KGaA: Lubeck, Germany, 2014. Available online: http://www.draeger.com/sites./PublishingImages/Products/cin_x-am_5600/UK/9046715_PI_X-am_5600_EN_110314_fin.pdf (accessed on 10 August 2015).

18. Picarro. *PICARRO Surveyor*; Picarro: Santa Clara, CA, USA, 2014; Available online: https://picarro.app.box.com/s/mtmyqr0k2kfotg2uf40z (accessed on 10 August 2015).

19. Picarro. *Picarro G2401 CO_2 + CO + CH_4 + H_2O CRDS Analyzer*; Picarro, Ed.; Picarro: Santa Clara, CA, USA, 2015.

20. QLDGov. Tapered Element Oscillating Microbalance. Available online: https://www.qld.gov.au/environment/pollution/monitoring/air-pollution/oscillating-microbalance/ (accessed on 20 August 2014).

21. ThermoScientific. Thermo Scientific TEOM 1405-DF. Available online: http://www.thermo.com.cn/Resources/200802/productPDF_3275.pdf (accessed on 10 August 2015).

22. Geotech. AQMesh Operating Manual. Available online: http://www.geotechuk.com/media/215152/aqmesh_operating_manual.pdf (accessed on 10 August 2015).

23. LAvionJaune. Ultra-Light, Standalone Lidar System for UAVs (Laser Scanner, IMU, RTKGPS, Processing Unit). Available online: http://yellowscan.lavionjaune.com/data/leafletYS.pdf (accessed on 10 August 2015).

24. Sihler, H.; Kern, C.; Pöhler, D.; Platt, U. Applying light-emitting diodes with narrowband emission features in differential spectroscopy. *Opt. Lett.* **2009**, *34*, 3716–3718.

25. Kern, C.; Trick, S.; Rippel, B.; Platt, U. Applicability of light-emitting diodes as light sources for active differential optical absorption spectroscopy measurements. *Appl. Opt.* **2006**, *45*, 2077–2088.

26. Choi, S.; Kim, N.; Cha, H.; Ha, R. Micro Sensor Node for Air Pollutant Monitoring: Hardware and Software Issues. *Sensors* **2009**, *9*, 7970–7987.

27. Thalman, R.M.; Volkamer, R.M. Light Emitting Diode Cavity Enhanced Differential Optical Absorption Spectroscopy (led-ce-doas): A Novel Technique for Monitoring Atmospheric Trace Gases. *Proc. SPIE* **2009**, *7462*.

28. Khan, A.; Schaefer, D.; Roscoe, B.; Kang, S.; Lei, T.; Miller, D.; Lary, D.J.; Zondlo, M.A. Open-path greenhouse gas sensor for UAV applications. In Proceedings of the 2012 Conference on Lasers and Electro-Optics (CLEO), San Jose, CA, USA, 6–11 May 2012; pp. 1–2.

29. Malaver, A.; Gonzalez, F.; Motta, N.; Depari, A.; Corke, P. Towards the Development of a Gas Sensor System for Monitoring Pollutant Gases in the Low Troposphere Using Small Unmanned Aerial Vehicles. In Proceedings of Workshop on Robotics for Environmental Monitoring, Sydney University, Sydney, Australia, 11 July 2012.

30. Neumann, P.P.; Hernandez Bennetts, V.; Lilienthal, A.J.; Bartholmai, M.; Schiller, J.H. Gas source localization with a micro-drone using bio-inspired and particle filter-based algorithms. *Adv. Robot.* **2013**, *27*, 725–738.

31. Bennetts, V.H.; Lilienthal, A.J.; Neumann, P.P.; Trincavelli, M. Mobile robots for localizing gas emission sources on landfill sites: Is bio-inspiration the way to go? *Front. Neuroeng.* **2011**, *4*, 735–737.

32. Watai, T.; Machida, T.; Ishizaki, N.; Inoue, G. A Lightweight Observation System for Atmospheric Carbon Dioxide Concentration Using a Small Unmanned Aerial Vehicle. *J. Atmos. Ocean. Technol.* **2005**, *23*, 700–710.

33. Brown, J.; Taras, M. *Remote Gas Sensing of SO₂ on a 2D CCD (Gas. Camera)*; Resonance LTD: Barrie, ON, Canada, 2008.

34. Lega, M.; Napoli, R.M.A.; Persechino, G.; Kosmatka, J. New techniques in real-time 3D air quality monitoring: CO, NOₓ, O₃, CO₂, and PM. In Proceedings of the NAQC 2011, San Diego, CA, USA, 7–11 March 2011.

35. Lega, M.; Kosmatka, J.; Ferrara, C.; Russo, F.; Napoli, R.M.A.; Persechino, G. Using Advanced Aerial Platforms and Infrared Thermography to Track Environmental Contamination. *Environ. Forensics* **2012**, *13*, 332–338.

36. Saghafi, A.; Day, S.; Fry, R.; Quintanar, A.; Roberts, D.; Williams, D.; Carras, J.N. Development of an Improved Methodology for Estimation of Fugitive Seam Gas. Emissions from Open Cut Mining. Available online: http://www.acarp.com.au/abstracts.aspx?repId=C12072 (accessed on 10 August 2015).

37. Gonzalez, F.; Castro, M.P.; Narayan, P.; Walker, R.; Zeller, L. Development of an autonomous unmanned aerial system to collect time-stamped samples from the atmosphere and localize potential pathogen sources. *J. Field Robot.* **2011**, *28*, 961–976.

38. Gonzalez, L.F.; Castro, M.P.; Tamagnone, F.F. Multidisciplinary design and flight testing of a remote gas/particle airborne sensor system. In Proceedings of the 28th International Congress of the Aeronautical Sciences, Optimage Ltd., Brisbane Convention & Exhibition Centre, Brisbane, QLD, Australia, 23 September 2012; pp. 1–13.

39. Malaver, A.; Motta, N.; Corke, P.; Gonzalez, F. Development and Integration of a Solar Powered Unmanned Aerial Vehicle and a Wireless Sensor Network to Monitor Greenhouse Gases. *Sensors* **2015**, *15*, 4072–4096.

40. Reed, W.R. *Significant Dust Dispersion Models for Mining Operations*; Department of Health and Human Services: Pittsburgh, PA, USA, September 2005.

41. Stockie, J.M. The Mathematics of Atmospheric Dispersion Modeling. *SIAM Rev.* **2011**, *53*, 349–372.

42. Visscher, A.D. An Air Dispersion Modeling Primer. In *Air Dispersion Modeling*; John Wiley & Sons, Inc.: Hoboken, NJ, USA, 2013; pp. 14–36.

43. Roldán, J.J.; Joossen, G.; Sanz, D.; del Cerro, J.; Barrientos, A. Mini-UAV Based Sensory System for Measuring Environmental Variables in Greenhouses. *Sensors* **2015**, *15*, 3334–3350.

44. Haas, P.; Balistreri, C.; Pontelandolfo, P.; Triscone, G.; Pekoz, H.; Pignatiello, A. Development of an unmanned aerial vehicle UAV for air quality measurements in urban areas. In Proceedings of the 32nd AIAA Applied Aerodynamics Conference; American Institute of Aeronautics and Astronautics, Atlanta, GA, USA, 16–20 June 2014.

45. Skycam. Swamp Fox UAV. Available online: http://www.kahunet.co.nz/swampfox-uav.html (accessed on 13 June 2014).

46. Budde, M.; ElMasri, R.; Riedel, T.; Beigl, M. Enabling Low-Cost Particulate Matter Measurement for Participatory Sensing Scenarios. In Proceedings of the 12th International Confrence on Moile and Ubiquitous Multimedia MUM, Lulea, Sweden, 2–5 December 2013; ACM: Lulea, Sweden, 2013; p. 19.

47. M.Chardon, C.; Trefois, C. *Standalone Sketch to Use with a Arduino Fio and a Sharp Optical Dust Sensor GP2Y1010AU0F*; Creative Commons: San Francisco, CA, USA, 2012.

48. Fiume, M.M. *Safety Assessment of Talc As Used in Cosmetics*; Cosmetic Ingredient Review: Washington, DC, USA; 12; April; 2013.

49. Klingler, G.A. *Digital Computer Analysis of Particle Size Distribution in Dusts and Powders*; Aerospace Research Laboratories, Office of Aerospace Research, United States Air Force: Wright-Patterson Air Force Base, OH, USA, 1972.

50. CASA. Civil Aviation Safety Regulations 1998. *Unmanned Air and Rockets*; Australian Government ComLaw: Canberra, Australia, 1998.

CHAPTER 6

Occurrence and Concentrations of Toxic VOCs in the Ambient Air of Gumi, an Electronics-Industrial City in Korea

Sung-Ok Baek *, Lakshmi Narayana Suvarapu and Young-Kyo Seo

Department of Environmental Engineering, Yeungnam University, Gyeongsan 712-749, Korea

ABSTRACT

This study was carried out to characterize the occurrence and concentrations of a variety of volatile organic compounds (VOCs) including aliphatic, aromatic, halogenated, nitrogenous, and carbonyl compounds, in the ambient air of Gumi City, where a large number of electronics industries are found. Two field monitoring campaigns were conducted for a one year period in 2003/2004 and 2010/2011 at several sampling sites in the city, representing industrial, residential and commercial areas. More than 80 individual compounds were determined in this study, and important compounds were then identified according to their abundance, ubiquity and toxicity. The monitoring data revealed toluene, trichloroethylene and acetaldehyde to be the most significant air toxics in the city, and their major sources were mainly industrial activities. On the other hand, there was no clear evidence of an industrial impact on the concentrations of benzene and formaldehyde in the ambient air of the city. Overall, seasonal variations were not as distinct as locational variations in the VOCs concentrations, whereas the within-day variations showed a typical pattern of urban air pollution, *i.e.,* increase in the morning, decrease in the afternoon, and an increase again in the evening. Considerable decreases in the concentrations of VOCs from 2003 to 2011 were observed. The reductions in the ambient concentrations were confirmed further by the Korean PRTR data in industrial emissions within the city. Significant decreases in the concentrations of benzene and acetaldehyde were also noted, whereas formaldehyde appeared to be almost constant between the both campaigns. The decreased trends in the ambient levels were attributed not only to the stricter regulations for VOCs in Korea, but also to the voluntary agreement of major companies to reduce the use of organic solvents. In addition, a site planning project for an eco-friendly industrial complex is believed to play a contributory role in improving the air quality of the city.

KEYWORDS

VOCs; BTEX; HAPs; ambient air; electronics industry; Gumi city

1. INTRODUCTION

Over the past two decades there has been a rapid increase in urbanization and industrialization in Korea. With this has come a dramatic increase in the number of manufacturing facilities, residences, and office buildings, together with increases in both the number and density of motor vehicles. As the total area of South Korea is very small, most urban areas are densely populated and new towns have developed rapidly in the vicinity of industrial complexes. Thus, industrial emissions and motor vehicles are believed to be the major causes of ambient air pollution in most Korean cities [1].

Recently, emissions of hazardous air pollutants (HAPs), particularly toxic volatile organic compounds (VOCs), from large industrial complexes have been of great concern in Korea [2,3]. Gumi City is a typical industrial city in Korea with a population of approximately 450,000. The Gumi National Industrial Complexes (GNIC) is often called the Korean Silicon Valley and the Mecca of the Korean electronics industry. GNIC plays a key role in the South Korean economy, being responsible for a major proportion of the country's exports. More than 1700 companies with approximately 90,000 workers are located in the GNIC. Among the industries in the city, 84% are electronic companies, 6% are petro-chemical, 4% are non-metal, 3% are textile and 3% are other industries [4]. The electronic industry is occasionally regarded as an "environment-friendly" or "stack-free" business because it consumes very little fossil fuels and industrial water. However, in general, the electronic industries use a variety of organic and/or inorganic solvents in their manufacturing processes [1]. As a result, the fugitive emissions of toxic chemicals volatilized from the industrial processes may cause unknown air quality problems in such a large industrial city [5]. Therefore, monitoring of the ambient air quality with respect to toxic VOCs is an important task in view of the health of workers and residents living around the industries [6].

Although questions have been raised from the general public and NGOs in Gumi City regarding the quality of ambient air [5,7], the air quality data that is available for this city has not been adequate to provide accurate information on the real-life situation to policy makers. This is largely because the air quality data available in the city focuses mostly on criteria pollutants, such as particulate matter, sulfur dioxide and nitrogen oxides, whereas little attention has been given to non-criteria pollutants, such as VOCs. In Korea, the first and only ambient air quality standard associated with the VOCs appeared in 2007 for benzene, which is 1.5 ppb as an annual average [1]. Before developing a solid scheme for the effective management of air quality problems, scientific information on at least three factors should be provided, which are abundance, ubiquity and toxicity of a wide range of target pollutants.

Environmental implications of VOCs can be summarized into two aspects, *i.e.*, one is that many VOCs (particularly aliphatic hydrocarbons such as olefins) play an important role as the precursors of secondary air pollutants, such as ozone, aldehydes and organic aerosols [8], and the other is that some VOCs (particularly aromatics and halogenated ones) can have adverse effects on human health [8]. As an example, benzene is considered one of the most important VOCs because of its carcinogenicity. Benzene and formaldehyde are classified as group 1 carcinogens (proven carcinogen to humans) by the International Agency for Research on Cancer [9]. Exposure to other VOCs, such as toluene, ethylbenzene and trichloroethylene, may also have adverse health effects on humans [10].

High levels of ozone in the summertime are now very common phenomena everywhere in Korea, due to the VOCs and nitrogen dioxide from vehicle emissions and industrial sources [1]. On the other hand, the ozone problem in Gumi City has been less severe than other urban and industrial areas in Korea [11]. Therefore, if there is an air quality problem associated with VOCs in the city, the nature of the problem can be inferred to be related to health implications rather than the ozone issues, because a large amount of organic solvents are used in this city. According to the Korea Pollutant Release and Transfer Registers (PRTR) Data for 2009 [12], the top five VOCs released into the air of the city are toluene (65.4 ton/year), trichloroethylene (53.8 ton/year), xylenes (37.3 ton/year), methyl ethyl ketone (23.0 ton/year), and tetrachloroethylene (21.9 ton/year), comprising more than 99% of the VOCs emissions identified from industrial sources within the city.

Over the past few decades, a considerable number of studies have been conducted on atmospheric VOCs in urban and industrial areas [13,14,15,16,17,18,19,20,21,22,23,24,25,26]. Most studies focused on vehicle emissions in large urban areas [13,14,15,16,17] or the formation of secondary air pollutants as photochemical byproducts [18,19]. Regarding industrial VOCs, several field studies have been reported for different types of industries, such as petrochemical, oil refinery, chemical processing, steel industries, and painting processes [20,21,22,23,24,25,26,27]. On the other hand, little is known about the occurrence and concentrations of VOCs associated with the electronic industry.

The main aim of this study was to characterize the occurrence and concentrations of a wide range of VOCs including aliphatic, aromatic, nitrogenous, halogenated, and carbonyl compounds in the ambient air of Gumi City. The temporal and spatial variations of ambient levels of VOCs were investigated by field monitoring in several sites in the city covering four seasons from 2003 to 2004 (first campaign), and from 2010 to 2011 (second campaign). More specific objectives of this study were: (i) to provide quantitative information on the concentrations of toxic VOCs in residential, commercial and industrial areas in the city; (ii) to investigate the extent to which industrial sources influence the air quality in non-industrial areas in the city; and (iii) to assess the impact of environmental policies on the reduction of VOCs emissions by investigating the long term trends of the VOCs concentrations in the city.

2. EXPERIMENTAL SECTION

2.1. Description of Sampling Sites

To characterize the atmospheric VOCs in Gumi City, two field campaigns were carried out at 7 year intervals, *i.e.*, the first was from December 2003 to November 2004 and the second was from April 2010 to March 2011. In the first campaign, ambient air samples were collected at five sites in the city, *i.e.*, three industrial, one residential and one commercial site, as shown in Figure 1. Industrial site I was located in the center of the oldest and the largest complex; thus particular attention was given to this site. Industrial site II was in a mixed zone of half industrial and half commercial/residential areas. Nevertheless, it was named as an industrial site because big companies, such as LG and Samsung, are located near the sampling site; hence considerable impacts of industrial sources are expected. Industrial site III was located within the smallest complex in a valley, which was isolated geographically from the other complexes. This site was excluded for the second campaign because many companies in this complex had moved to a new complex developed in the outskirts of the city (information on the locations of the GNIC can be found in a website [4]). Instead, the frequency of sampling at other sites was increased during the second campaign. The residential site was located in a new town, being surrounded by a large number of apartment buildings. Finally, the commercial site was in a traditional old town, where many shopping centers and office buildings were concentrated. All sampling sites were well prepared for air sampling purposes in terms of electricity and vandalism, because these sites all belong to the National Air Quality Monitoring Network in Korea.

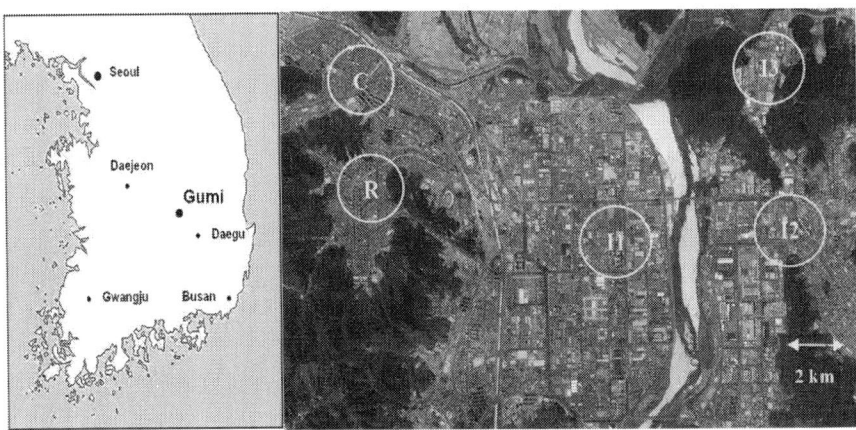

Figure 1. Locations of Gumi City in Korea and sampling sites in the city: I1—industrial site I, I2—industrial site II, I3—industrial site III; R—residential site; C—commercial site.

2.2. Sampling Period, Frequency and Duration

The sampling periods, frequencies and duration for the two campaigns are summarized in Table 1. For both campaigns, air sampling was conducted in four seasons at each site, *i.e.*, spring, summer, fall, and winter. In each season, sampling was carried out over seven consecutive days in each site. One exception is industrial site I, where samples were taken every six days throughout the year for the first campaign (but not in the second campaign). This site was identified as the most important site, as mentioned earlier.

During the first campaign, all the VOCs samples were collected manually in duplicate, and the average of the two measurements was regarded as one data-point. Each sampling was carried out for 150 min and three times per day, *i.e.*, morning (09:00~11:30), afternoon (13:00~15:30), and evening (18:30~21:00). On the other hand, single samples were collected automatically during the second campaign, and the sampling frequency was increased to six times per day (00:00~04:00, 04:00~08:00, and so on). For carbonyl compounds, the sampling frequency and duration were same as the VOC samples for the first campaign, while only two samples per day were collected for the second campaign, *i.e.*, one in the morning (09:00~11:00) and the other in the afternoon (13:00~15:00). As a result, 507 and 668 samples for VOCs (476 and 212 for carbonyls) were collected during the first and second campaigns, respectively.

Table 1. Sampling sites and periods for the first and the second campaigns.

	2003/2004 Campaign	2010/2011 Campaign
Sampling site *	3 industrial sites (I, II & III)	2 industrial sites (I & II)
	1 residential site	1 residential site
	1 commercial site	1 commercial site
Sampling period **	Winter: 13~19 January 2004	Spring: 21~27 May 2010
	Spring: 9~15 April 2004	Summer: 3~9 August 2010
	Summer: 13~19 August 2004	Fall: 15~21 October 2010
	Fall: 15~21 October 2004	Winter: 5~11 January 2011
Sampling method	Manual sampling	Automatic sampling
Sampling flowrate	150 mL/min for VOCs	100 mL/min for VOCs
	1 L/min for carbonyls	1 L/min for carbonyls
Sampling duration	2.5 h (150 min) for VOCs	4 h (240 min) for VOCs
	2.5 h (150 min) for carbonyls	2 h (120 min) for carbonyls
Sampling frequency	3 per day for VOCs and carbonyls	6 per day for VOCs and 2 per day for carbonyls
Number of samples	507 for VOCs and 476 for carbonyls	668 for VOCs and 212 for carbonyls

* In the first campaign, industrial site I was selected as a supersite, and industrial site III was excluded in the second campaign. ** In the first campaign, sampling was carried out every 6 days at industrial site I from December 2003 to November 2004.

2.3. Sampling and Analytical Methods

The protocol of the VOCs measurement methods used in this study was in principle similar to the USEPA TO-17 [28]. For the 2010–2011 campaign, the VOC samples were collected at a flow rate of 100 mL/min on stainless steel adsorbent tubes (1/4" × 9 cm), using low flow rate pumps equipped with a mass flow controller (Flec pump, Chematec, Roskilde, Denmark) and sequential automatic tube samplers (STS 25, Perkin Elmer, Beaconsfield, UK and MTS32, Markes, Llantrisant, UK). Each adsorbent tube was packed with 120 mg of Carbograph 2 (20/40 mesh) in the front position and 280 mg of Carbograph 1 (40/60 mesh) in the back position. Before sampling, the adsorbent tubes were preconditioned with helium as a carrier gas at 250 °C for 2 h using a thermal conditioner (TC-20, Markes). The preconditioned tube was kept in a 50 mL vial with a PTFE lined cap. The analysis of VOCs was performed using a GC/MS (HP6890/5973, Hewlett-Packard, Wilmington, DE, USA) with an automatic thermal desorption apparatus (Unity/Ultra, Markes). During the thermal desorption process, the eluted VOCs from the sampling tube were transferred to a cold trap (packed with 12 mg of Tenax TA and 47 mg of Carbotrap) at 300 °C and a flow rate of 50 mL/min for 10 min. Subsequently, the cold trap was heated rapidly from −15 °C to 320 °C, and maintained at that temperature for 5 min. The VOCs were then injected onto a capillary column (Rtx-1, 0.32 mm × 105 m × 1.5 μm, Restek, Bellefonte, PA, USA). The initial temperature of the GC oven was set to 50 °C for 10 min, then increased to 250 °C at a rate of 5 °C/min. The valves of the thermal desorber and transfer line were maintained at 180 °C. The carrier gas (helium) flow rate in the column was 1.4 mL/min (15 psi) and the outlet split flow of the thermal desorber was 10 mL/min.

For the 2003–2004 campaign, the VOCs were collected by drawing air through a stainless steel sampling tube (1/4″ × 9 cm) containing 100 mg of Carbotrap-C (equivalent to Carbograph 2) backed up with 300 mg of Carbotrap (equivalent to Carbograph 1) at a flow rate of 150 mL/min. The analysis of VOCs was performed using an automatic thermal desorption unit (ATD-400, Perkin Elmer) connected to the GC/MS. The GC column and operating conditions of the analytical system for the first campaign were virtually the same as those for the second campaign.

For both campaigns, carbonyl compounds were collected on DNPH-silica cartridges (LpDNPH S10L, Supelco, Bellefonte, PA, USA) at a flow rate of 1 L/min. An ozone scrubber (Supelco) was placed in front of the DNPH cartridge. The carbonyls were extracted with 3 mL of acetonitrile, and then analyzed by HPLC with UV detection at 360 nm (Shimadzu HPLC system, Kyoto, Japan). Details of the carbonyl sampling and analytical methods used in this study can be found elsewhere [7,29].

2.4. Target Analytes and Preparation of Standard Samples

In this study, the target VOCs for the two campaigns were slightly different due to a difference in the standard mixtures used for calibration. During the first campaign, two types of gas standards were used, *i.e.*, a TO-14A 41 Component Mix (1 ppm, Restek, Bellefonte, PA, USA) and a mixture of 57 ozone precursors (1 ppm, Restek). Some of VOCs were duplicated in the two mixtures and some very volatile VOCs, such as $C_2 \sim C_3$ hydrocarbons could not be measured using the sampling and analytical methods of this study. As a result, a total of 83 individual VOCs were finally selected as the target analytes. For the second campaign, an EPA TO-15/17 Calibration Mix (1 ppm, Supelco) containing 62 components were used. Among the 62 compounds, four very volatile VOCs (propylene, ethanol, Freon 12, and chloromethane) were excluded from the determination because these compounds were found to be collected inefficiently by the adsorbent tubes used in this study. Methyl ethyl ketone (MEK) was included in the 62 Mix, but this compound was determined by HPLC to be a carbonyl compound. Acrylonitrile was not included in the 62 Mix, but this compound was calibrated with the 41 Mix standards. In addition, for some important VOCs, which were unavailable in the gas standard mixtures, a liquid standard mixture was prepared with individual liquid standards (Sigma-Aldrich Co., St. Louis, MO, USA). These were naphthalene, N,N-dimethylformamide (DMF), epichlorohydrin, nitrobenzene, aniline, phenol, 2-methoxyethanol, 2-ethoxyethaol, and 2-ethoxyethylacetate. These 9 VOCs are of concern in Korea, as being included among 48 priority air toxics [3]. As a result, a total of 67 compounds were determined for the second campaign. A self-manufactured spiking apparatus [30,31] was used to prepare the standard samples by spiking known amounts of the gas standard mixture into pre-conditioned tubes. The liquid standard samples were prepared by spiking the standard into cleaned adsorbent tubes using a packed column injector ofthe GC at 250 °C and a helium flow rate of 100 mL/min [32]. In addition to these compounds, 15 carbonyl compounds were determined for both campaigns using a standard mixture (TO11/IP-6A Aldehyde/Ketone-DNPH Mix., Supleco) [33].

2.5. Quality Control and Quality Assurance

An evaluation of duplicate precision for the VOC samples analyzed by the adsorption and thermal desorption method is essential because replicate analysis is practically impossible for such samples. According to the TO-17, a criterion of 30% for duplicate precision was recommended for the sorbent-based sampling of VOCs. In this study, the mean duplicate precision (MDP) was within 30% for the majority of the target VOCs, and the within-a-day repeatability appeared to be less than 10% overall, while the between-days repeatability was less than 20%. The method detection limits (MDLs) for each target VOC were evaluated according to the USEPA guidelines [34], and the MDLs were estimated to be 0.01~0.05 ppb, depending on the individual compounds. More details on the performance of the VOC sampling and analytical methods used in this study can be found elsewhere [30,31,32,33]. Until now, no standard reference materials (SRM) are available for VOC samples. In this study, as an alternative approach to evaluate the accuracy of VOC measurements, inter-lab comparison studies were carried out with a "third party" laboratory at the Korea Research Institute of Standards and Science (KRISS) by sharing parts of duplicate samples ($n = 26$) [5]. The results of the inter-lab comparisons showed that the MDPs for toluene and trichloroethylene were 27.5% and 19.2%, respectively, while benzene was 32.1%. More volatile compounds, such as dichloromethane, appeared to be less accurate (50.7%). However, most of target VOCs showed a MDP of 30% or lower.

3. RESULTS AND DISCUSSION

In this study, a very wide range of VOCs were determined. The monitoring data will be discussed with respect to their frequencies of detection, locational, and temporal variations. A particular emphasis will be given on the comparison of concentrations in "industrial" and "non-industrial" areas to evaluate industrial impacts on the ambient levels of VOCs in Gumi city. Finally, data from the two campaigns will be compared to investigate a long term trend in the VOC concentrations in the city.

3.1. Occurrence of VOCs in the Ambient Air of Gumi City

In Gumi City, the most ubiquitous VOCs in ambient air appeared to be benzene, toluene, m,p-xylenes, formaldehyde, acetaldehyde, and MEK because they have been detected in every sample during both campaigns. The frequencies of detection for individual compounds are summarized in Table 2. In this study, the detection frequency is defined as the percentage of the number of samples over the MDL for the total number of effective samples. For the first campaign, 32 compounds among 98 target compounds were detected in a more than 50% frequency, while 24 compounds have not been detected in any sample. During the second campaign, the detection frequencies of 26 among 82 target analytes were more than 50%, while 39 compounds were detected in less than 5% of the total samples. In this paper, therefore, the discussion will focus on a number of selected VOCs with respect to their abundance, ubiquity and toxicity.

3.2. Spatial Variations of VOCs Concentrations

In order to investigate locational distributions of VOCs within Gumi City, the concentration data at each sites for the selected VOCs from the first and the second campaigns are summarized in Table 3 and Table 4, respectively. A wide range of concentrations were documented. For example, the toluene and trichloroethylene levels during the first campaign ranged from 0.40 ppb to 50.78 ppb, and from "less than MDL" to 38.11 ppb, respectively. The highest concentration of 77.71 ppb was observed for acetaldehyde in the first industrial site of the first campaign. On the other hand, the ambient levels and variations of VOCs during the second campaign appeared to be considerably lower than in the first campaign, indicating that there have been many changes during the seven year period not only within the city, but in the governmental policies for improving the air quality in Korea. This issue will be discussed more intensively in the later part of this paper.

Table 2. Frequencies of detection for the target VOCs in the ambient air of Gumi city.

Detection Frequency	First Campaign (2003/2004)	Second Campaign (2010/2011)
100%~95%	benzene, toluene, *m,p*-xylene, pentane, 2-methylpentane, formaldehyde, acetaldehyde, acetone, methyl ethyl ketone	benzene, toluene, ethylbenzene, *m,p*-xylenes, hexane, vinylacetate, methyl tert-butyl ether, *o*-xylene, formaldehyde, acetaldehyde, acetone, methy ethyl ketone
75%~95%	isopentane, ethylbenzene, isobutane, hexane, dichloromethane, methylcyclopentane, propionaldehyde, trichloroethylene	ethyl acetate, carbon tetrachloride, heptane, 1,2,4-trimethylbenzene, methyl isobutyl ketone, cyclohexane, propionaldehyde

Detection Frequency	First Campaign (2003/2004)	Second Campaign (2010/2011)
50%~75%	3-methylpentane, butane, o-xylene, 1-butene, heptane, decane, nonane, dodecane, 3-methyl- hexane, m,p-ethyltoluene, 1,2,4-trimethylbenzene, crotonaldehyde, iso-valeraldehyde	naphthalene, trichloroethylene, styrene, 1,3,5-trimethylbenzene, crotonaldehyde, iso-valeraldehyde
25%~50%	2-methylhexane, *n*-octane, 2,3-dimethylbutane, methylcyclohexane, styrene, Freon 11, cyclohexane, butyraldehyde	Freon 113, 2-propanol, phenol, 4-ethyltoluene, 1,2-dichloropropane, Freon 11, tetrachloroethylene, 1,1,1-trichloroethane, N,N-dimethylformamide
5%~25%	tetrachloroethylene, 1-pentene, isoprene, 2-pentene, *trans*-2-butene, cyclopentane, 3-methylheptane, 1,1,1-trichloroethane, *cis*-2-pentene, 2-methyl- heptane, 1,3,5-trimethylbenzene, 1,2,3-trimethyl benzene, 1,2-dichloropropane, 2,2,4-trimethyl pentane, *o*-ethyltoluene, 2,3-dimethylpentane, p-diethylbenzene, carbon tetrachloride, 2,4-dimethylpentane	1,2-dichloroethane, chloroform, 2-methoxyethanol, dichloromethane, tetrahydrofuran, 1,4-dioxane, carbon disulfide, butyraldehyde

| 0%~5% | cis-2-butene, 2,3,4-trimethylpentane, chloroform, n-propylbenzene, 1,3-butadiene, 2,2-dimethyl-butane, Freon 12, 1-hexene, Freon 113, 1,4-dichlorobenzene, chlorobenzene, 1,2,4-trichloro-benzene, chloroethane, isopropylbenzene, 1,2-dichlorobenzene, acrylonitrile, 1,1,2,2-tetra-chloroethane, 1,1-dichloroethene, 1,2-dichloro-ethane, Freon 114, vinyl chloride, bromomethane, 1,1-dichloroethane, acrolein, cis-1,2-dichloro-ethylene, 1,2-dibromoethane, trans-1,3-dichloro-propene, 1,1,2-trichloroethane, cis-1,3-dichloro-propene, 1,3-dichlorobenzene, m-diethylbenzene, bromodichloromethane, bromoform, dibromo-chloromethane, trans-1,2-dichloroethylene, valeraledehyde, 1,2,3-trichlorobenzene, hexaaldehyde, o,m,p-tolualdehyde, 2,5-dimethyl-benzaldehyde | 2-ethoxyethanol, 1,3-butadiene, chlorobenzene, acrylonitrile, Freon 114, 1,1-dichloroethene, 1,2-dichlorobenzene, trans-1,2-dichloro-ethylene, Freon 12, 2-ethoxyethylacetate, benzyl chloride, 1,2,4-trichlorobenzene, 2-hexanone, bromomethane, 1,1-dichloroethane, bromodichloromethane, 1,1,2,2-tetrachloro-ethane, 1,3-dichlorobenzene, 1,4-dichloro-benzene, vinyl chloride, chloroethane, bromoform, cis-1,2-dichloroethylene, cis-1,3-dichloropropene, trans-1,3-dichloro-propene, 1,1,2-trichloroethane, dibromo-chloromethane, 1,2-dibromoethane, hexachloro-1,3-butadiene, anilin, epichlorohydrin, nitrobenzene, acrolein, n-valeraldehyde, hexaaldehyde, o,m,p-tolualdehyde, 2,5-dimethyl-benzaldehyde |

For the first campaign, toluene appeared to be the most abundant (5.60 ppb as a mean of total data, $n = 507$) among the VOCs group, followed in order by n-pentane (2.19 ppb), trichloroethylene (1.34 ppb) and m,p-xylenes (0.88 ppb). Among the carbonyl group, acetaldehyde (4.44 ppb, n = 476) was the most abundant compound, followed by formaldehyde (3.55 ppb), acetone (3.31 ppb) and MEK (2.74 ppb). Most of these compounds are strongly associated with industrial manufacturing processes, while formaldehyde in the urban atmosphere is commonly derived from both indoors and outdoors [35]. MEK, another well-known organic solvent used in industrial settings, is also emitted as a combustion product from motor vehicles, landfills and so forth [36].

Overall, the most abundant VOC for the second campaign was also toluene (2.06 ppb, $n = 668$), followed by ethyl acetate (0.65 ppb), m,p-xylenes (0.38 ppb) and benzene (0.35 ppb). For carbonyl compounds, the mean concentration (2.21 ppb, $n = 212$) of acetaldehyde decreased to a half level of the first campaign, while formaldehyde (3.46 ppb) almost the same as before. On the other hand, such comparison of the rankings for the two campaigns may not be appropriate because of the difference in the target compounds for each campaign.

Although the overall rankings of individual compounds in each site were similar, the levels of some VOCs and carbonyls varied widely between the sampling sites, indicating the effects of local emission sources on the sites. For the first campaign, the mean concentrations of toluene, acetaldehyde, and trichloroethylene appeared to be much higher in industrial sites I and II than in the other sites (Table 3), whereas saturated hydrocarbons, such as pentane, decane and nonane, were detected not only more frequently but at higher levels in the industrial sampling site III. Interestingly, a large municipal landfill is located in the vicinity of this site. Thus, the increased levels of alkanes were estimated to be affected by the VOC emissions from the landfill [37,38].

Table 3. Concentrations (in ppb) of selected VOCs in Gumi city during 2003/2004.

VOCs	Industrial Site I (n = 179)		Industrial Site II (n = 82)		Industrial Site III (n = 82)		Residential Site (n = 82)		Commercial Site (n = 82)	
	Mean ± SD [1]	Max [2]	Mean ± SD	Max	Mean ± SD	Max	Mean ± SD	Max	Mean ± SD	Max
n-Butane	0.39 ± 0.44	2.21	0.40 ± 0.56	2.93	0.68 ± 0.76	3.28	0.24 ± 0.28	1.87	0.30 ± 0.32	1.49
iso-Pentane	0.78 ± 0.67	4.51	0.62 ± 0.52	1.99	1.67 ± 1.30	7.79	0.47 ± 0.35	1.63	0.55 ± 0.38	2.14
n-Pentane	1.90 ± 2.51	26.16	1.09 ± 1.23	6.71	7.39 ± 7.86	38.99	0.45 ± 0.38	1.74	0.43 ± 0.30	1.41
Dichloromethane	1.24 ± 1.27	8.16	0.63 ± 0.68	3.47	0.31 ± 0.30	1.28	0.55 ± 0.59	3.82	0.72 ± 0.65	3.88
2-Methylpentane	0.81 ± 0.73	6.10	0.59 ± 0.49	2.83	0.82 ± 0.51	4.16	0.43 ± 0.27	1.41	0.52 ± 0.31	1.36
3-Methylpentane	0.34 ± 0.39	3.27	0.25 ± 0.22	0.97	0.42 ± 0.40	2.52	0.17 ± 0.16	0.90	0.22 ± 0.19	1.10
n-Hexane	0.53 ± 0.62	5.92	0.53 ± 0.49	2.17	1.21 ± 1.31	8.49	0.36 ± 0.20	0.93	0.39 ± 0.32	2.29
Benzene	0.54 ± 0.28	1.64	0.68 ± 0.33	1.88	0.69 ± 0.31	1.91	0.60 ± 0.27	1.65	0.67 ± 0.33	1.71
Carbon tetrachloride	0.07 ± 0.03	0.20	0.08 ± 0.02	0.13	0.05 ± 0.01	0.09	0.08 ± 0.02	0.13	0.06 ± 0.02	0.11
Trichloroethylene	3.12 ± 5.29	38.11	0.49 ± 0.47	2.04	0.47 ± 0.65	3.60	0.20 ± 0.25	1.37	0.31 ± 0.49	2.57
Toluene	5.50 ± 5.12	37.12	7.93 ± 8.69	50.78	8.95 ± 6.32	32.05	2.82 ± 1.96	9.54	2.43 ± 1.70	8.39
Tetrachloroethylene	0.13 ± 0.35	3.89	0.08 ± 0.06	0.37	0.24 ± 0.28	1.75	0.04 ± 0.04	0.30	0.05 ± 0.03	0.22
Ethylbenzene	0.54 ± 0.66	5.77	0.54 ± 0.70	5.33	0.50 ± 0.32	1.93	0.28 ± 0.21	1.77	0.34 ± 0.35	2.55
m,p-Xylenes	1.10 ± 1.29	9.17	0.96 ± 0.97	5.99	1.00 ± 0.58	2.51	0.57 ± 0.55	2.91	0.48 ± 0.34	1.98
Styrene	0.23 ± 0.34	3.30	0.10 ± 0.07	0.35	0.31 ± 0.36	1.59	0.05 ± 0.04	0.40	0.05 ± 0.06	0.39
o-Xylene	0.30 ± 0.33	2.21	0.28 ± 0.29	1.84	0.31 ± 0.19	0.76	0.16 ± 0.15	0.72	0.14 ± 0.10	0.63
1,2,4-TMB [3]	0.18 ± 0.21	1.39	0.20 ± 0.20	1.06	0.31 ± 0.24	1.08	0.11 ± 0.10	0.49	0.13 ± 0.11	0.56
n-Decane	0.43 ± 0.48	3.25	0.31 ± 0.54	4.20	1.91 ± 1.90	8.85	0.14 ± 0.15	0.80	0.18 ± 0.21	1.39
n-Nonane	0.25 ± 0.41	3.22	0.22 ± 0.37	2.56	1.49 ± 2.24	16.60	0.11 ± 0.11	0.62	0.15 ± 0.16	0.90
Formaldehyde *	3.40 ± 1.66	9.10	3.86 ± 2.01	12.45	3.87 ± 1.90	10.31	3.46 ± 1.77	11.10	3.36 ± 1.62	7.85
Acetaldehyde *	7.35 ± 9.19	77.71	4.22 ± 3.22	15.69	2.52 ± 1.90	9.72	2.34 ± 1.21	5.43	2.05 ± 0.95	5.45
Acetone *	3.28 ± 1.56	11.06	3.59 ± 1.49	9.07	4.27 ± 2.70	15.61	2.74 ± 1.36	7.45	2.68 ± 1.63	13.38
MEK * [4]	2.40 ± 1.97	12.98	3.32 ± 2.64	13.28	4.81 ± 4.74	28.13	1.80 ± 1.12	4.79	1.71 ± 1.25	6.75
Propionaldehyde *	0.80 ± 0.90	4.11	1.04 ± 1.36	7.25	1.01 ± 1.18	6.19	1.21 ± 1.39	5.33	1.03 ± 1.12	5.61
Crotonaldehyde *	0.31 ± 0.31	2.06	0.36 ± 0.44	1.85	0.61 ± 0.69	4.12	0.10 ± 0.12	0.65	0.17 ± 0.30	2.09

[1] Standard deviation; [2] Maximum; [3] 1,2,4-Trimethylbenzene; [4] Methyl ethyl ketone. * The number of samples for the carbonyl compounds are 172, 78, 77, 75, and 74 for the industrial sites #1, #2, #3, residential, and commercial site, respectively.

For the second campaign, as shown in Table 4, there is also a clear tendency for some VOCs of higher levels at industrial sites than the other sites. According to the Korea PRTR Data [12], the major compounds used in industries located in Gumi City are toluene, ethylbenzene, m,p-xylenes, trichloroethylene, isopropyl alcohol, ethyl acetate, and styrene. All these VOCs were found generally at higher levels in industrial sites. On the other hand, the mean concentrations of benzene and methyl tert-butyl ether (MTBE) appeared to be similar at all four sites, suggesting that their major source is vehicle emissions. Although the annual average concentrations of formaldehyde for industrial site I and II were higher than that of the non-industrial sites, there was no statistically significant difference between the sites. This suggests that formaldehyde is emitted from a variety of sources, such as motor vehicles, furniture, insulation materials, and combustion, which are ubiquitous both indoors and outdoors in an urban area [7]. In addition, formaldehyde can be formed in the atmosphere as a byproduct of photochemical reactions [8,19]. DMF is a solvent used extensively in the textile industries for the dying process, and it was found to be higher at industrial site II than in the other sites because this site is surrounded by many textile industries. Although a few studies reported the DMF data in occupational settings [39], there is little information on the ambient levels for this compound.

Table 4. Concentrations (in ppb) of selected VOCs in Gumi City during 2010/2011.

VOCs	Industrial Site I ($n = 168$)		Industrial Site II ($n = 168$)		Residential Site ($n = 168$)		Commercial Site ($n = 164$)	
	Mean ± SD [1]	Max [2]	Mean ± SD	Max	Mean ± SD	Max	Mean ± SD	Max
Isopropyl alcohol	0.71 ± 1.91	18.84	0.23 ± 0.52	3.62	0.03 ± 0.08	0.58	0.06 ± 0.26	2.14
Dichloromethane	0.03 ± 0.12	0.88	0.09 ± 0.40	3.51	0.01 ± 0.04	0.29	0.05 ± 0.26	2.18
MTBE [3]	0.18 ± 0.14	0.66	0.25 ± 0.26	1.22	0.25 ± 0.31	2.21	0.26 ± 0.27	1.54
Ethyl acetate	1.25 ± 1.71	10.30	0.79 ± 1.20	8.61	0.24 ± 0.32	2.09	0.30 ± 0.80	6.94
n-Hexane	0.23 ± 0.15	0.97	0.26 ± 0.19	1.42	0.26 ± 0.38	4.01	0.24 ± 0.22	1.24
Vinyl acetate	0.31 ± 1.02	2.09	0.29 ± 1.22	2.26	0.23 ± 0.75	1.31	0.24 ± 0.52	1.01
Benzene	0.32 ± 0.18	0.81	0.34 ± 0.21	0.98	0.32 ± 0.21	0.97	0.41 ± 0.46	2.75
Carbon tetrachloride	0.07 ± 0.17	0.26	0.06 ± 0.10	0.16	0.07 ± 0.05	0.12	0.07 ± 0.07	0.19
Trichloroethylene	0.41 ± 0.46	2.70	0.27 ± 0.33	1.83	0.06 ± 0.10	0.71	0.05 ± 0.07	0.47
MIBK [4]	0.39 ± 0.46	3.09	0.37 ± 0.48	3.88	0.07 ± 0.07	0.39	0.07 ± 0.10	0.84
Toluene	2.98 ± 2.02	10.30	3.02 ± 2.77	15.39	1.14 ± 0.97	5.68	1.11 ± 1.18	8.81
Tetrachloroethylene	0.01 ± 0.03	0.11	0.03 ± 0.05	0.31	0.01 ± 0.01	0.07	0.01 ± 0.03	0.15
Ethylbenzene	0.33 ± 0.30	1.94	0.36 ± 0.35	1.97	0.15 ± 0.19	2.13	0.14 ± 0.11	0.81
m,p-Xylenes	0.56 ± 0.66	6.51	0.51 ± 0.46	2.44	0.23 ± 0.31	2.62	0.20 ± 0.15	1.02
Styrene	0.08 ± 0.09	0.57	0.07 ± 0.09	0.61	0.01 ± 0.02	0.10	0.02 ± 0.02	0.11
o-Xylene	0.18 ± 0.23	2.26	0.19 ± 0.19	1.33	0.08 ± 0.12	0.96	0.07 ± 0.05	0.34
1,2,4-TMB [5]	0.08 ± 0.16	1.76	0.07 ± 0.07	0.43	0.06 ± 0.13	1.00	0.04 ± 0.03	0.13
N,N-DMF [6]	0.16 ± 0.28	2.00	0.34 ± 0.58	3.31	0.04 ± 0.10	0.51	0.03 ± 0.08	0.70
Naphthalene	0.03 ± 0.03	0.15	0.03 ± 0.02	0.10	0.03 ± 0.02	0.11	0.03 ± 0.02	0.09
Formaldehyde *	3.77 ± 1.85	9.74	3.67 ± 1.68	7.59	3.17 ± 1.55	7.27	3.20 ± 1.63	8.41
Acetaldehyde *	2.58 ± 1.65	7.49	3.32 ± 1.70	8.05	1.43 ± 1.32	5.20	1.51 ± 1.22	6.00
Acetone *	0.49 ± 0.45	2.82	0.50 ± 0.65	3.12	0.35 ± 0.30	1.29	0.36 ± 0.31	3.42
MEK * [7]	1.05 ± 1.55	4.30	0.93 ± 0.85	4.73	0.55 ± 0.32	1.80	0.48 ± 0.40	2.13
Propionaldehyde *	0.02 ± 0.10	0.31	0.02 ± 0.12	0.31	0.02 ± 0.05	0.23	0.02 ± 0.04	0.25
Crotonaldehyde *	0.06 ± 0.15	0.66	0.07 ± 0.08	0.49	0.03 ± 0.05	0.22	0.02 ± 0.06	0.34

[1] Standard deviation; [2] Maximum; [3] Methy ter-butyl ether; [4] Methyl iso-butyl ketone; [5] 1,2,4-Trimethylbenzene; [6] Dimethylformamide; [7] Methyl ethyl ketone. * The number of samples for the carbonyl compounds is 53 for each site.

3.3. Comparison of VOCs Concentrations in Industrial and Non-Industrial Areas

To investigate the impacts of industrial sources on individual compounds in residential and commercial areas, the measured VOCs and carbonyl data were divided into two groups, *i.e.*, industrial and non-industrial groups. In other words, all the measured data from the industrial sites were pooled as a group, while the data from the residential and commercial sites was placed into another group. Comparisons of the two groups for the first and the second campaigns are illustrated in Figure 2 and Figure 3, respectively.

Figure 2 shows cumulative probabilities of the concentration data of selected VOCs for the two groups. Two distinct patterns can be found in Figure 2. Toluene, trichloroethylene, acetaldehyde, and MEK showed much higher levels in the industrial data group than the non-industrial group, whereas no statistically significant differences ($\alpha = 0.05$) were found for benzene and formaldehyde between the two groups. This indicates that the major sources of VOCs with higher levels in Figure 2 were mostly industrial emissions. On the other hand, there was no evidence of potential impacts of industrial sources on the occurrence of benzene and formaldehyde in the ambient air of Gumi City.

As shown in Figure 3, the data from the second campaign also showed very similar patterns to the first campaign, even though the concentrations of the individual compounds are different from the previous survey. Distinct differences in the levels of toluene, ethylbenzene, xylenes, trichloroethylene, acetaldehyde, and MEK were noted between the two groups, indicating all these compounds are strongly associated with industrial emissions. The mean concentration of trichloroethylene in the industrial data group appeared to be almost 7 times higher than in the non-industrial group, whereas toluene, ethylbenzene and xylenes were 2~3 times higher. Acetaldehyde and MEK also showed approximately 2 times higher levels. In contrast, no significant differences in benzene and MTBE were noted between the two groups. As mentioned earlier, the Korean PRTR data also confirmed the non-use of benzene and MTBE in the industries in Gumi City [12]; thus the main source of benzene and MTBE in this city is believed to be vehicle emissions instead of industrial activities.

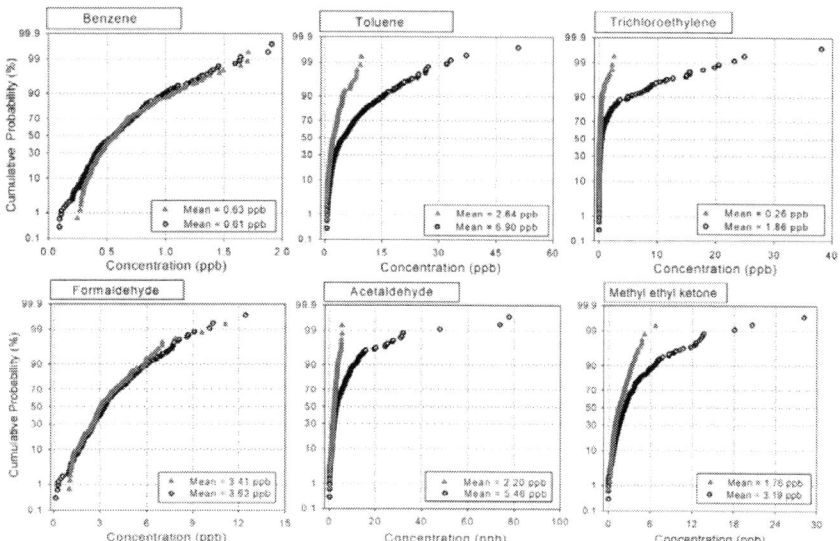

Figure 2. Comparison of the concentration distributions of selected VOCs in industrial (circle dot) and non-industrial areas (triangle dot) for the first campaign during 2003/2004.

Acetaldehyde is a notorious mal-odorant but it has not been reported in the Korean PRTR, because this compound is not a common solvent or a raw material in the industry. Nevertheless, the high levels of acetaldehyde in Gumi City have been of concern for a long time due to the public complaints of foul smells; this issue was discussed in a previous paper [7]. Alcohols, such as methanol and isopropanol, can play a role as a precursor of acetaldehyde emitted from industrial processes [40,41]. In fact, it was reported that a large amount of those alcohols are being used in the electronic industries in Gumi industrial complexes [7,12].

3.4. Temporal Variations of VOCs Concentrations

Seasonal variations in the ambient levels of VOCs can be influenced by a number of factors, including source variations, fuel consumption, chemical reactivity, meteorology, and the location and time of sampling. In the first campaign, seasonal variations of VOCs did not appear to be significantly different from each other. Therefore, seasonal data for industrial site I, which showed not only higher levels in most cases, but also more frequently measured than other sites, were demonstrated in Figure 4 for selected VOCs, as a typical example. The data from the second campaign were presented in Figure 5, where seasonal average concentrations of the selected VOCs in four sites were compared each other.

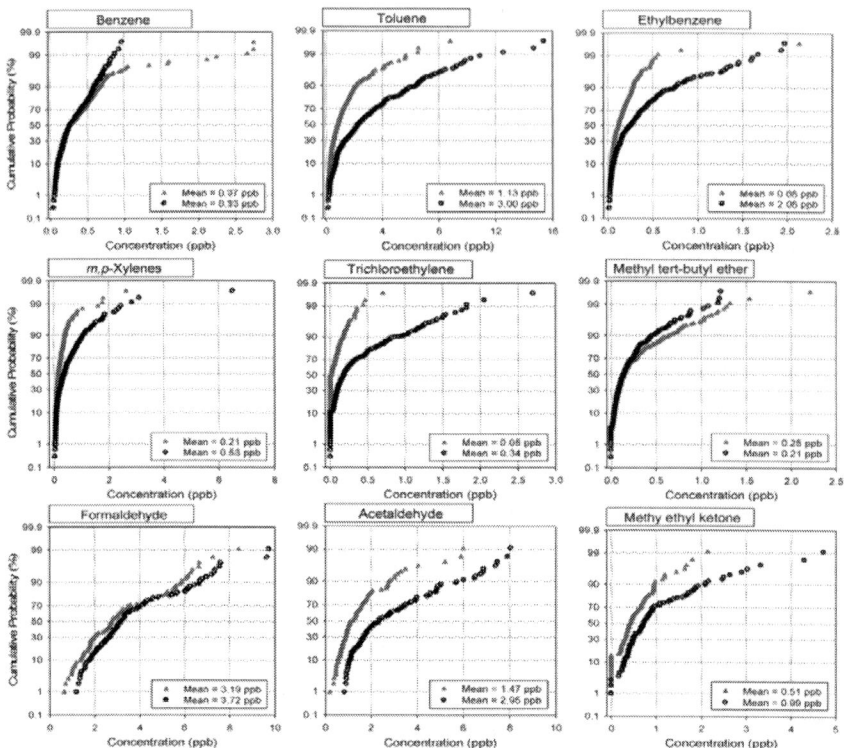

Figure 3. Comparison of the concentration distributions of selected VOCs in industrial (circle dot) and non-industrial areas (triangle dot) for the first campaign during 2010/2011.

Examination of Figure 4 and Figure 5 showed that the seasonal patterns of benzene, formaldehyde, acetaldehyde and other VOCs were all different from each other. First of all, the benzene concentrations increased during the winter and fall seasons in both campaigns. In typical urban areas, where industrial emissions are ignored, increased levels of VOCs might be generally expected during the colder seasons than warmer seasons [8]. Although elevated temperatures in the summer will obviously increase the evaporation of VOCs, the decay or removal of VOCs through photochemical reactions will be more significant during the summer. In addition, the ambient VOCs levels tend to increase due to air stagnation during the cold season. VOCs emissions from cold start vehicles account for a large proportion of the total VOCs emissions from motor vehicles equipped with catalytic converters, particularly during the winter [8,38]. Therefore, the increased winter benzene levels observed in this study suggest that catalysts may not remove sufficient VOCs from the exhaust under stagnant winter conditions in Korea.

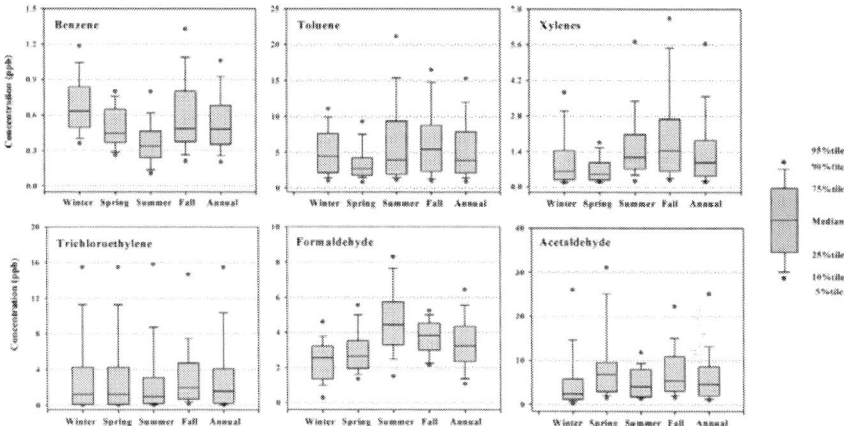

Figure 4. Seasonal variations of the selected VOCs at industrial site I during 2003/2004.

On the other hand, other VOCs associated with industrial emissions, such as toluene, ethylbenzene, xylenes and trichloroethylene, showed relatively fewer variations throughout the year, but were generally higher in fall than in the other seasons. Assuming that the industrial activity in Gumi City is constant throughout a year, the relatively higher concentrations during the fall season might be due to meteorological conditions. The meteorological conditions during both campaigns are summarized in Table 5. The data was obtained from the Gumi Regional Meteorological office, which is located near the commercial sampling site (within a distance of 500 m from the site C in Figure 1). Interestingly, the average wind speed for the sampling period for the fall season was 1.2 m/s for both campaigns, which was the lowest among the four seasons in 2010, and lower than winter and spring, but similar to the summer in 2004. No precipitation was noted during this period. Therefore, such elevated levels of industrial VOCs in the fall might be caused by the limited dilution, no scavenging, and poor dispersion in the atmosphere.

In contrast to benzene, the formaldehyde levels appeared to be the highest in summer and lowest in winter (Figure 4 andFigure 5), showing a typical "high-in-summer and low-in-winter" pattern, which has been reported elsewhere [35,36,42,43]. The increased concentrations of formaldehyde in summer were attributable not only to the increased volatile emissions due to the higher temperature, but also to the formation of secondary pollutants as a byproduct of the photochemical reactions in summer [40,43,44]. Acetaldehyde, methyl ethyl ketone and acetone are also known to be emitted as secondary pollutants of photochemical processes [45]. In this study, however, the concentrations of these carbonyl compounds did not vary noticeably from season to season, indicating that these compounds are largely associated with local emission sources with a variety of independent and irregular industrial activities.

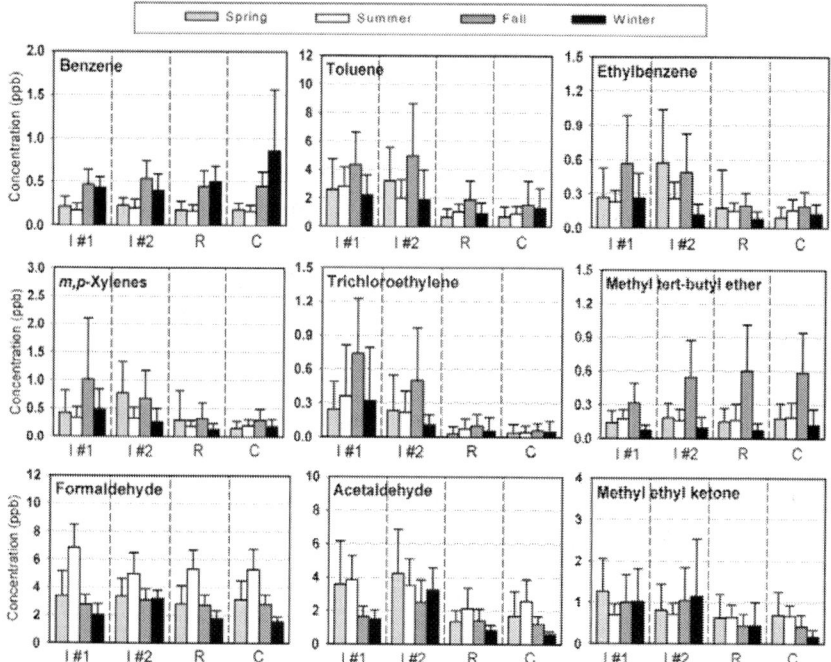

Figure 5. Seasonal variations of the selected VOCs at four different sites during 2010/2011 (the error bar indicates a standard deviation). I #1: industrial sit 1, I #2: industrial site #2, R: residential site, C: commercial site.

During the first campaign, to investigate diurnal variations in the VOCs levels, the samples were taken three times per day, *i.e.*, in the morning, afternoon and evening. More detailed diurnal variations were investigated during the second campaign by collecting six samples consecutively in a day. Figure 6 gives an example of the daily variations of the VOCs measured at industrial site II. This site was selected as an example because it is located in a mixed zone of industrial, residential and commercial areas. The concentrations of VOCs during a day varied, depending on the sites and compounds, but it was generally noticed that the levels of most VOCs, except formaldehyde, increased in the morning, decreased in the afternoon, and then increased again in the evening. The rise and fall in the VOCs levels during a day is apparently due not only to the atmospheric stability but also to traffic volumes in the Gumi area. In general, the traffic volume increases in the morning and evening. In contrast, the mixing height increases in the afternoon, but decreases in the morning and evening. In most cases, however, the concentrations of formaldehyde increased during the afternoon, particularly in summer. This suggests that there is a contributive portion of the secondary formation of formaldehyde through photochemical reactions in the air, as mentioned previously in this paper.

Table 5. Meteorological conditions during the first and the second campaigns in Gumi City.

		2003/2004 Campaign			
Sampling Period	Season	Average temp. (°C)	Wind Speed (m/s)	Prevailing Wind Direction	Rain (mm)
13–19 January	Winter	1.6	2.0	W	1.0
9–15 April	Spring	20.2	1.4	W	0.0
13–19 August	Summer	26.2	1.0	WSW	56.5
15–21 October	Fall	17.1	1.2	NW	0.0
		2010/2011 Campaign			
Sampling Period	Season	Average temp. (°C)	Wind Speed (m/s)	Prevailing Wind Direction	Rain (mm)
21–27 May	Spring	17.8	1.4	WNW	50.5
3–9 August	Summer	28.3	1.5	WNW	57.0
15–21 October	Fall	14.3	1.2	WNW	0
5–11 January	Winter	−3.0	2.8	WNW	0

3.5. Comparison of Concentrations of VOCs in 2003–2004 and 2010–2011 in Gumi City

An assessment of spatial and temporal variations of the VOCs concentrations in Gumi City suggests that there might have been significant changes in air quality in the city during the 7 year period. To evaluate the long term trends in the levels of VOCs and carbonyls in the city, site-by-site comparisons were carried out for commonly measured compounds in both campaigns, and the results are presented in Table 6. For comparison, the % reduction was calculated for each compound of concern from the data in Table 3 and Table 4, which is defined as a percentage of the difference between the two campaigns relative to the data from the first campaign. Table 6 shows that in most cases, the concentrations of VOCs and carbonyls have decreased considerably in the second campaign compared to the first campaign. The only exception is formaldehyde, which showed no significant reductions during the seven year period. Other VOCs and carbonyls appeared to be reduced in the range of 21.3%~98.2%, depending on the sampling sites. The largest reduction was found for dichloromethane, which was 93.6% on city-wide average, whereas other chlorinated VOCs, such as trichloroethylene and tetrachloroethylene, were reduced to approximately 25% of that measured in the first campaign. Aromatic VOCs, such as toluene, xylenes and ethylbenzene, also showed 43.0%~55.4% reductions, indicating that the concentrations of these compounds decreased to almost half of those measured in the first campaign.

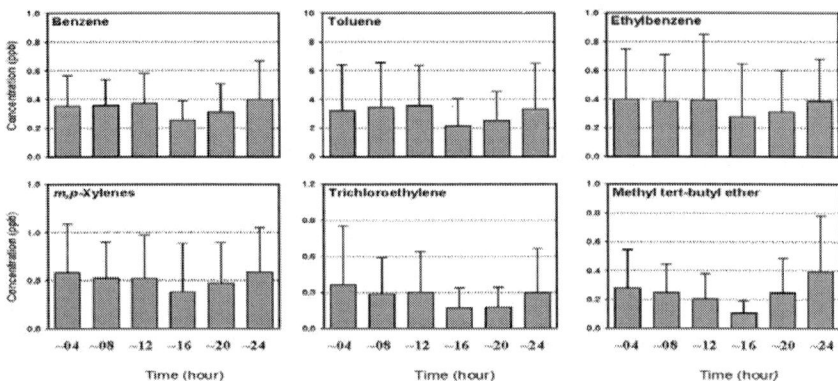

Figure 6. Diurnal variations of the 4 hourly average concentrations of selected VOCs at industrial site II during the 2010/2011 campaign (the error bar indicates a standard deviation).

The decreased trend for the organic solvents was confirmed further using the Korea National PRTR database [12]. According to the PRTR data, 305 tons of toluene were emitted into the Gumi atmosphere during 2004, whereas 106 tons were released in 2010, showing a 65.2% reduction. This study showed that the ambient concentrations of toluene decreased by 45.8% and 61.9% in the industrial sites I and II, respectively. A similar trend was also observed in the case of xylenes (as a sum of *m*-, *p*- and *o*-), *i.e.*, a 44.0% reduction in the emission data and a 48.3% reduction in the ambient concentrations. In case of trichloroethylene, however, the present study found a greater reduction (65.9% as an average of two industrial sites) than the PRTR database, which is 52.6%. Confirmation of the reduced emissions for the other compounds is not possible due to a lack of information in the 2004 PRTR database. Nevertheless, this study clearly shows that a reduction of the industrial emissions of VOCs and other organic solvents, such as acetone and MEK, resulted in decreased concentrations in the ambient atmosphere of the industrial areas, which led to decreased levels of those compounds in the air of the residential and commercial areas in Gumi City.

Table 6. Estimation of the reductions of the average concentrations (in ppb) of selected VOCs in the 2010/2011 campaign relative to the 2003/2004 campaign in Gumi city.

	% Reduction by Mean Concentrations in 2010/2011 Relative to 2003/2004				
	Industrial Site I	Industrial Site II	Residential Site	Commercial Site	Total Mean ± SD
Dichloromethane	97.6	85.7	98.2	93.1	93.6 ± 5.8
n-Hexane	56.6	50.9	27.7	38.5	43.4 ± 15.5
Benzene	40.7	50.0	46.7	38.8	44.1 ± 5.2
Trichloroethylene	86.9	44.9	70.0	83.9	71.4 ± 19.1
Toluene	45.8	61.9	59.6	54.3	55.4 ± 7.1
Tetrachloroethylene	92.3	62.5	75.0	80.0	77.5 ± 12.3
Ethylbenzene	38.9	33.3	46.4	58.8	44.4 ± 11.0
m,p-Xylenes	49.1	46.9	59.6	58.3	53.5 ± 6.4
Styrene	65.2	30.0	80.0	60.0	58.8 ± 21.0
o-Xylene	40.0	32.1	50.0	50.0	43.0 ± 8.7
1,2,4-Trimethylbenzene	55.6	65.0	45.5	69.2	58.8 ± 10.6
Formaldehyde	−10.9	4.9	8.4	4.8	1.8 ± 8.6
Acetaldehyde	64.9	21.3	38.9	26.3	37.9 ± 19.5
Acetone	85.1	86.1	87.2	86.6	86.2 ± 0.9
Methyl ethyl ketone	56.3	72.0	69.4	71.9	67.4 ± 7.5

Very high levels of acetaldehyde in industrial site I (Table 3) have been of concern in Gumi City, and a previous study [7] suggested the necessity of intensive supervision and surveillance for this compound as one of the top priority pollutants in the city. Interestingly, it was reported that the number of public complaints of foul smells in Gumi City had reduced considerably in 2009 compared to previous years [46], which was attributed to the decreased concentrations of acetaldehyde in industrial areas, as shown in Table 6. The mean concentration of benzene, a group 1 carcinogen, was reduced by approximately 44.1% on average in the four sites.

The considerable reductions in the concentrations of VOCs during a period between the two campaigns can be attributed to many reasons. First of all, governmental regulations for VOCs and HAPs have been strengthened during this period [1]. In Korea, the number of HAPs was 25 until 2007, but 9 VOCs and a group of polycyclic aromatic hydrocarbons (PAHs) were added to the list in 2008. Among the 35 HAPs, 24 chemicals (or a class of chemicals) are organic compounds. The emission standards for HAPs have been strengthened steadily over the last 20 years. According to the Korean air quality preservation law, if an industry deals with chemicals included in the HAPs list, very strict obligations are applied to the company for permission, monitoring, reporting, installation and operation of control facilities, *etc.* In addition to governmental policies, voluntary agreements between major companies and the Ministry of Environment to reduce the use of organic solvents appear to play a contributory role in reducing VOC emissions [1]. The introduction of advanced technology to control VOCs emissions in major industries also appear to contribute to the decrease in the concentrations of atmospheric VOCs in the city. Finally, the improvement in air quality is likely to be associated with the incidental effects of a site planning project implemented by Gumi city government. The project has been started in 2005 and is still on-going to redevelop the old industrial complex as an "eco-friendly" complex. Many old companies have been moved to a new industrial complex in the outskirts of the city (approximately 10 to 20 km away in northeast direction from the old complexes), and the empty spaces are being developed as a neighborhood park [4].

4. CONCLUSIONS

This paper reported the results of two field monitoring campaigns at a seven year interval in Gumi City, which was developed specifically for the electronic industry in Korea in the 1970s. More than 80 individual compounds were measured in this study, and then important compounds were identified with respect to their concentrations, frequencies of detection, and toxicities. The monitoring data showed that toluene, trichloroethylene and acetaldehyde are the most significant toxic VOCs in Gumi City, and their major sources are mainly industrial activities. On the other hand, there was no clear evidence of the industrial impacts on the concentrations of benzene and formaldehyde in the ambient air of the city. The ambient concentrations of benzene appeared to be similar at all sampling sites, suggesting that its major source is vehicle emissions. Formaldehyde also showed a similar pattern to benzene, but an additional contribution of secondary formation in the atmosphere to the ambient levels was observed in the summertime. Overall, the seasonal variations were not as distinct as locational variations in the VOCs concentrations, whereas the diurnal variations showed a typical pattern of urban air pollution, *i.e.*, increase in the morning, decrease in the afternoon and an increase again in the evening.

In most cases, there were considerable reductions in the VOCs concentrations from 2004 to 2011 in Gumi City. Citywide average reductions of 55.4%, 48.3% and 71.4% for the concentrations of toluene, xylenes and trichloroethylene, respectively, were estimated. The reductions in the ambient concentrations for the three compounds were confirmed by the Korean PRTR data for industrial emissions within the city during the same period. Significant reductions in concentrations of benzene and acetaldehyde were also observed, whereas formaldehyde appeared to be relatively constant during the period. The decreased trends in the concentrations of VOCs and carbonyls were attributed not only to the stricter regulations on VOCs in Korea, but also to the voluntary cooperation and introduction of advanced technology to control VOCs emissions in major industries. In addition, a site planning project by Gumi City for redeveloping the old industrial complex as an eco-friendly complex is likely to play a contributory role in improving the air quality of the city.

ACKNOWLEDGMENTS

This study was supported by the 2013 Yeungnam University Research Grants. Authors wish to appreciate valuable assistances in the field monitoring for graduate students. We also acknowledge that some parts of field monitoring works were supported by Gumi City Government and National Institute of Environment Research.

AUTHOR CONTRIBUTIONS

S.O. Baek conceived and designed this study; Y.K. Seo contributed to the field works; L.N. Suvarapu investigated some literatures and contributed to writing a draft; S.O. Baek and Y.K. Seo analyzed the data; S.O. Baek wrote the final manuscript.

REFERENCES

1. Korea Ministry of Environment. *White Paper*; Korea Ministry of Environment: Seoul, Korea, 2014; p. 589. (In Korean).

2. Korea Ministry of Environment (KMOE). *Report of Air Pollution Status on Hazardous Air Pollutants in Korea*; KMOE: Seoul, Korea, 2000; p. 156.

3. Park, J.M. Management and emissions of hazardous air pollutants from air sources. In Proceedings of the International Symposium on ROK-USA Collaboration for Policies of Air Quality Management, 72–93, Seoul, Korea, 29 November 2011.

4. Gumi City. Available online: http://english.gumi.go.kr/open_content/main_page (accessed on 2 February 2015).

5. Baek, S.O. Characterization of volatile organic compounds in the ambient air of a large industrial city in Korea. In Proceedings of the Symposium on

Air Quality Measurement Methods and Technology, Chapel Hill, NC, USA, 4–6 November 2008.

6. Bruhl, R.J.; Linder, S.H.; Sexton, K. Case study of municipal air pollution policies: Houston's air toxic control strategy under the White Administration, 2004–2009. *Environ. Sci. Technol.* **2013**, *47*, 4022–4028.

7. Seo, Y.K.; Baek, S.O. Characterization of carbonyl compounds in the ambient air of an industrial city in Korea. *Sensors* **2011**, *11*, 949–963.

8. Field, R.A.; Goldstone, M.E.; Lester, J.N.; Perry, R. The sources and behaviour of tropospheric anthropogenic volatile organic compounds. *Atmos. Environ.* **1992**, *26A*, 2983–2996.

9. International Agency for Research on Cancer (IARC). *IARC Monographs on the Evaluation of Carcinogenic Risks to Humans*; International Agency for Research on Cancer (IARC): Lyon, France, 2012; Volume 100 F.

10. WHO. *Guidelines for Air Quality*; WHO: Geneva, Switzerland, 2000; p. 190.

11. Korea Ministry of Environment. *Environmental Statistics Yearbook*; Korea Ministry of Environment: Seoul, Korea, 2014; p. 710. (In Korean).

12. Korea Ministry of Environment. *PRTR Information System*; Korea Ministry of Environment: Seoul, Korea, 2010.

13. Lung, C.H.; Horng, T.J.; Yu, C.S.; Hsiung, L.K.; Yi, M.S. VOC concentration profiles in an ozone non-attainment area: A case study in an urban and industrial complex metroplex in southern Taiwan. *Atmos. Environ.* **2007**, *41*, 1848–1860.

14. Hsieh, L.T.; Yang, H.H.; Chen, H.W. Ambient BTEX and MTBE in the neighborhoods of different industrial parks in Southern Taiwan. *J. Hazard. Mater.* **2006**, *A128*, 106–115.

15. Jia, C.; Foran, J. Air toxics concentrations, source identification, and health risks: An air pollution hot spot in southwest Memphis, TN. *Atmos. Environ.* **2013**, *81*, 112–116.

16. Hoque, R.R.; Khillare, P.S.; Agarwal, T.; Shridhar, V.; Balachandran, S. Spatial and temporal variation of BTEX in the urban atmosphere of Delhi, India. *Sci. Total Environ.* **2008**, *392*, 30–40.

17. Jia, C.; Batterman, S.; Godwin, C. VOCs in industrial, urban and suburban neighborhoods, Part 1: Indoor and outdoor concentrations, variation, and risk drivers. *Atmos. Environ.* **2008**, *42*, 2083–2100.

18. Czader, B.H.; Byun, D.W.; Kim, S.T.; Carter, W.P. A study of VOC reactivity in the Houston-Galveston air mixture utilizing an extended version of SAPRC-99 chemical mechanism. *Atmos. Environ.* **2008**, *42*, 5733–5742.

19. Villanueva, F.; Tapia, A.; Notario, A.; Albaladejo, J.; Martinez, E. Ambient levels and temporal trends of VOCs, including carbonyl compounds, and ozone at Cabaneros National Part border, Spain. *Atmos. Environ.* **2014**, *85*, 256–265.

20. Simpson, I.J.; Marrero, J.E.; Batterman, S.; Meinardi, S.; Barletta, B.; Blake, D.R. Air quality in the industrial heartland of Alberta, Canada and potential impacts on human health. *Atmos. Environ.* **2013**, *81*, 702–709.

21. MaCarthy, M.C.; Aklilu, Y.A.; Brown, S.G.; Lyder, D.A. Source apportionment of volatile organic compounds measured in Edmonton, Alberta. *Atmos. Environ.* **2013**, *81*, 504–516.

22. Tiwari, V.; Hanai, Y.; Masunaga, S. Ambient levels of volatile organic compounds in the vicinity of petrochemical industrial area of Yokohoma, Japan. *Air Qual. Atmos. Health* **2010**, *3*, 65–75.

23. Liu, P.W.G.; Yao, Y.C.; Tsai, J.H.; Hsu, Y.C.; Chang, L.P.; Chang, K.H. Source impacts by volatile organic compounds in an industrial city of southern Taiwan. *Sci. Total Environ.* **2008**, *398*, 154–163.

24. Kume, K.; Ohura, T.; Amagai, T.; Fusaya, M. Field monitoring of volatile organic compounds using passive air samplers in an industrial city in Japan. *Environ. Pollut.* **2008**, *153*, 649–657.

25. Roukos, J.; Riffault, V.; Locoge, N.; Plaisance, H. VOC in an urban and industrial harbor on the French North Sea coast during two contrasted meteorological situations. *Environ. Pollut.* **2009**, *157*, 3001–3009.

26. Ramirez, N.; Cuadras, A.; Rovira, E.; Borrull, F.; Marce, R.M. Chronic risk assessment of exposure to volatile organic compounds in the atmosphere near the largest Mediterranean industrial site. *Environ. Int.* **2012**, *39*, 200–209.

27. Celebi, U.B.; Vardar, N. Investigations of VOC emissions from indoor and outdoor painting processes in shipyards.*Atmos. Environ.* **2008**, *42*, 5685–5695.

28. USEPA. *Compendium Method TO-17: Determination of Volatile Organic Compounds in Ambient Air*; USEPA: Washington, DC, USA, 1999; p. 49.

29. USEPA. *Compendium Method TO-11A: Determination of Formaldehyde in Ambient Air using Adsorbent Cartridge Followed by High Performance Liquid Chromatography (HPLC)*; USEPA: Washington, DC, USA, 1999; p. 51.

30. Seo, Y.K.; Suvarapu, L.N.; Cho, B.Y.; Baek, S.O. A study on optimal combination of adsorbents for sampling of C4~C10 hydrocarbons in ambient air and analysis with thermal desorption gas chromatography coupled with mass spectrometry. *Asian J. Chem.* **2014**, *26*, 5283–5290.

31. Seo, Y.K.; Suvarapu, L.N.; Baek, S.O. Monitoring of volatile organic compounds at Gyeongju: A historical and tourist place in South Korea. *Asian J. Chem.* **2014**, *26*, 2493–2499.

32. Baek, S.O.; Jenkins, R.A. Characterization of trace organic compounds associated with aged and diluted sidestream tobacco smoke in a controlled atmosphere—Volatile organic compounds and polycyclic aromatic hydrocarbons.*Atmos. Environ.* **2004**, *38*, 6583–6599.

33. Baek, S.O.; Jenkins, R.A. Performance evaluation of simultaneous monitoring of personal exposure to environmental tobacco smoke and volatile organic compounds. *Indoor Built Environ.* **2001**, *10*, 200–208.

34. USEPA. *Determination and Procedure for the Determination of the Method Detection Limit, Code of Federal Regulations, Part 136, Appendix B*; USEPA: Washington, DC, USA, 1990; p. 537.

35. Feng, Y.; Wen, S.; Chen, Y.; Wang, X.; Lü, H.; Bi, X.; Sheng, G.; Fu, J. Ambient levels of carbonyl compounds and their sources in Guangzhou, China. *Atmos. Environ.* **2005**, *39*, 1789–1800.

36. Pang, X.; Mu, Y. Seasonal and diurnal variations of carbonyl compounds in Beijing ambient air. *Atmos. Environ.* **2006**,*40*, 6313–6320.

37. Kim, K.H.; Choi, Y.J.; SunWoo, Y.; Baek, S.O.; Jeon, E.C.; Hong, J.H. The emissions of major aromatic VOC as landfill gas from urban landfill sites in Korea. *Environ. Monit. Assess.* **2006**, *118*, 407–422.

38. Sing, H.B.; Salas, L.; Viezee, W.; Ferek, R. Measurement of volatile organic chemicals at selected sites in California.*Atmos. Environ.* **1992**, *26A*, 2929–2946.

39. Wrbitzky, R. Liver function in workers exposed to *N,N*-dimethylformamide during the production of synthetic textiles. *Int. Arch. Occup. Environ. Health* **1999**, *72*, 19–25.

40. Altshuller, A.P. Production of aldehydes as primary emissions and from secondary atmospheric reactions of alkenes and alkanes during the night and early morning hours. *Atmos. Environ.* **1993**, *27*, 21–32.

41. Northway, M.J.; Gouw, J.A.; Fahey, D.W.; Gao, R.S.; Warneke, C.; Roberts, J.M.; Flocke, F. Evaluation of the role of heterogeneous oxidation of alkenes in the detection of atmospheric acetaldehyde. *Atmos. Environ.* **2004**, *38*, 6017–6028.

42. Grutter, M.; Flores, E.; Andraca-Ayala, G.; Báez, A. Formaldehyde levels in downtown Mexico City during 2003.*Atmos. Environ.* **2005**, *39*, 1027–1034.

43. Tanner, R.L.; Meng, Z. Seasonal variations in ambient atmospheric levels of formaldehyde and acetaldehyde. *Environ. Sci. Technol.* **1984**, *18*, 723–726.

44. Moussa, S.G.; el-Fadel, M.; Saliba, N.A. Seasonal, diurnal and nocturnal behaviors of lower carbonyl compounds in the urban environment of Beirut, Lebanon. *Atmos. Environ.* **2006**, *40*, 2459–2468.

45. Liu, Y.; Yuan, B.; Li, X.; Shao, M.; Lu, S.; Li, Y.; Chang, C.C.; Wang, Z.; Hu, W.; Huang, X.; *et al.* Impact of pollution controls in Beijing on atmospheric oxygenated volatile organic compounds (OVOCs) during the 2008 Olympic Games: observation and modeling implications. *Atmos. Chem. Phys.* **2015**, *15*, 3045–3062.

46. Korea Ministry of Environment (KMoE). *Present Status of Designated area for Odor Control, KMoE*; Korea Ministry of Environment (KMoE): Seoul, Korea, 2009.

CHAPTER 7

Regional Air Quality of the Nigeria's Niger Delta

Precious N. Ede, David O. Edokpa[*]

Institute of Geosciences and Space Technology, Rivers State University of Science and Technology, Port Harcourt, Nigeria

ABSTRACT

There is no systematic attempt to evaluate the air quality of any settlement in the Niger Delta region over a long period. Records of air quality data for this study were generated through secondary sources from impact assessment of facilities aimed at implementing air quality regulations on the environment. Suspended particulate matter in the region's atmosphere ranged from 40 mg/m³ in Brass to 98 mg/m³ in Port Harcourt. Carbon monoxide concentrations were highest in Mbiama (191 mg/m³). Nitrogen dioxide concentration was highest in Bonny (187 mg/m³), and sulphur dioxide concentrations ranged from 19 mg/m³ in Ukwugba to 90 mg/m³ in Port Harcourt. Total hydrocarbon ranged from 78 mg/m³ in Odukpani to 192 mg/m³ in Nchia. Carbon dioxide ranged from 400 ppm in Buguma to 450 ppm in Port Harcourt. The most abundant of the VOCs is benzene and toluene. Ethylene was detected only in one station at concentration of 0.1 mg/m³ which was negligible. The most abundant of the metals was zinc, which was present at above 2 mg/m³ in most of the study settlements. In remote settlements like Buguma and Emuoha, some of the metals were not detected at all. In some instances, short-term limits for the pollutants exceeded WHO standards. The need for stakeholders in the region to articulate initiatives that support quality environmental practices was emphasized as laws pertaining to air quality regulations which are weak and less enforceable.

KEYWORDS

Niger Delta, Regional Air Quality, Emissions from Petroleum

1. INTRODUCTION

The Niger Delta is the southernmost part of Nigeria where all the petroleum exploration and production takes place and the generated income used to service at least 80% of the national budget. The implication is that industries are attracted to the region because of the presence of cheap and easy access to oil and gas. The spin-off is a region whose economy is always booming irrespective of situations elsewhere in the country. Unlike some parts of the world where there are cogent debates on energy options to optimize environmental sustainability, the Nigerian state is preoccupied with how to harness its petroleum endowment adequately to fund infrastructure and other aspirations of its teeming population. Nigeria is a member of the non-Annex 1 countries with no clear-cut limit on how much greenhouse gas it can emit, according to the Kyoto Protocol; this further promotes a negative emissions profile. The continuous flaring of associated gas known for the emission of all major air contaminants for instance; due to lax national regulations have caused Nigeria to be reputed as one of the highest. Globally, the volume of gas flared between 1996 and 2006 (during which time awareness of the detrimental impact of flare emissions on the global climate grew) remained relatively constant, ranging between 150 and 170 billion cubic meters (BCM). Nigeria's share of the total volume is approximately 24.1 BCM of gas; in comparison, the U.S. flares 2.8 BCM during the same time period [1] [2] . As a result of the associated activities with the exploration and production of oil and gas, the atmosphere of the Niger Delta is the primary sink into which the emissions are deposited. Rapid settlement and population growth in Nigeria as a whole and significant economic activities prompted by the huge petroleum industry in the Niger Delta are enough reasons to associate the area with diminished air quality. This widely acknowledged fact has led to series of studies and publications; most notable of which are the World Bank [3] study of the region and the Niger Delta Human Development Report by UNDP [4] that narrate paradoxes of poverty in the midst of plenty. The Environmental Assessment of Ogoniland by UNEP [5] delivered a catalogue of devastation due to oil pollution in Ogoniland, which was one of many ethnic Niger Delta settlements. These reports illustrate the level of pollution in the Niger Delta and in this paper the air quality of the area will be discussed in the context of the prevailing economic activities.

2. SOURCES OF EMISSIONS THAT IMPACT THE REGIONAL AIR QUALITY

There are several sources of pollutants that affect the air quality of the Niger Delta region and some may be natural. The anaerobic condition of the Niger Delta swamps is ideal for the natural production of atmospheric contaminants like ammonia and methane and this has been documented in Funtua near Brass, Bayela State [6] and the swamps around Port Harcourt [7] . Anthropogenic sources include economic and industrial activities especially in large urban centres and industrial complexes; and the best known process that impact on the air quality of the region is the petroleum industry. An estimation of emissions using the emission factors from domestic sources in the Niger Delta which includes domestic lighting (using generators) and domestic cooking (using firewood and kerosene) constituted sources of significant emission of pollutants as the estimates show 70 kt/year of CO; 50 kt/year of NO_x; 3 kt/year of PM_{10}; 2.4 kt/year of SO_2; 60 t/year of VOC; 5.7 mt/year of CO_2 and 2 kt/year of CH_4 released from domestic emissions [8] . A profile of gaseous emissions admitted to by one of the oil companies in Table 1 is illustrative. The sources of air pollutants from petroleum production include crude oil and condensate spills, gas flares, and vapours from storage, processing and transportation facilities.

There are over 14 public power plants in operation or under construction located in Afam I and II, Eleme, Port Harcourt, Sapele, Ughelli, Kwale, Gbaran, Aba, Omoku, Ikot Abasi, Odukpani, Alaoji and Azura-Edo. Many oil companies and heavy industries use private gas turbines to power their production facilities and residential quarters; for instance, the petrochemical plant in Eleme has four gas turbines and the liquefied natural gas complex in Bonny has ten gas turbines. Several new power plants are proposed in the near future, although these plants utilize natural gas regarded as relatively clean, they nevertheless emit noxious gases. The petroleum- based industries in the region include three refineries, two petrochemical plants and a 6-train gas liquefaction plant [11] that are still expanding, just as another fertilizer plant is under construction in Eleme by Idorama.

The heavy industries are notably the Aladja steel complex, the Ikot Abasi aluminium smelting company, and the Onne fertilizer plant. Typical sources of air pollutants in the Niger Delta are categorized in Table 2.

The Nigerian oil economy has fuelled the growth of cities in the Niger Delta. The most prominent of these cities is Port Harcourt, but there are others like Warri, Bonny, Eket and Omoku whose growth and recent sprawl are decidedly consequent upon the regional oil boom. In a spatially constrained island city like

Bonny where large petroleum based liquefied gas plant and crude oil export terminal have led to severe urban congestion, the environment and air quality is invariably undermined [13] . Metropolises such as Port Harcourt are important for the large number of vehicle traffic and industries emissions. A myriad of human activities like construction sites, machine operation and maintenance also release emissions that impact air quality. Owning a power generating set has become a necessity in both the rural and urban areas of the region because public power supply in Nigeria is grossly unreliable [14]. As a result, all the major oil producing companies have undertaken to generate their operational power needs independently, through gas turbines. These generating sets are key sources of emissions into the atmosphere.

Table 1. Gas emissions in SPDC (Shell Nigeria) operations (000 tons).

Pollutant	1999	2000	2001	2002	2003	2004
CO_2	18.10	21,838	22,489	15,467	18,798	19,798
CH_4	86.5	98.4	111.6	72.8	87.0	90.7
VOCs	135.3	160.2	183.3	100.4	117.2	122.2
Gas flared	6458	7693	7909	5222	6385	6611
SO_x	1.5	1.7	1.8	1.1	1.1	1.1
NO_x	20.1	17.8	27.3	22.3	23.1	21.9
HCFC/CFC	2650	3459	1901	2960	1198	2403

Source: [9] [10].

We have briefly overviewed the sources of emissions that impact on the air quality of the region. The location of a specific activity-form is likely to be associated with the potential air quality of its immediate vicinity and even beyond [15] [16] . The presence of heavy industries raises the level of atmospheric contamination in the region; likewise the sprawling urban centres and traffic congestion.

The air quality of an environment is tied to the atmospheric processes of the location. The air over the Niger Delta in terms of the weather systems does attenuate and sometimes enhance emissions dispersion; how they interact is noteworthy. Modelling of emissions to determine pollution transport in the region have been conducted by [15] - [17] . Small scale rural vocations such as bush burning as part of preparations for farming and the cottage processing of farm produce like oil palm also generate significant emissions [18] .

3. METHODOLOGY

There is no systematic attempt to evaluate the air quality of any settlement in the Niger Delta region over prolonged period. The region is host to many industries and rapidly growing cities, capable of impinging on good air quality. Records of air quality data in the region are sourced mainly through discrete efforts for impact assessment of facilities aimed at implementing various regulations on the environment. The greatest drawback of existing information in characterizing the regional air quality of Nigeria's delta region is that they are not continuous. A good regional air quality record should be based on a network of permanent stations with monitoring schedule. Moreover, reference to data generated to meet regulatory requirements may underplay actual levels of emissions present in the atmosphere because the polluter who generated the data intends to portray a seeming clean bill of health.

Table 2. Emission sources from human activities in the Niger Delta.

Category	Emission Source	Nature of Pollutant	Impact
Petroleum	a) Gas flaring b) Oil spill vapour c) Venting d) Fugitives	Heat radiation noxious emissions, Hazardous air pollutants (HAPs)	Thermal conduction, air pollution, noise, air pollution
Heavy industries	a) Cooling towers b) Separators c) Boilers d) Burners e) Venting	HC & VOC vapours HAPs, acid precursors, flue gas, particulates	Air pollution, vibration, acid rain
Other industries	a) Foundries b) Solvents c) Vapours d) Fuels e) Feedstock	Noxious gases, HAPs	Odour, health effects like carcinogenic
Power plants	Over six gas fired thermal plants exist.	HC, noxious gases, CO_2, particulates	Greenhouse-gas, air pollution
Automobiles, marine vessels and machines	a) Exhausts b) Leaks c) Wearing of tyres breaks, etc.	HC, noxious gases, CO_2, particulates, PAHs	Bodily irritation, smoke, noise
Other activities	a) Construction b) Agriculture c) Domestic sources	Methane, dust and particulates, noxious gases, etc.	Visibility, dust, etc

Source: [12].

During impact assessment, control measurements are taken to demonstrate how projects' location air quality differs from the general atmosphere. These controls can be assumed to reflect background air quality status of the region, except as already stated, they are not continuous over a significant period. It is from such records that this appraisal of air quality in the Niger Delta is based; therefore, data included here are not tied to any particular emission point nor are diurnal and seasonal variations captured. Although available records may be short on time, they are robust in the number of pollutants included. It has become quite routine for studies to include the primary pollutants like carbon monoxide (CO), oxides of nitrogen (NO_x), sulphur compounds (SO_x), lead and suspended particulates; sometimes too heavy metals and hydrocarbon species, such as the BTEX (benzene, toluene, ethylene and xylene) are also monitored. A few researchers are also linking air quality measurements to meteorological conditions and atmospheric modelling of emissions. The settlements in this report are based on the criteria of data accessibility to the author, although, these settlements were not the primary objective for generating the data. The data pertain to the settlements only in the sense that control measurements were extrapolated for them. The settlements included in this report are Ahoada, Bonny, Brass, Buguma, Ukwugba, Eket, Nchia, Emuoha, Mbiama, Odukpani, Omoku, Owasa, Port Harcourt, Sapele, Ughelli and Warri. These settlements straddle six states in the region as well as the ecological and climatological subdivisions of the Niger Delta, whose population estimated in the census of 2006 by NPC [19] was over 21 million.

The pollutants are grouped into three: the first category is mainly the primary pollutants emitted by industrial plants, automobiles and combustion processes. The emissions are carbon monoxide (CO) nitrogen dioxide (NO_2), sulphur dioxide (SO_2), total hydrocarbons (HC) and suspended particulate matter (SPM). Carbon dioxide (CO_2), a natural constituent of the atmosphere and a non-pollutant, but a greenhouse-gas is also included because of its growing profile in current environmental discourse. Carbon dioxide is not necessarily an air pollutant because it does not undergo secondary reactions, except as weak carbonic acid when mixed with moisture in the atmosphere. As a greenhouse gas, CO_2 is important along with methane (CH_4) and chlorofluorocarbons, which is why it is included in the list of common air pollutants in this instance.

The second category are the hydrocarbon species particularly, the BTEX group. This category has the least number of measurement records. The third category comprises trace metals presence in the atmosphere of the region. Measurements presented in this report are compared to standards to ascertain whether they are within regulatory limits.

Instrumentation for Data Measurement

The data utilised in this study was sourced through secondary sources from impact assessment of facilities aimed at implementing air quality regulations on the environment and the various methods used are summarized in Table 3.

4. RESULTS AND DISCUSSION

Results of measurement of emissions in the Niger Delta are summarised in Figure 1. Particulate load in the region's atmosphere ranged between 40 mg/m^3 in Brass to 98 mg/m^3 in Port Harcourt. The range of values for particulates depends on the time of the day, season, intensity of human activities and proximity to emission sources. All the settlements with high SPM such as

Owasa (79 mg/m^3), Nchia (75 mg/m^3) and Mbiama (73 mg/m^3) are located on busy highways, heavy industrial plants, numerous oil fields and gas flare points, or where construction activities are in progress. In terms of surface hydrology of the delta, the settlements with the highest suspended particulates are located on dryer lands and are accessible by road as well. It was estimated that the flaring of gas alone in the petroleum industry releases over 6000 tons of particulates into the atmosphere in the Niger Delta, annually [23] . [24] calculates that particulates from gas flares in Nigeria have a settling velocity of 1.15×10^8 m·s^{-1}. This is indicative that fractions of the particles are aerosols which drift like atmospheric air molecules. The removal process is therefore through wet deposition instead of gravitational settling. The particles are captured by water droplets present in the atmosphere and are removed by a process called washout if insoluble.

Table 3. Summary of methods used to measure emissions.

Parameter	Method
Sulphur dioxide	TCM/Pararosaniline—SO$_2$ is absorbed from the air in a potassium tetrachloromecurate and formaldehyde in the following amounts by controlling the flow rate of sample and reagents. A pararosaniline methyl sulphuric acid dye is formed. The absorbance of the coloured solution is determined at about 550 nm in a spectrophotometer. Concentrations in the range of <25 to 1000 µg/m^3 can be measured by this method [20]. This method requires relatively simple apparatus. It is essentially specific to SO$_2$ since all known interference are minimised or eliminated, samples are relatively stable after collection. It has been used widely as a reference method and is covered by an international standard [21].
Nitrogen dioxide	This method is intended for the manual determination of NO$_2$ in the atmosphere in the concentration range of a few to about 9400 µg/m^3 (5 ppm) for sampling periods of up to two hours and flow rate of up to 0.6/min. The principle of this method involve reacting NO$_2$ with diazolizing-coupling reagents (Sulphanilic acid and N-(1-naphthyl)-ethylene diamine dihydrochloride) to produce a deeply coloured azo dye whose intensity is measured spectrophotometrically. It is sensitive enough to detect low concentration of NO$_2$. This method has been used extensively in the United States and Europe, it has been tested by many workers and is highly recommended [20] [22].
Carbon dioxide	Colorimetric chemical sensors—This detects emissions at their permissible exposure limit after some minutes of exposure. The sensors consist of disposable array of cross-responsive nanoporous pigments whose colours are changed by diverse chemical interaction with analytes. CO$_2$ levels could range up to 10,000 ppm on the detector.
Carbon monoxide	Non-dispersive infrared—Non-Dispersive Infra-Red (NDIR) detectors are the industry standard method of measuring the concentration of carbon oxides. The constituent gas in a sample will absorb some infra-red at a particular frequency. By shining an infra-red beam through a sample cell (containing CO), and measuring the amount of infra-red absorbed by the sample at the necessary wavelength, a NDIR detector is able to measure the volumetric concentration of CO in the sample.
Hydrocarbons (total & BTEX)	An ultraviolet-visible spectrophotometer with automatic cell driver and spectrum design system equipped with quartz cells was used. With the spectrophotometer, the amount of a known chemical substance (concentrations) can be determined by measuring the intensity of light detected depending on the range of wavelength of light source.
Suspended Particulate Matter (SPM)	Gravimetric (non-destructive)—The sample is taken through continuous filtration of ambient air on glass fibre filtering material with a capturing capacity more than 99.5% and flow rate of 33 - 55 cm·s^{-1}. The filter head is turned with the open side down, at a distance of 1.5 - 3.0 m above the surface. The sampling time was for 24 hours. The sampling frequencies correspond to the character of the sampling site locations. The amount of sample captured on the filter (in µg) is determined gravimetrically as a difference between the weight of the filter prior to and after the exposure.
Heavy metals	Atomic absorption spectroscopy (AAS)—This method is very reliable and simple to use as it can analyze over 62 elements through atomic identification and quantification. Metals absorb ultraviolet lights in their elemental form when they are excited by heat and each metal has a characteristics wavelength that will be absorbed. The AAS instrument looks for a particular metal by focusing a beam of UV light at a specific wavelength through a flame and into a detector.

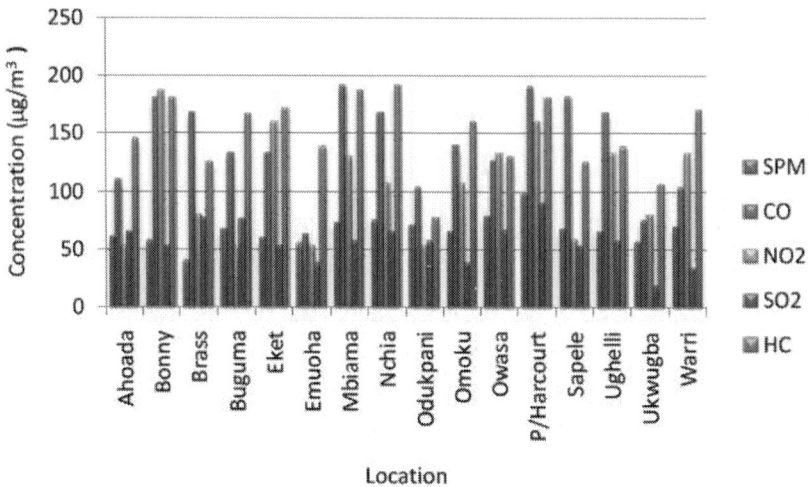

Figure 1. Concentration of major air pollutants in the region.

The main source for washout is impaction and interception, where a rain droplet removes particles in the layer below the cloud on its way down to the surface. If the particles are soluble they go into solution with rainwater. These observations are in line with [25] and Canada National Ambient Air Quality Objectives [26] .

Most petroleum related flare facilities onshore in Nigeria are horizontal and point to a surface pit surrounded by a bund wall. The stack height is therefore zero. But due to buoyancy forces acting on the emission, the effective plume rise can be as much as 10 m [27] . With an effective stack height of 10 m the residence time is estimated at 8.70×10^8 s by [24] ; thus, the possibility of these particles being carried to distant places by wind. Assuming a mean wind speed of 3 m·s^{-1}, particulates from flares can be transported and dispersed at 2.61×10^6 km in the downwind direction. Therefore in one hour, pollutants will be transported 10.8 km. This also supports calculations by [16] that pollutants are transported to distant areas far from their sources within the region.

Carbon monoxide concentrations were highest in Mbiama (191 mg/m^3). Although Mbiama is a small town, it is however split into two halves by the busy East-West highway. Long distant trailer drivers use the small town as a stopover place; so at all times, there are heavy duty vehicles steaming on the roadside. Spot measurement of CO in Mbiama easily spike above 36,614 mg/m^3 (20 ppm). This is a hazard to the periodic and daily markets that attract large number of traders from far flung areas of the Niger Delta. Emuohia, Ukwugba and Ahoada had the least concentration of CO. All CO measurement in this study exceeded short-term limits and the reason may be due to the prevalence of too many smoking vehicles on Nigeria roads.

Nitrogen dioxide concentration was highest in Bonny (187 mg/m^3). The least concentrations of NO$_2$were at Ahoada, Buguma and Odukpani (53 mg/m^3). In Owasa, Ughelli and Warri the most important means of mobility is 2-stroke engine motorbikes, which is likely responsible for the 133 mg/m^3 recorded. All sulphur dioxide measurements in this study were below short-term regulatory limits for 1 and 24 h for WHO (Table 4).

Sulphur dioxide is the criterion emission detected with the least concentrations in the regions atmosphere. It ranged between 19 mg/m^3 in Ukwugba to 90 mg/m^3 in Port Harcourt. Sulphur dioxide in the atmosphere arises from the combustion of fuels containing sulphur. Relatively, the petroleum obtained in Nigeria has negligible sulphur content; although a large part of the fuel supplied in the country are imported. Depending on the sources of such fuels, the sulphur content may be high. There are no publications on that, but the acrid smell of some imported fuels justifies the conclusion that their sulphur contents are higher than those produced in domestic refineries.

Total hydrocarbon in the atmosphere ranged from 78 mg/m^3 in Odukpani to 192 mg/m^3 in Nchia. The Niger Delta region as a whole is replete with petroleum production and processing facilities. There are many oil spill sites in the region and most waterways exhibit oil films on the surface. As much as 35% - 65% of spill components can evaporate into the atmosphere within weeks [28] and this is possible with the predominantly light crudes of Niger Delta, under a perennially humid-hot climate. Settlements like Bonny and Nchia had high concentrations of hydrocarbon in the atmosphere because export terminals, petroleum refineries and liquefied natural gas processing facilities are present in them. The sources of hydrocarbon in the Niger Delta region are fairly ubiquitous. There are hundreds of petroleum production and processing facilities within a few miles of each other in the region from where hydrocarbons are vented. In addition, crude oil spills, incessant petroleum product pipeline breakage, associated gas flares, emissions from automobiles, power generators and machines are other sources of hydrocarbons in the region's atmosphere.

The most abundant of the BITEX group is benzene and toluene. Benzene ranged from 0.2 - 0.3 mg/m^3 and it was present in all the locations. Toluene ranged from 0.1 - 0.5 mg/m^3 and it was also present in all the locations monitored. Xylene was the next in abundance and it ranged between 0.1 - 0.2 mg/m^3. Ethylene was detected only in one station at concentration of 0.1 mg/m^3. It is justifiable to conclude that ethylene presence in the regional atmosphere is negligible. Their importance to air quality studies is that they are carcinogenic [32] .

Carbon dioxide ranged between 400 ppm Buguma to 450 ppm in Port Harcourt. These values are similar to background concentrations of CO_2 [33] , although measurements taken near oil processing facilities like gas flares regularly rise above 450 ppm. Carbon compounds in the atmosphere are partly as a result of the burning of fossil fuel and the relevant compounds in this regard are carbon dioxide and carbon monoxide. Carbon dioxide rarely go into reaction in the atmosphere, its importance lies in the fact that it is one of the greenhouse gases, capable of increasing the overall heat balance of the earth. Since the turn of the century, background levels of carbon dioxide have increased to 400 ppm [33] .

Table 4. Emissions tolerance limits and standards for ambient air quality.

Pollutants	Nigeria (FMEV)		National Air Quality Standard for Nigeria	WHO Guidelines and World Bank Standards
	Long-Term Tolerance Limits 24 Hours (mg/m³)	Short-Term Tolerance Limits 30 min (mg/m³)		
Suspended particulate matter	0.015	0.5	250 µg/m³ (24 h) 600 µg/m³ (1 h)	60 - 90 µg/m³ (annual mean) 150 - 230 µg/m³ (24 h)
Carbon dioxide	10 vol. %	-		
Carbon monoxide	1.0	0.5	10 µg/m³ (1 h) 20 µg/m³ (8 h)	60 µg/m³ (24 h), 10 µg/m³ (8 h), 30 µg/m³ (1 h)
Nitrogen dioxide	0.085	0.085	75 - 113 µg/m³ (24 h)	40 - 50 µg/m³ (annual), 150 (8 h), 200 µg/m³ (24 h)
Sulphur dioxide	0.05	0.5	0.01 (24 h) 0.1 (1 h)	50 µg/m³ annual, 350 µg/m³ (1 h), 125 µg/m³ (24 h)
Benzene	0.8	0.2	-	1 µg/m³
Toluene	0.6	0.6	-	1 mg/m³ (odour) 7.5mg/m³
Ethylene	5.0	5.0	-	-
Xylene	-	0.2	-	-
Hydrocarbon (total)	2.0	5.0	160 µg/m³ (3 h)	160 µg/m³
Cadmium	0.003	0.01	-	10 - 10 µg/m³, 0.04 µg/m³
Chromium	0.001	0.0015	-	1 µg/m³
Lead	0.005	0.002	-	0.5 - 1 µg/m³ (annual)
Manganese	0.01	0.03	-	1 µg/m³ (annual)
Vanadium	0.002	-	-	1 µg/m³ (24 h)

Sources: [29]-[31].

Trace metals in the atmosphere are sometimes categorized as hazardous air pollutants (HAPs) along with volatile organic compounds (VOCs). Five metals were detected, namely; cadmium, chromium, lead, vanadium, zinc and manganese. Arsenic, barium, nickel and mercury were investigated but not detected. The most abundant of the metals was zinc, which was present at above (2.0 mg/m³) in all the settlements included in this study.

In remote settlements like Buguma and Emuoha some of the metals were not detected at all. Metals enter the atmosphere from several sources including the crust, automobile emissions (especially in the fuel and brake linings) and crude oil processing. Figure 2 shows trace metal measurements in the atmosphere of the region. Lead in the atmosphere of the region has been traced to its presence in some gasoline sold in Nigeria (0.74 mg/l), and described as one of the highest in the world by the [3] . Its accumulation in human tissues can affect mental function and may result in other health effects. The significant concentration of zinc in these areas of the Niger Delta may indicate the presence of trace metal pollution from oil exploration activities done in the flow stations close to study areas as well as cases of oil spills in the areas [34] [35] . [36] has reported the concentrations of trace metals in some oil producing locations from Brass and Bonny within the Niger Delta region. A study conducted by [37] , show a significant levels of zinc in Port Harcourt (22.91 ± 6.26 μg/m³) in the atmosphere. Industries which may be responsible for zinc concentration in the atmosphere in Port Harcourt include Pipeline Products and Marketing Company, PPMC at Eleme, Port Harcourt Refinery Company at Alasa, Eleme, Eleme Petrochemical Industry, etc. [38] . These industries together with regular petroleum exploration and exploitation activities may be responsible for the release of trace metals in the atmosphere which is being transported by meteorological factors from Port Harcourt to other part of the Niger Delta [37] .

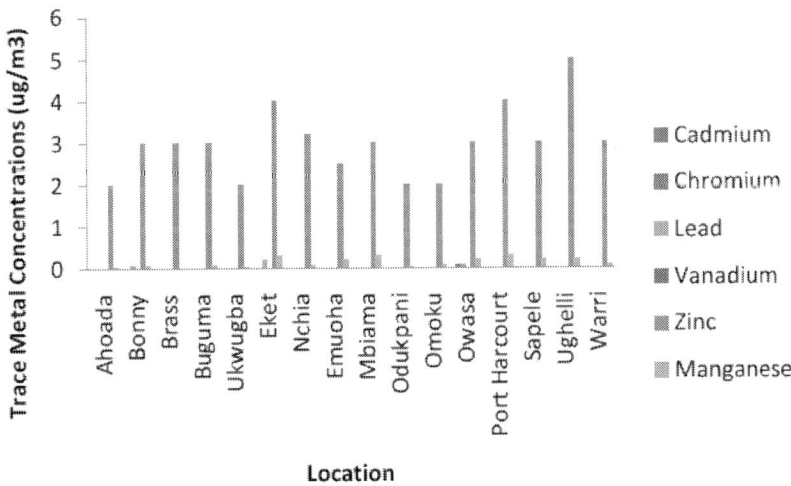

Figure 2. Concentration of trace metals in the region's atmosphere.

5. CONCLUSION

Laws on environment, as they pertain to air quality regulations, are presently very weak in Nigeria and existing environmental standards do not include specific provisions for all possible emission sources. Stakeholders at all level in the country have to articulate initiatives that support quality environmental practices. Specific air pollution and environmental standards ought to be more stringent targeting sources like factories, incineration and vehicle emissions. There should be air quality standards for designated areas based on the kind of activities that are prevalent. Standards easily degenerate into disputes if clear-cut protocols are not instituted to guide practice and enforcement on mutually agreeable methods. Pioneering the use of air quality modeling in Niger Delta as obtainable in some countries appears to be appropriate because it is a versatile instrument for regulatory controls. Traffic flow in large cities should be enhanced at all times because high mobility produces turbulence that aids dispersion, just as steaming automobiles in a traffic jam emit pollutants that can fumigate under certain weather conditions. There should be permanent air monitoring stations all over the region to provide air quality alerts and spot places where activity changes should be prescribed in order to improve air quality.

REFERENCES

1. Elvidge, C.D., Baugh, K.E., Tuttle, B.T., Howard, A.T., Pack, D.W. and Erwin, E.H. (2007) A Twelve-Year Record of National and Global Gas Flaring Volumes Estimated Using Satellite Data. Final Report to the World Bank, 107 p.

2. JINN (2010) Gas Flaring in Nigeria: An Overview. Justice in Nigeria Now.http://justiceinnigerianow.org/gas-flaring

3. World Bank (1995) Defining an Environmental Development Strategy for the Niger Delta. Vol. 1, Industry and Energy Operations, West Central Africa Department.

4. UNDP (2006) Niger Delta Human Development Report. United Nations Development Programme, Abuja, 299 p.

5. UNEP (2011) Environmental Assessment of Ogoniland. United Nations Environmental Programme, Nairobi, 262 p.

6. Initiates (1999) Gas Emission Study/Investigation at Fantuo Community. NAOC, Port Harcourt, 65 p.

7. Obunwo, C.C., Ede, P.N. and Jegede, M. (2009) Estimation of Methane from Two Mangrove Habitats in Port Harcourt of the Niger Delta Area, Nigeria. International Journal of Chemistry, 1, 62-65.

8. Fagbeja, M.A., Hill, J., Chatterton, T., Longhurst, J. and Akinyede, J. (2013) Residential-Source Emission Inventory for the Niger Delta—A

Methodological Approach. Journal of Sustainable Development, 6, 98-120. http://dx.doi.org/10.5539/jsd.v6n6p98

9. SPDC (2004) 2003 People and Environment Annual Report. 45 p.

10. SPDC (2005) 2004 People and Environment Annual Report. 33 p.

11. NLNG (2002) Nigeria LNG Plant Overview. Training Department, Bonny.

12. Ede, P.N. (1998) Pollution and the Rivers State Environment. Nigerian Research Review, 1, 81-89.

13. Ede, P.N., Edokpa, D. and Israel-Cookey, C. (2011) The Relative Contribution of an LNG Plant Emission to the Regional Air Quality of Nigeria. Journal of the Nigerian Society of Chemical Engineers, 26, 1-11.

14. Ede, P.N. and Oriji, I.B. (2013) Emissions from Private Power Generating Equipment in Port Harcourt, Nigeria. Nature and Science, 11, 59-64.

15. Sonibare, J.A. and Ede, P.N. (2009) Potential Impacts of Integrated Oil and Gas Plant on Ambient Air Quality. Energy and Environment, 20, 331-345.http://dx.doi.org/10.1260/095830509788066394

16. Ede, P.N., Edokpa, D.O. and Ayodeji, O. (2011) Aspects of Air Quality Status of Bonny Island, Nigeria Attributed to an LNG Plant. Energy & Environment, 22, 891-909.http://dx.doi.org/10.1260/0958-305X.22.7.891

17. Edokpa, D.O. and Ede, P.N. (2013) Challenge of Associated Gas Flaring and Emissions Propagation in Nigeria. Academia Arena, 5, 28-35.

18. Ede, P.N., Obunwa, C.C. and Nlerumchi, S.C. (2010) Air Quality Studies around Some Local Palm Oil Mill Plant at the Northern Fringes of the Niger Delta Area, Nigeria. Journal of Chemical Society of Nigeria, 35, 6-10.

19. NPC (2007) Legal Notice on Publication of 2006 Census Report. National Population Commission, Lagos.

20. WHO (1976) Selected Methods of Measuring Air Pollutants. WHO Offset Publication No. 24, E, Geneva.

21. ISO 6767 (1990) Ambient Air Determination of the Mass Determination of Sulphur Dioxide Tetrachloromercurate (TCM)/Pararosaniline Method.

22. ISO 6768 (1985) Ambient Air Determination of the Mass Concentration of Nitrogen Dioxide Modified Griess Saltzman Method.

23. Ede, P.N., Seiba, I.H. and Igwe, F.U. (2006) Combustion Efficiency Determined for Selected Flare Points in the Niger Delta Area. African Journal of Environmental Pollution and Health, 5, 19-25.

24. Seiba, I.H. (2006) The Production and Removal from the Air of Particulates in Associated Gas Flares in Nigeria. Unpublished Master's Thesis, Institute of Geosciences and Space Technology, Rivers State University of Science and Technology, Port Harcourt, 121 p.

25. Guttorp, P. (1986) The Boundary Layer.

26. Canada National Ambient Air Quality Objective for Particulate Matter (Executive Summary) (2003) Physical and Chemical Characteristics.http://www.hesc.go.ca/hecssesc/airquality/publications/particulate/matter-exec-summa

27. Ede, P.N. (1995) An Analysis of the Atmospheric Impact of Gas Flaring in Rivers State. Unpublished Master's Thesis, Department of Geography, University of Port Harcourt, Choba, 102 p.

28. Asthana, D.K. and Asthana, M. (2003) Environment: Problems and Solutions. S. Chand & Company, New Delhi, 434 p.

29. Federal Ministry of Environment, FMEv (1991) Guidelines and Standards for Environment Pollution Control in Nigeria. FEPA, Lagos, 250 p.

30. WHO (2006) WHO Air Quality Guidelines—Global Update 2005: Summary of Risk Assessment. The World Health Organisation, Geneva, 22 p.

31. World Bank (1995) Initial Draft of Industrial Pollution Prevention and Abatement Handbook. World Bank, Environment Department, Washington DC.

32. Khitoliya, R.K. (2004) Environmental Pollution Management and Control for Sustainable Development. S. Chand and Company, New Delhi, 309 p.

33. IPCC (2014) Summary for Policymakers. In: Edenhofer, O., Pichs-Madruga, R., Sokona, Y., Farahani, E., Kadner, S., Seyboth, K., et al., Eds., Climate Change 2014, Mitigation of Climate Change, Cambridge University Press, Cambridge, United Kingdom and New York, 9.

34. Obi-Iyeke, G.E. (2014) Trace Metal Dynamics in Some Leafy Vegetables Consumed in Warri, Niger Delta Region of Nigeria. International Journal of Research and Reviews in Applied Sciences, 18, 279-284.

35. Omofonmwan, S.I. and Odia, L.O. (2009) Oil Exploration and Conflict in the Niger Delta Region of Nigeria. Journal of Human Ecology, 26, 25-30.

36. Nwadinigwe, C.A. and Nwaorgu, O.N. (1999) Metal Contaminants in Some Nigerian Well-Head Crudes: Comparative Analysis. Journal of the Chemical Society of Nigeria, 24, 118-121.

37. Uno, U.A., Ekpo, B.O., Etuk, V.E., Etuk, H.S. and Ibok, U.J. (2013) Comparative Study of Levels of Trace Metals in Airborne Particulates in Some Cities of Niger Delta Region of Nigeria. Environmental and Pollution, 2, 110-121.

38. Ekweozor, I.K.E., Dambo, W.B. and Daka, E.R. (2003) Zinc and Cadmium Levels in Crassostrea gasar from the Lower Bonny Estuary, Nigeria. Journal of Nigerian Environmental Society, 1, 31-40.

CHAPTER 8

Seasonal Influence on the Ambient Air Quality in Al Jahra City for Year 2010

Raslan Alenezi[1*], Bader Al-Anzi[2], Abdallah Abusam[3], Aamir Ashfaque[4]

[1]*Department of Chemical Engineering, College of Technological Studies, Public Authority Applied Education and Training, Adailiya, Kuwait*

[2]*Department of Environment Technology and Management, College for Women, Kuwait University, Kuwait City, Kuwait*

[3]*Water Technologies Department, Water Resources Division, Kuwait Institute for Scientific Research, Safat, Kuwait*

[4]*Gulf Paper Manufacturing Company, Fahaheel, Kuwait.*

ABSTRACT

Eight primary criteria air contaminants were measured continuously for the year 2010 to evaluate ambient air quality in Al Jahra, which is one of the oldest and busiest cities in the state of Kuwait. The state of the art instrumentation was used to record the pollutants concentration to ppb levels maintaining quality control and quality assurance. Hourly base data for Non-Methane Hydrocarons (NM-HC), CH_4, CO, CO_2, O_3, SO_2, NO_2 and Particulate Matter (PM_{10}) were analyzed for year 2010. Meteorological parameters contributing to air pollution, such as (temperature, solar intensity, wind speed and direction) have also been considered. The effect of winter and summer seasonal changes on pollutant concentration levels were analyzed to identify the most probable sources for the application of the futuristic mitigation methods for pollution abatement. The obtained results consistently suggest that the foregoing pollutant concentration levels are higher in winter than summer due to poor dispersion and shallow inversion layer with the exception of O_3, CO_2 and PM_{10}. However, all of the pollutant concentrations are below the allowable standards limits except for NM-HC.

KEYWORDS

Pollutants; Al Jahra; Seasonal Changes; Monitoring Station

1. INTRODUCTION

Air is one of the major components of the environment that has been abused by human reckless behavior due to renaissance and prosperity as part of urbanization causing what is known today as air pollution. Air pollution has become a burdensome and an international problem threatening the existence of life due to its adverse impact on all living organisms, as well as facilities. Air pollution comprises outdoor and indoor air pollution, which was rated by experts as high risk to human health. According to World Health Organization (WHO) 1 of 6 people lives in polluted urban area, which is more than 1.1 billion people [1]. Outdoor air pollutants come mostly from two main sources. First source is natural source such as volcanic eruption and forest fire. Second source is industrial source such as burning fossil fuels in motor vehicles and power and industrial plants. Such urbanization processes have contributed significantly to the deterioration of air quality by increasing pollution levels in urban areas. As one might expect, humans have been producing increasing amounts of pollution as time has progressed, they now account for the majority of pollutants released into the air.

Monitoring the level of air pollutant is vital for their control and for decision makers. Therefore, the concentrations of air pollutants in Kuwait are constantly measured and monitored by number of fixed Air Quality Monitoring Stations, (AQMS) belonging to Kuwait-Environment Public Authority (KU-EPA). Mainly air quality in Kuwait is affected from burning fossil fuel in power plants, traffic, oil activities and petroleum refineries.

The area of air quality in the State of Kuwait has been discussed by many researchers [2-8]. Wahab and Bouhamra [9] used a mobile Air Pollution Monitoring Laboratory, (APML) to study air pollution in a residential area of Kuwait, which was affected by road traffic increase at an unimaginable scale. They reported that the levels of Nitrogen Oxides (NO_x) and Non-Methane Hydrocarons (NM-HC) exceeded the proposed ambient air quality standard for residential areas in Kuwait. Also, AbdulWahab [3] has studied two cases of air pollution in the industrial area in the Sultanate of Oman and an urban residential area in Kuwait and come up with the same conclusion for NM-HC and NO_x.

Ettouney et al., [2] have analyzed air pollution and meteorological data for two locations in Kuwait; Jahra and Umm Al Hyman over a period of 4 years (2001-2004). There evaluations for the data were calculated to obtain annual hourly averages and annual 1-h maxima for each year for each pollutant as well as metrological parameter. This way, the impacts of the seasonal variation (winter, spring, summer and autumn) of these pollutants and metrological parameters on the air quality of these urban areas in different seasons have been masked. In the winter and summer seasons there is substantial difference in the metrological parameters such as ambient temperature, humidity, solar radiations and wind speed that have influence the dispersion of air pollutants. Alenezi and Ashfaque [10] reported that high concentrations of different air pollutants are observed in winter season due to prevalent meteorological conditions. Moreover, Ettouney et al., [2] have presented the annual hourly average for the wind speed and wind direction, which is inaccurate values and misleading. They should have analyzed wind speed vector taking into consideration the wind speed values and direction resolving into their respective horizontal component (along x-axis) and vertical component (along y-axis) and later resultant has to be calculated as an appropriate estimate of the magnitude and direction of wind velocity for the specified period of time, however they calculated the arithmetic mean instead.

Al-Bassam and Popov [11] performed statistical analyses on the variation of selected primary pollutants that are usually emitted by automobiles. Their study was carried out for 11 days in February 2006. Data were measured near a private school using a mobile air monitoring station. A computer dispersion model (CALINE4) was also used in their study, after validation, to predict the CO concentrations as function of number of cars in the vicinity of the school. Results showed high levels of NM-HC almost all of the times above the Kuwaiti EPA limit, while CO and Nitrogen Dioxide (NO_2) concentrations found to be below the KU-EPA limits. The model results suggest that 30% reduction in CO is achievable if number of cars is decreased by half. However, this model should be tested to include pollutants of concern such as NM-HC, which has exceedance over 75% of the time from KU-EPA limits. Also, 11 days in the entire year is not enough to provide an accurate and more realistic results, and hence further studies must be performed for long duration to include seasonal effect.

The state of Kuwait is located in the northeastern corner of Arabian Peninsula and one of the Gulf Cooperation Council, GCC countries, which is mainly depending on oil as natural resource. Kuwait is a major exporter of crude oil, which is equal to 2.6 million barrels per day (mbbld). Kuwait locally produces heavy fuel oil that has been used for power generation that contains about 4% sulphur [8]. In Kuwait, motor vehicles and buses are the only means of road transportation [11]. Kuwait environment is exposed to air pollution augmented by extreme weather conditions. Kuwait is characterized by a typical desert type of weather with long summer spells and high frequency of dust storms, arid periods, and humid conditions. In the summer season, high unlimited emissions of various air pollutants such CO, carbon dioxide (CO_2), methane (CH_4), NM-HC, NO_x and Particulate Matters (PM_{10}) resulted as the power plants operated at their full capacity to relinquish the thirst of Heating Ventilation and Air Conditioning, (HVAC) of all buildings.

Due to its size, location (proximity to main pollutants sources) and large population, Al Jahra city was selected as the study area. This work is based on the quantitative analysis of air pollution data reflecting the ambient air quality of Al Jahra being the largest governorate in the state of Kuwait. The exceedances of the air pollutants threshold values are calculated during the period of study. Also, the diurnal profiles and the seasonal variations of these air pollutants are studied and the relationships between each other were investigated. The year of 2010 was divided into two main seasons (winter and summer) furthermore, these seasons are divided into three months each, starting from winter (January to March) and summer season (July to September). Similar diurnal trend of concentration levels at different times in the day were observed for both seasons. The meteorological parameters (temperature, wind speed and direction and solar intensity) were also considered in the discussion of the results.

Locations and Measurements

Kuwait map (**Figure 1**) shows the location of the Al Jahra residential area relative to Kuwait city and other urban areas in Kuwait. According to the satellite image of Kuwait, Al Jahra district is surrounded by several utility industries, Northern oil fields, powers and desalination plants, wastewater treatment plant and free-ways connecting it to the rest of Kuwait Cities and neighbouring countries. As shown in **Figure 1** and wind rose plot (**Figure 14**), Al Jahra area is located on the path of the predominant north-westerly wind that blows most of the times from this direction throughout the year transporting these pollutants to the city. Kuwait municipality's 2003 records describe Al Jahra as a relatively run down residential suburb of central Kuwait, covering total area (11,230) km^2. Al Jahra has a total population of (269,915), housed in (34,755) residential blocks [12]. The district is situated on the major free-way, 6th ring road at the north end.

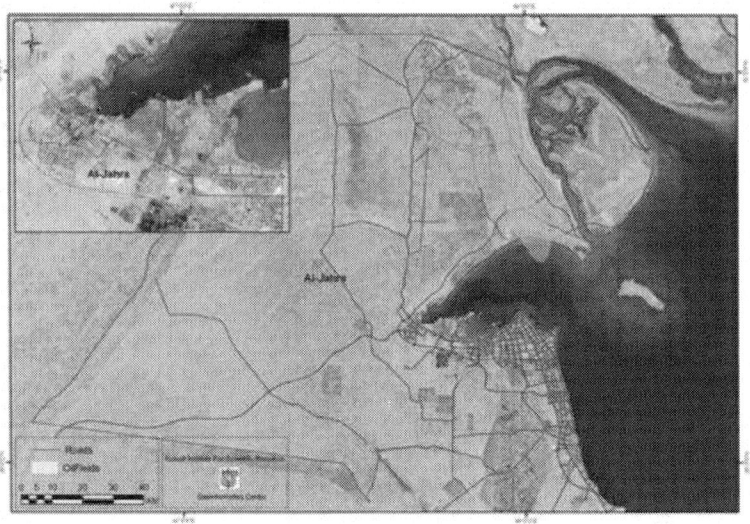

Figure 1. Al Jahra residential area with respect to neighboring areas in Kuwait.

The air quality data in Kuwait were collected using a fixed AQMS operated by KU-EPA, which measures continuously each 5 minutes the concentration level of various pollutants. This monitoring station is located above the main polyclinic in the middle of the residential area. The sampling site is selected on the basis of availability of power, security and elevated position.

2. RESULTS AND DISCUSSIONS

The ambient air quality data for two extreme seasons (summer and winter) as hourly average concentrations and meteorological conditions were recorded utilizing monitoring stations fixed by KU-EPA at selected sites throughout the state of Kuwait. The pollutants measured were CO, CH4, PM_{10}, NM-HC, SO_2 and NO_2. The collected meteorological conditions include wind speed and wind direction, ambient temperature and solar intensity. All of the data were recorded from a station placed 10 m above Al Jahra polyclinic located in the centre of commercial districts of the city. Figures 2-15 depict the behaviour of the measured pollutant concentrations and meteorological conditions in Al Jahra city during the designated seasons in year 2010.

2.1. Air Pollutants

2.1.1. Nitrogen Dioxide

Figure 2 presents seasonal average concentrations of NO_2 for each hour in year 2010. NO_2 concentrations for winter season were consistently higher than summer values by about (50%) due to many reasons. The foremost cause for this behaviour is summer weather being extremely hot, humid and dusty forcing most of residents, especially foreigners, to travel to their respective countries for summer vacations. About 67% of the population in the state of Kuwait comprises expatriates as reported by ministry of planning in Kuwait (http://unstats.un.org/unsd/statcom/_seminar/kuwait.pdf). This in turn reduces human consumption of fossil fuel used in HVAC and vehicles, which decreases NO emission and hence NO_2formation. Also summer season is characterized as windy and dusty that dilutes and disperses the pollutants evenly decreasing the concentrations of pollutants in the air. The maximum concentration level was recorded during evening time at 19:00 hr for winter season, which is equal to 63.02 ppb (seasonal average). This concentration is in concordance with the results reported by Ettouney et al., [2] for years 2001 to 2004 where the highest yearly concentration was 46 ppb in winter season. The NO_2 peak values were calculated for each year and plotted in **Figure 3** that showed a linear trend, depicting the influence of population increase associated with increase in number of automobiles and power consumption. The minimum NO_2 concentration was 9.9 ppb measured at 15:00 in the summer season, where most of the residents stayed indoor avoiding the extreme temperature (approaching 50°C) during this hour.

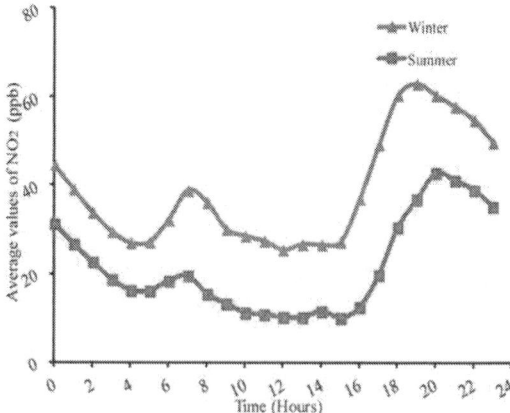

Figure 2. Hourly average NO_2 concentrations for winter and summer seasons.

2.1.2. Carbon Monoxide

The diurnal CO hourly seasonal average concentrations are plotted in **Figure 4**. Winter CO pattern exhibits two maxima; the first one at about at 7:00 am early morning with concentration of 1.22 ppm and the second one was at about 22:00 hr late evening with 1.62 ppm. This is attributed to heavy road traffic caused by people heading to work, school, shopping and other commercial activities. Generally, winter values are higher than those measured during the summer season about (80%); however there are times especially at low values where CO concentrations during the summer season exceed those recorded in winter season. This is partially caused by drivers (chauffeurs) who keep their cars running to maintain inner car temperatures when they are waiting in the parking area during the summer for quite sometimes to avoid hot weather. CO concentration, unlike NO_2, is not showing any trend due to its long decay life about 7 days that forces constant background concentration which does not follow any typical behaviour.

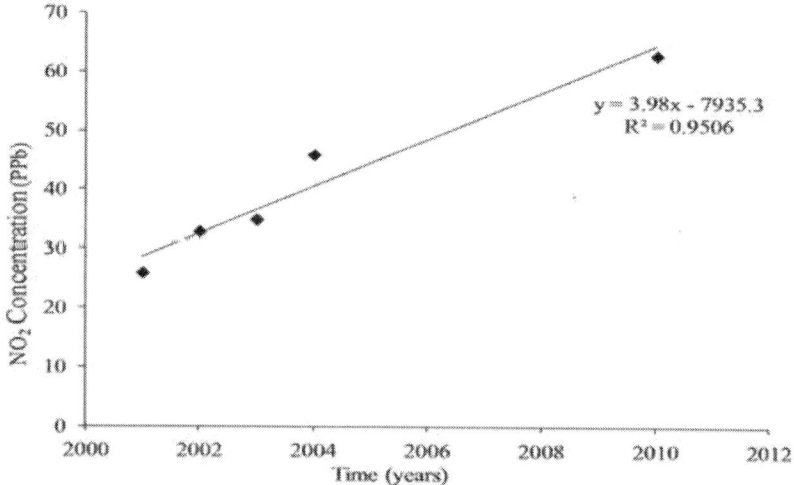

Figure 3. Maximum hourly average concentrations of NO_2 for various years.

2.1.3. Non-Methane Hydrocarbons and Methane

Diurnal variations for NM-HC and methane concentrations during year 2010 are plotted in Figures 5 and 6, respectively for winter and summer seasons. The concentration of NM-HC gas is most of the time above the standards set by Kuwait Environmental Public Authority, KEPA, (0.24 ppm average value from

6:00 to 9:00 am). The seasonal variation of these gases is different from NO_2 and CO concentrations, showing no significant change in the concentrations of NM-HC between winter and summer seasons (less than 0.1 ppm) due to consistent emission from their sources petroleum related activities (oil production, transportation, refining, dispensing and consumption). However, for early morning and late afternoon hours higher values during winter season are recorded reaching maximum concentrations of 0.48 and 0.56 ppm respectively. Methane concentration is plotted in **Figure 6**, which exhibits almost constant concentration levels for both seasons with greater values recorded during winter season. NM-HCs are among primary pollutants responsible of harmful ozone formation in troposphere with the association of NO_x and sunlight as O_3 precursors. This coincides with high values measured for NO_2 and ozone concentrations during afternoon time.

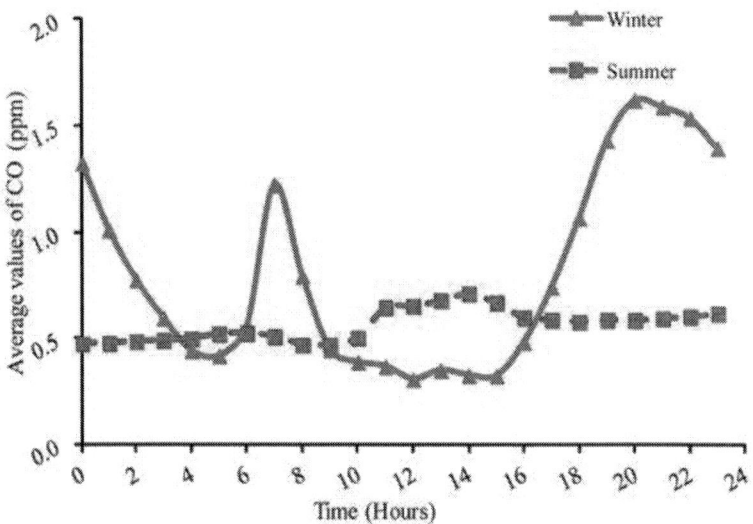

Figure 4. Hourly average CO concentrations for winter and summer seasons.

2.1.4. Carbon Dioxide

Unlike previous results, **Figure 7** shows that summer season measures higher concentrations of CO_2 than winter season. This is clearly due to full capacity operation of desalination and power plants units during summer season to reduce the deficit between demand and supply created by increase in urbanization and industrial development. The same figure shows that average ground level concentration of CO_2 is almost the same throughout the year and equal to 381 ppm except for early morning from 6 - 8 am where slight change has been observed, however it is still noticeably higher than winter values. CO_2 concentration has showed an increasing trend like most of the other places in the world.

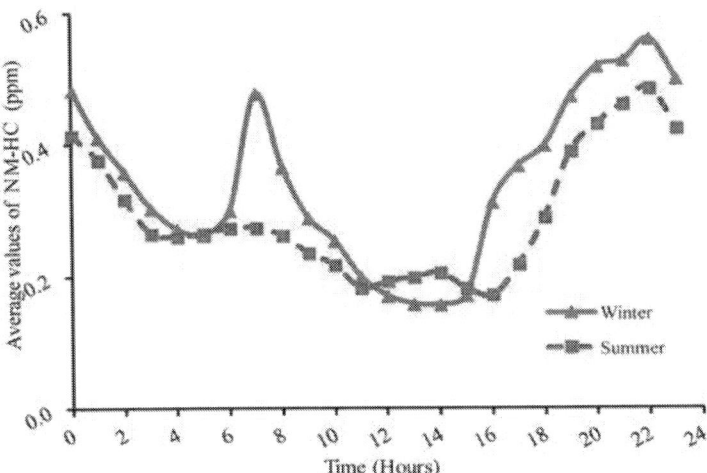

Figure 5. Hourly average NM-HC concentrations for winter and summer seasons.

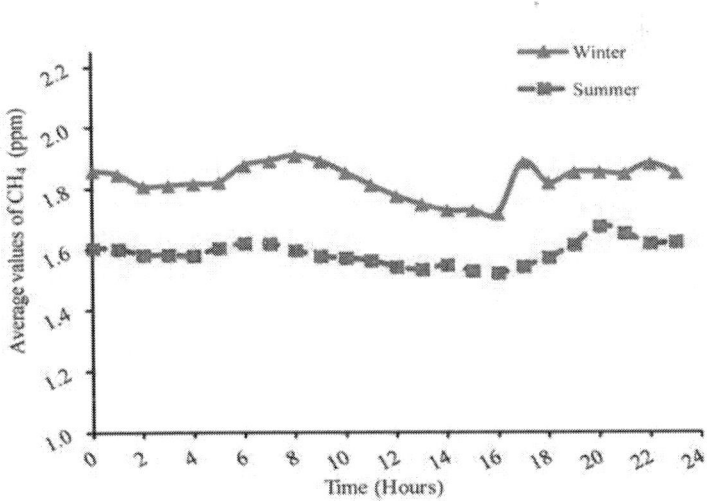

Figure 6. Hourly average CH_4 concentrations for winter and summer seasons.

2.1.5. Ozone

Figure 8 represents seasonal average ground level hourly concentration of ozone. It is obvious that there is a distinct maximum recorded during afternoon time in summer season and persisted for about two hr (16 to 18 hr). Assessment of O_3 accumulation rate from 8:00 to 18:00, revealed two phases: a slow and a fast one, with different durations in each of the two seasons, due to different levels of precursors. Before the sunrise, auto emissions (particularly NO) actually break down ozone present in the atmosphere and after sunrise, ozone concentrations will gradually increase because of the photodissociation of (NO_2). Also, the constituen of automobile exhaust and the combination of the resultant atomic oxygen (O) and molecular oxygen (O_2) already present in the atmosphere [13,14]. This corresponds to low concentrations of NO_x such as NO_2 (**Figure 9**). Similar trend of O_3 concentration was observed by Ettouney et al., [2] with NO_x associated with high temperature (solar UV radiation), certifying O_3 generation based on its precursors strength. Therefore, concentration of O_3 increases during afternoon time until it reaches a maximum value displayed by hump in **Figure 8** and then drops down to reach a minimum value at about midnight, whereas, NO_2 concentrations show exactly the opposite behaviour during the same time as shown in **Figure 9**. These results agree with reported values of Abdul-Wahab [3] that shows a decreasing trend in ozone concentration with an increase in the concentrations of NO_x. They also suggested that O_3 increases with SO_2, however a negative correlation obtained for temperature greater than 27°C.

2.1.6. Particulate Matters (PM₁₀)

Kuwait has one of severe arid climate in the world, where hot summer season persists for over six months a year with low annual rainfall of about 110 mm/yr and short winter time lasting for almost three months. There are natural dusty events throughout the year but more severe in summer due to Indian monsoon. **Figure 10** shows the concentration variation of PM_{10} for winter and summer seasons; summer season PM_{10} concentrations are higher than winter seasonal values. It is clear that seasonal average hourly PM_{10} concentrations during summer season fluctuates between maximum value of 248.8 µg/m3 recorded at dawn time (3:00 am) and minimum value of 157.8 µg/m3 recorded at early afternoon (13:00). Low rate of precipitation and desertification aggravated land and soil erosion facilitating the dust to be carried by strong wind during the summer season forming the main natural source responsible for local PM_{10} concentration in Kuwait. The prevalent wind is 80% of the time blowing from North-west direction over the desert plans of Mesopotamia as shown in **Figure 10**. Other artificial source that recently became a factor is construction of new residential houses in Sa'ad Albduallah city across the street from Al Jahra and other development north of the Al Jahra area.

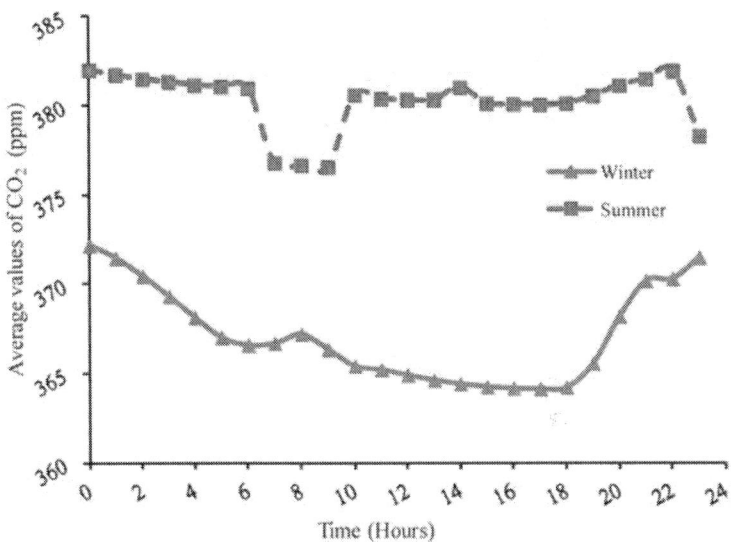

Figure 7. Hourly average CO_2 concentrations for winter and summer seasons.

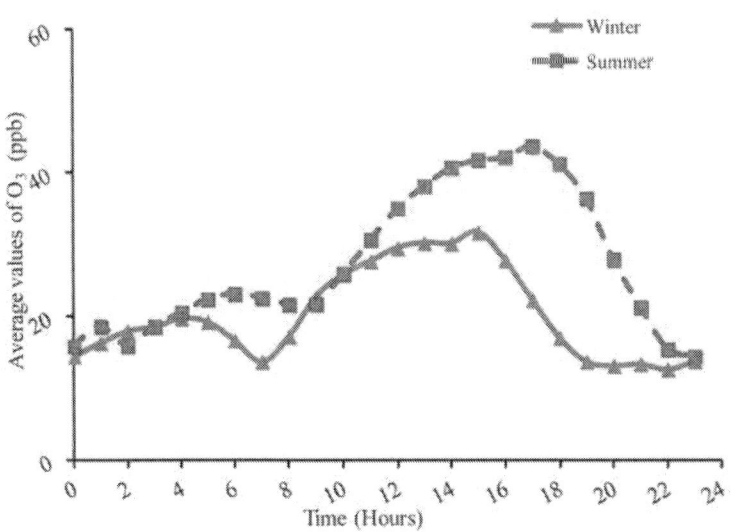

Figure 8. Hourly average O_3 concentrations for winter and summer seasons.

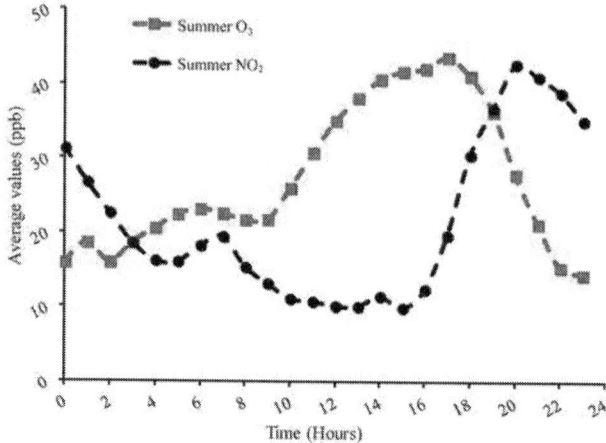

Figure 9. Hourly average O_3 and NO_2 concentrations for winter and summer seasons.

2.1.7. Sulfur Dioxide

As shown in **Figure 11**, higher concentrations of SO_2 during winter were observed with multiple maxima measured for most of the day (10 hr - 20 hr). Similar trends were reported for the same city by Ettouney et al. [2] and M. S. Al-Rashidi et al., [15] for years 2001 to 2004 with variable magnitudes. Interestingly enough, the values recorded in the current study is less than those reported by Ettouney et. al. [2] for year 2001 and higher than the rest of the years (2002-2004). This was due to use of heavy fuel oil containing 4% sulphur in year 2001 in power industry and later gas oil and crude oil with low sulphur contents were used [8].

2.2. Meteorological Conditions in Al Jahra

The meteorological conditions play an important role in pollutant dispersion affecting the ground level concentrations in Al Jahra residential areas. In Kuwait, winters are characterized with cold and damp weather accompanied with occasional precipitation of rain that reduces the temperature even further to reach 1°C during late night and dawn times. Wind conditions in winters are calm and stable, which causes poor pollutants dispersion and hence increases their concentrations in the air. Summer is dry and harsh with maximum day temperatures occasionally slightly exceeding 50°C, which makes it almost intolerable without significant air conditioning. Winds are considerably turbulent and are predominantly blowing from northwest direction. Turbulent wind along with high inversion layer during the summer helps to dilute pollutants concentrations by increasing their dispersion rate.

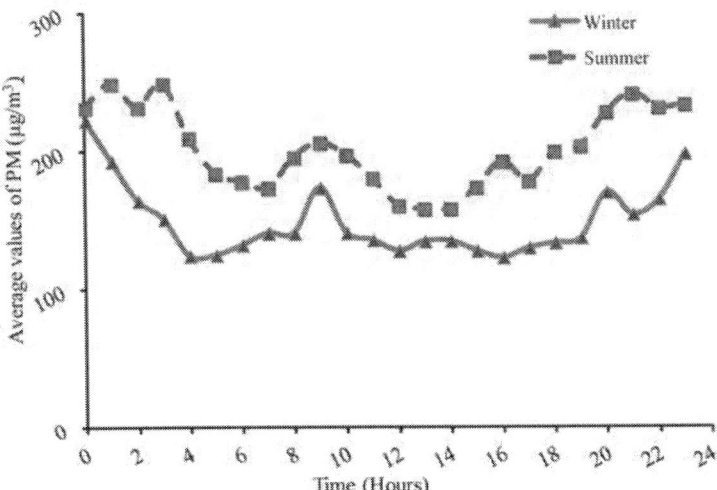

Figure 10. Hourly average PM_{10} concentrations for winter and summer seasons.

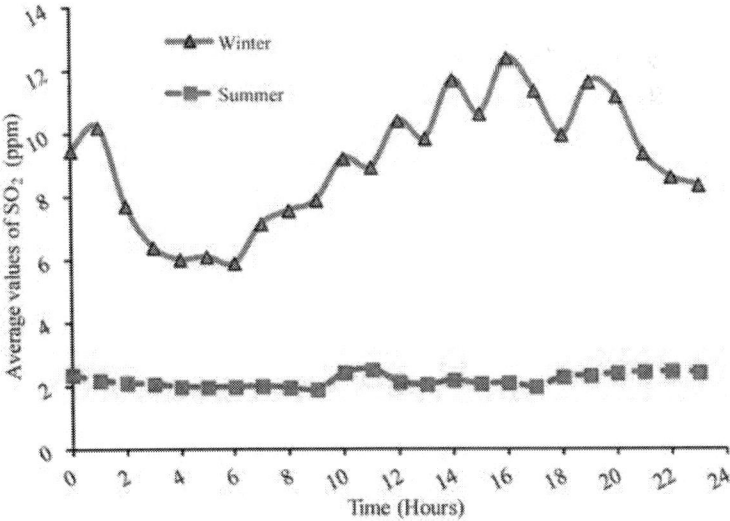

Figure 11. Hourly average SO_2 concentrations for winter and summer seasons.

2.2.1. Solar Intensity

Figure 12 shows the average solar intensity variations during winter and summer seasons in year 2010. The highest maximum hourly value of solar intensity is recorded during the summer season on the 3rd of July at 11:00 hr and equal to 774.1 w/m². The average seasonal solar intensity for the entire summer season is 193.71 ± 0.89 σ w/m², where standard deviation (σ) is equal to 227.470 w/m². In the winter season, the hourly value of solar intensity is found to be 728.5 w/m² recorded on 27th of March at 12:00 hour. The average seasonal solar intensity for the entire period of winter is 151.5 ± 0.85 σ w/m², where standard deviation (σ) is equal to 204.7 w/m².

2.2.2. Temperature Variations

As stated previously, the weather in Kuwait is harsh due to long summer season with high temperature reaching up to 50°C (**Figure 13**). The highest temperature value was 50°C recorded on 16th of July at 15:00 hr. The average seasonal temperature in summer season is 39.5 ± 0.85 σ, °C, where standard deviation (σ) is 4.65°C. While in the winter season, the maximum temperature was 40.4°C measured on 16th of March at 14:00 hr and the average seasonal temperature is 20.1 ± 0.78 σ°C, where standard deviation (σ) is 5.77°C.

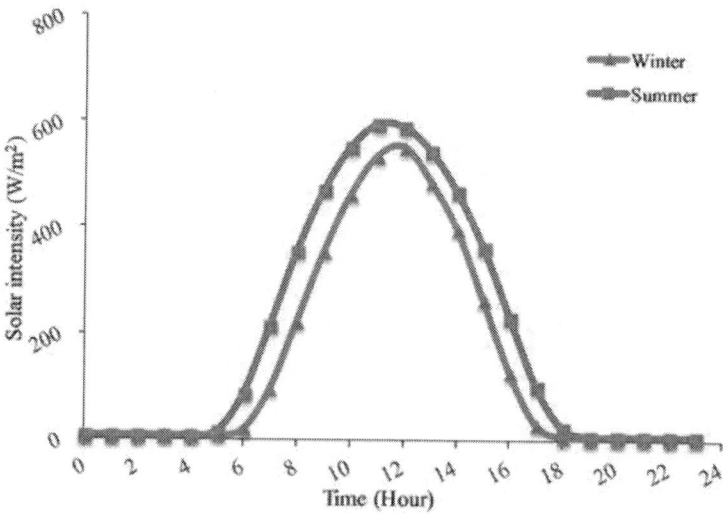

Figure 12. Average solar intensity variations during winter and summer seasons in year 2010.

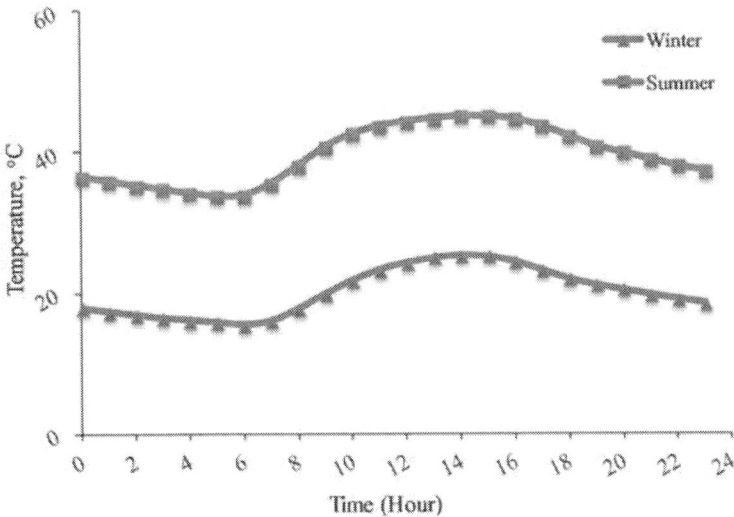

Figure 13. Average temperature variations during winter and summer seasons in year 2010.

Summer season is also characterized with high winds causing severe dust storms that disperse pollutants emissions. In winter, high ground level concentrations of pollutants are mainly influenced by more stable meteorological conditions.

2.2.3. Wind Speed and Direction

One of the main factors influencing pollutants ground level concentration is wind speed and direction. Therefore, **Figure 14** is generated to show the wind rose for Al Jahra district during the summer season of year 2010, which indicates that the prevailing wind is from Northwestern direction.

Reviewing **Figure 14**, one can clearly see that the highest hourly based wind speed is equal to 6.94 m/sec measured on 4th of September at 04:00 hr, with 300.5° bearing North-western direction, reflecting the dominant influence of wind direction.

Figure 15 shows the wind rose for winter season. The maximum observed hourly average value is 6 m/sec on 18th of March at 13:00 hour, with 143.75° bearing southeastern direction, blowing from Arabian Gulf.

3. CONCLUSIONS AND RECOMMENDATIONS

Assessment of air quality was carried out on data collected from Al Jahara monitoring station for year 2010 to investigate the impact of fossil fuel sources on hourly average pollutant concentrations for winter and summer seasons as precaution measurement for air quality assurance. All of the pollutants under study show higher concentration during winter season except for CO_2, O_3, and PM_{10}, whereas, hourly average concentration for NM-HC and PM_{10} exceeded the permissible limits. It is obvious from the present work that the ground level concentration of O_3 depends on NO_2concentrations and temperature (O_3 precursors). Data from present and previous studies show increasing pattern of NO_2, which exhibits a linear relationship with a time slope of 4 (ppb year-1).

Present results are accurate and reflect the actual pollutant concentrations as they vary each hour due to unstable seasonal meteorological conditions. Meteorological results revealed that higher values were obtained for all conditions (parameters) during the summer season, which significantly influenced the pollutant concentrations.

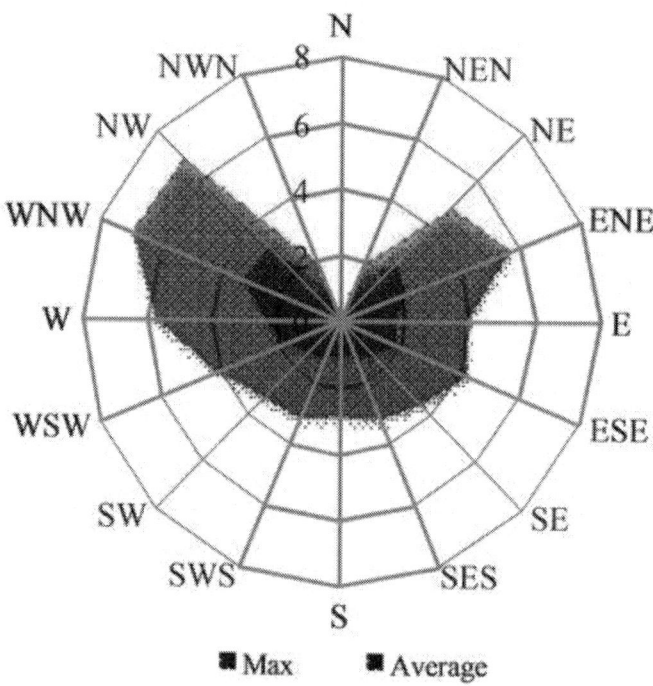

Figure 14. Winde rose for Al Jahra city for summer season.

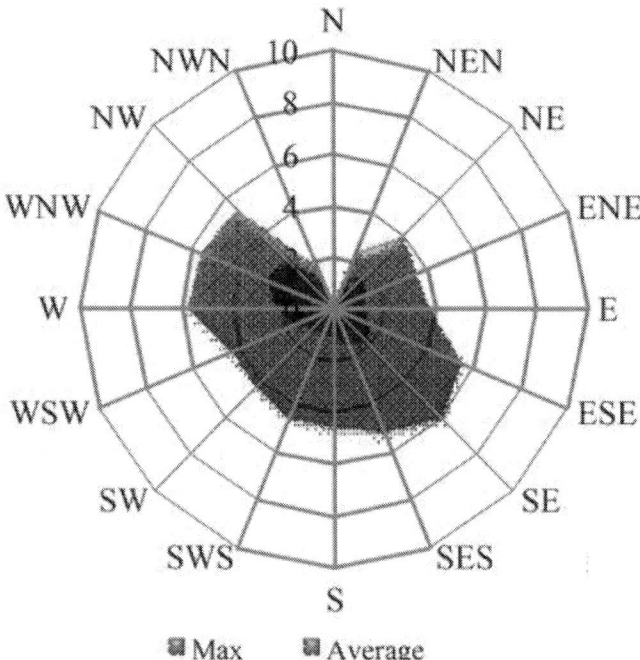

Figure 15. Winde rose for Al Jahra city for winter season.

A well studied long-term policy plan must be prepared and implemented soon to minimize the emission of such gases into the air to maintain air quality and healthy environment. This issue concerns GCC the most, which rely mainly on limited natural resources as fossil fuel. However, this resource has become less attractive because of the time constraints imposed by depletion and environmental degradation. Consequently, seeking other alternative sources of sustainable energy is imperative in order to survive and protect our environment such as biofuel, nuclear, solar, wind, hydro tidal and geothermal (renewable).

4. ACKNOWLEDGEMENTS

The authors want to thanks KU-EPA for providing air pollution data.

REFERENCES

1. S. A. Al-Haider and S. M. Al-Salem, "Outdoor Air Quality Data Analysis of Al-Mansoriah Residential Area (Kuwait): Air Quality Indices Results," Transactions on Ecology and the Environment, Vol. 116, 2008, pp. 189-196.

2. R. S. Ettouney, J. G. Zaki, M. A. El-Rifai and H. M. Ettouney, "An Assessment of the Air Pollution Data from Two Monitoring Stations in Kuwait," Toxicological and Environmental Chemistry, Vol. 92, No 4. 2010, pp. 655- 668.

3. S. A. Abdul-Wahab, "Two Case Studies of Air Pollution from Oman and Kuwait," International Journal of Environmental Studies, Vol. 66, No. 2, 2009, pp. 179-191.

4. A. A. Ramadan, M. Al-Sudairawi, S. Alhajraf and A. R. Khan, "Total SO_2 Emissions from Power Stations and Evaluation of Their Impact in Kuwait Using a Gaussian plume Dispersion Model," American Journal of Environmental Sciences, Vol. 4, No. 1, 2008, pp. 1-12.

5. G. Andria, G. Cavone and A. M. L. Lanzolla, "Modelling study for Assessment and Forecasting Variation of Urban Air Pollution," Measurement, Vol. 41, No. 3, 2008, pp. 222-229.

6. E. Al-Bassam and A. Khan, "Air Pollution and Road Traffic in Kuwait," WIT Press, Dresden, 2004, pp. 741- 750.

7. W. S. Bouhamra and S. A. Abdul-Wahab, "Description of Outdoor Air Quality in a Typical Residential Area in Kuwait," Environmental Pollution, Vol. 105, No. 2, 1999, pp. 221-229.

8. B. N. Al-Azmi, V. Nassehi and A. Khan, "SO_2 and NO_x Emissions from Kuwait Power Stations in Years 2001 and 2004 and Evaluation of the Impact of These Emissions on Air Quality Using Industrial Sources Complex Short-Term (ISCST) Model," Water, Air, & Soil Pollution, Vol. 203, No. 203, 2009, pp. 169-178.

9. S. A. Abdul-Wahab and W. S. Bouhamra, "Diurnal Variations of Air Pollution from Motor Vehicles in Residential Area," International Journal of Environmental Studies, Vol. 61, No. 1, 2004, pp. 73-98.

10. R. Alenezi and A. Ashfaque, "Assessment of Ambient Air Quality in Al Jahra Governorate, for 2008," International Journal of Energy and Environment, Vol. 5, No. 4, 2011, pp. 582-591.

11. E. Al-Bassam, V. Popov and A. Khan, "Air Quality in the Vicinity of a Governmental School in Kuwait," Transactions on Ecology and the Environment, Vol. 116, 2008, pp. 237-246.

12. State of Kuwait, "Annual Statistical Abstract, 2008," General Statistic Office, State of Kuwait, 2009.

13. A.-N. Riga-Karandinos and C. Saitanis, "Comparative Assessment of Ambient Air Quality in Two Typical Mediterranean Coastal Cities in Greece," Chemosphere, Vol. 59, No. 8, 2005, pp. 1125-1136.

14. J. Seinfeld and S. Pandis, "Atmospheric Chemistry and Physics: From Air Pollution to Climate Change," Wiley Interscience, New York, 1998.

15. M. S. Al-Rashidi, V. Nassehi and R. J. Wakeman, "Investigation of the Efficiency of Existing Air Pollution Monitoring Sites in the State of Kuwait," Environmental Pollution, Vol. 138, No. 2, 2005, pp. 219-229.

CHAPTER 9

Artificial and Biological Particles in Urban Atmosphere

Chang-Jin Ma[1] and Mariko Yamamoto[1,2]

[1] *Department of Environmental Science, Fukuoka Women's University, Fukuoka, Japan*
[2] *Fukuoka Prefectural Munakata High school, Munakata, Japan*

1. INTRODUCTION

In urban areas including smaller towns, the presence of ambient artificial particulate matters (PM) (e.g., aged fine PM and asbestos) and biological particles has been one of the main causes of adverse effects on human health and air quality. Bioaerosols are airborne particles of biological origin (e.g., bacteria, fungi, pollen, viruses). Some biological contaminants trigger allergic reactions, including hypersensitivity pneumonitis, allergic rhinitis, and some types of asthma.

In Japan, there is peculiar annual variation in these ambient PM due to the sources as well as the changing weather. Most synergistic effects take place during spring because of windborne pollen from trees, especially Japanese cedar trees.

From 1949 to 1970, a large number of Japanese cedar trees were planted in many parts of Japanese Island because of a great demand for new housing after World War II. Planted cedar forests now cover 12% of Japan's total land area, which is more than 45,000 km^2. Vast amounts of pollen, especially from cedar trees, drifts over wide areas every spring [1]. As a result, more and more Japanese over the last 30 years have been affected by pollen allergies.

The air quality in urban area, especially its $PM_{2.5}$ (less than 2.5 μm in aerodynamic diameter) level, has become of increasing public concern because of its importance and sensitivity related to health risks [2]. This inhalable $PM_{2.5}$ is generally emitted from both natural (e.g., volcanic eruptions, forest fires, etc.) and anthropogenic sources (combustion processes in industrial sector and automobiles), and it can also be formed when gases react in the air. Inhalable particles, particularly $PM_{2.5}$, have been demonstrated to have the greatest impact on human health, visibility reduction, and solar radiation change, especially in densely populated urban areas.

Asbestos, a naturally occurring fibrous mineral, can be found naturally in the outdoor atmosphere and in some drinkable water, including water from natural sources around the world [3]. As asbestos had been a popular building material since the 1950s, it is still found in many buildings, including hospitals, schools and homes. Inhaling asbestos fibers is known to cause several serious and even fatal lung diseases. Studies have shown that the non-occupationally exposed population have 10,000 - 999,999 asbestos fibers in each gram of dry lung tissue, which translates into millions of fibers and tens of thousands of asbestos bodies in every person's lungs [4]. However, most building material products manufactured today do not contain asbestos. In the industrialized countries, asbestos was phased out of building products mostly in the 1970s, with most of the remainder phased out by the 1980s [5]. In 2006, the Japanese Ministry of Health, Labour and Welfare issued a final ruling banning most asbestos-containing products with the exception of 5 kinds of materials (e.g., some sealing materials). Those 5 unbanned materials were also banned eventually in 2011. Ever since its initial phase out in 2006 and permanent ban in 2011, it can still be found today in some older buildings and consumer goods. Ambient asbestos fibers will finally be lost in the air and eventually precipitate on the ground. These pieces of asbestos are likely to settle on the soil but can be re-released again into the atmosphere [6].

To assess the impact of both artificial and biological PM on the environment, including air quality, ecosystems, and human health, it is necessary to know its concentration, chemical composition, and the interplay among their components. The ambient outdoor PM in urban areas has seldom been evaluated with respect to both artificial and biological components. In light of this situation, we undertook a field campaign to evaluate the artificial and biological PM in an urban environment in Japan during springtime.

2. MATERIAL AND METHODS

2.1. Description Of Fukuoka Prefecture

Fukuoka Prefecture is located on Kyushu Island, Japan's third largest island, located southwest of the main island, Honshu. According to the latest estimates (June 1, 2013), its population and total area are 5,088,480 and 4,971 km^2, respectively [7].

Fukuoka Prefecture faces the sea on three sides, bordering on Saga, Oita, and Kumamoto prefectures and facing Yamaguchi Prefecture across the Kanmon Straits. Fukuoka Prefecture includes the two largest cities on Kyushu, Fukuoka and Kitakyushu, and much of Kyushu's industry. Fukuoka prefecture's main cities form one of Japan's main industrial centers, accounting for nearly 40% of the economy of Kyushu. Major industries include automobiles, transport equipment, electronic parts and machine, general machinery, iron and steel, semiconductors, steel, and food [8].

2.2. Sampling And Monitoring Of Ambient Pm

An intensive measurement of PM was conducted at four selected sites (A: 33.65°N; 130.45°E, B: 33.63°N; 130.42°E, C: 33.39°N; 130.26°E, D: 33.66°N; 130.40°E,) in the Fukuoka Prefecture. The location of each sampling site in Figure 1 is indicated by filled circles.

Site A has six-lane roads with heavy traffic conditions. Site B is a residential area and surrounded by a residential area without any known point sources. Site C is an industrial area with roughly 60 manufacturing companies including small-to-medium sized enterprises. Most of them are distributed within a radius of one kilometer from site C. Site D is a desolate area with resort beach without the influx of people, as our study was conducted during an off-season period.

An intensive measurement of PM was simultaneously conducted at four-sites for two days beginning on April 18, 2007 and ending in April, 2007. For the sampling of size-classified PM, four impactor samplers (Tokyo Dylec Co.) were simultaneously operated. Particles were collected directly on the filter arranged behind the jet-nozzles of the 1st stage of sampler. For the collection of particle samples, airflow was maintained at 20 l min^{-1}. The size fraction of PM and filter kind at each stage are coarse fraction (> 2.5 μm) at the first stage (a 47 mm diameter, non hole Nuclepore˙ polycarbonate filter) and fine fraction (PM$_{2.5}$) at a back-up stage (a 47 mm diameter, 0.01 μm hole Nuclepore˙ polycarbonate filter), respectively.

Although the influx value (i.e., deposition rate, grains cm^{-2}) is a widely used method for studies of airborne pollen, in this study, for the purpose of measuring the ambient concentration (grains m^{-3}), airborne pollen grains were also collected on the 1st stage of impactor samplers. The directly collected pollen grains on the natural filter surface are more easily observed by a SEM without filter pretreatment compared to those collected on a traditional plastic tape coated by adhesive.

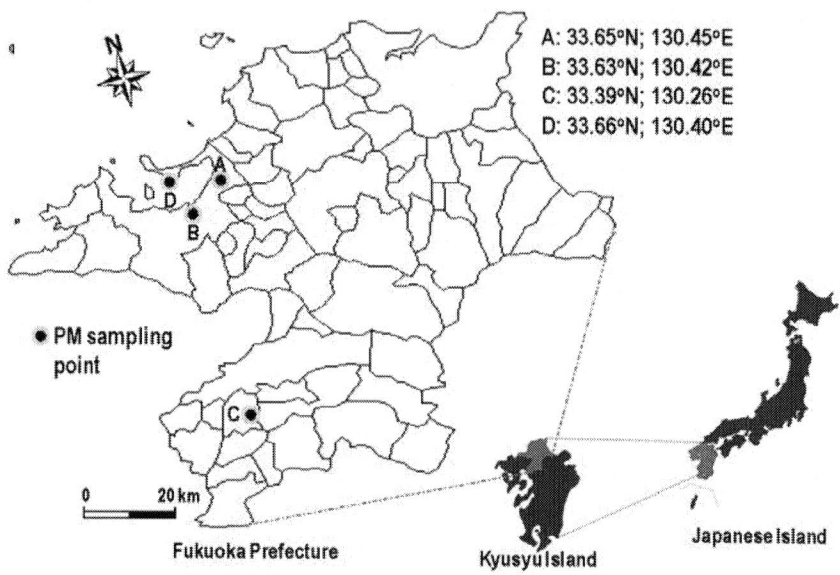

A: 33.65°N; 130.45°E
B: 33.63°N; 130.42°E
C: 33.39°N; 130.26°E
D: 33.66°N; 130.40°E

PM sampling point

0   20 km

Fukuoka Prefecture

Kyusyu Island

Japanese Island

Figure 1. Locations Of Sampling Stations For Pm In Fukuoka Prefecture

In order to assess PM$_{2.5}$ mass concentration, 4 Dust scan Scouts (Rupprecht & Patashnick Co. Model 3020) were simultaneously operated at each site. This PM$_{2.5}$ monitoring system makes use of near-forward light scattering to measure the concentration of particulate matter in ambient air. The light source is a safety-interlocked laser that operates at a wave length of 670 nm. The scattered light caused by the presence of particles is received by a sensor, forming the basis of the monitor's computations. During sampling period, there was no Asian dust event. The wind speed was measured at 2.4 - 4.7 m s^{-1}, with the temperature at around 11.2 - 19.4 °C and the average relative humidity at 56%.

2.3. Sample Pretreatment And Analysis

2.3.1. Sample Pretreatment

Figure 2 schematically illustrates the procedures of sample pretreatment and analysis. After sampling, filters were placed in sterilized airtight petridishes and stored in a refrigerator until analysis.

For the laboratory analysis, the filters capturing $PM_{2.5}$ were extracted with deionized water by ultrasonic treatment. And then the extracted water was filtrated through a 25 mm diameter Nuclepore filter with 0.08 μm pore size to separate into the soluble and insoluble fractions. After filtration, the filtrate was considered to be soluble fractions. Blank filters were handled in the same manner as the samples.

Meanwhile, the coarse particles (>2.5 μm) deposited on the first stage were progressed to single particle analysis for identification of asbestos and pollen. Samples for observation of pollen were coated with a very thin layer of platinum by a machine called a sputter coater.

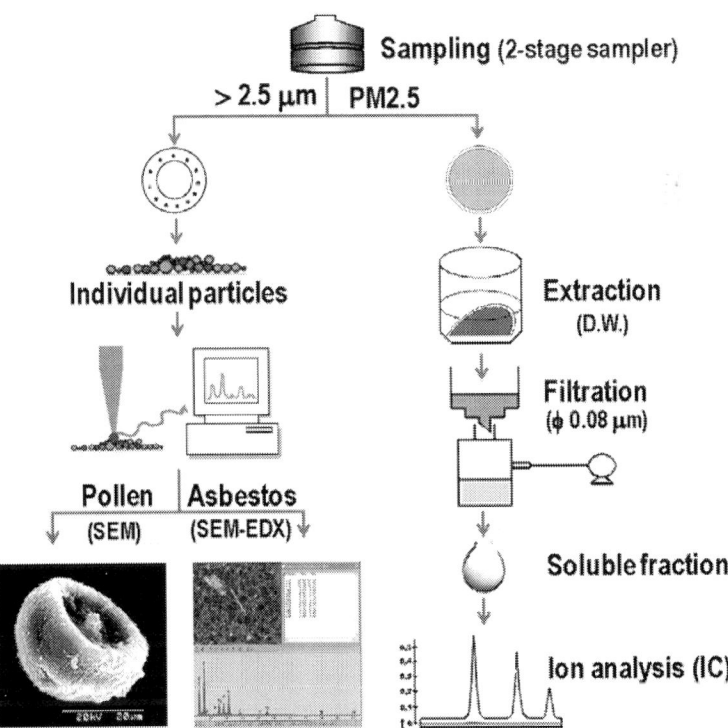

Figure 2. Procedures Of Sample Pretreatment and Analysis

2.3.2. Analysis Of Major Ionic Species In Pm₂.₅

In general, sulfate, nitrate, and ammonia had the greatest ambient concentrations in particles [9,10]. The concentrations of major ions in $PM_{2.5}$ (ammonium, nitrate, and sulfate) were determined by Ion Chromatography (IC) (Dionex DX-100). To attain the reliability of the analyzed data, the quality assurance and quality control (QA/QC) was conducted by analyzing a set of known standard species. The data obtained by 5-time repeated IC analyses were tested for precision by checking the relative standard deviation (% RSD, (SD / mean) x 100) of each concentration in standard solution (0.5, 2, and 5 mg/L). As a result, all three ionic species maintained low % RSD levels (i.e., ammonium: 4.63 - 7.59%, nitrate: 0.04 - 9.03%, and sulfate: 1.96 - 9.72%). This high reproducibility is a clear indication of a methodological soundness.

2.3.3. Identification Of Asbestos And Pollen

The most common method of identifying asbestos fibers in ambient PM is by polarized light microscopy. However, a Scanning Electron Microscopy (SEM) can be also usefully applied to the observation of asbestos. The advantage of using a SEM for asbestos identification is that it has better resolution through higher magnification and a greater depth of focus than polarized light microscopy. Most of all, a SEM equipped with an energy dispersive X-ray spectrometer (EDX) (i.e., SEM-EDX) can quantify the elemental components in asbestos. In Figure 3, the diagram of a SEM-EDX is schematically illustrated.

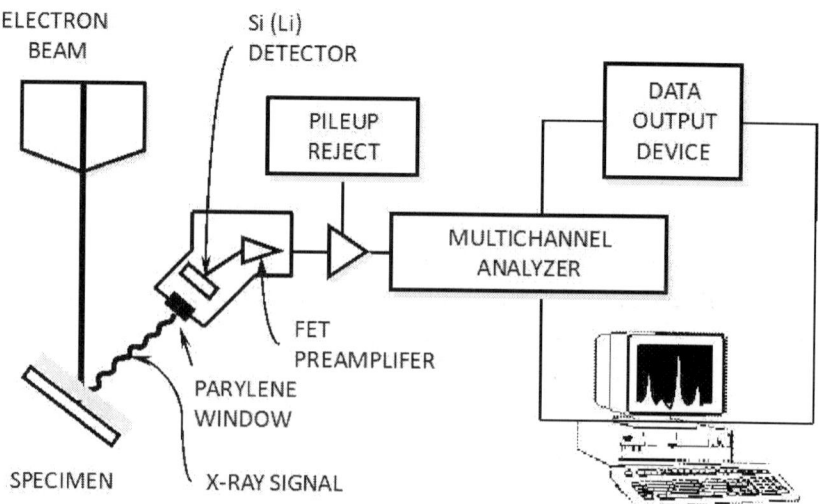

Figure 3. Schematic diagram of a SEM-EDX

Using an EDX and a computer system, information about the elemental properties of asbestos fibers can be gathered and graphed in their appropriate relative ratios. Computation of the exact ratios of the elemental compositions in asbestos fiber allows the researcher to distinguish not only one type of asbestos fiber from another but also asbestos fibers from non-asbestos fibers.

In this study, for the purpose of observing and analyzing the morphological and chemical properties of airborne asbestos and pollen grains, an SEM (JEOL JSM-5400) equipped with an EDX (Philips, EDAX DX-4) was employed. The samples were placed inside the SEM's vacuum column (10^{-6} Torr) through an air-tight door. Pollen species were also distinguished and countered under 3000 x magnification and 15 - 20 kV working conditions.

Calculation of the airborne asbestos fiber concentration on the filter sample was carried out using the following formula:

$$F_{con} = \frac{A_{eff} \times N_{total}}{a_{field} \times n_{field} \times V_{air}}$$

where F_{con} is airborne fiber concentration (fiber/L, f L^{-1}), A_{eff} is effective collecting area of filter (cm^2), N_{total} is the total number of fibers in an SEM field area, a_{field} is an SEM field area (cm^2), n_{field} is total number of fields counted on the filter, and V_{air} is total air volume (L) calculated by sample collection time (min) and pump flow rate (L/min).

3. RESULTS AND DISCUSSION

3.1. Spatial Distribution Of Pm $_{2.5}$ And Major Water-Soluble Ions

Air pollution, especially $PM_{2.5}$, has become a serious issue in East Asia, and there is rising public criticism regarding its effects. Concern in Japan is also increasing as winds transfer the pollution into domestic areas. Figure 4 shows the distribution of $PM_{2.5}$ monitoring data at four-sampling sites (i.e., a heavy traffic area (A), a residential area (B), an industrial area (C), and a desolate area (D)).

Site C, an industrial area, showed the heighest $PM_{2.5}$ (65.3 µg m^{-3}) followed by site A (35.3 µg m^{-3}), site B (22.0 µg m^{-3}), and site C (11.3 µg m^{-3}). As might be expected, the highest $PM_{2.5}$ level was monitored at site C where there is a compact mass of manufacturing companies. Among the four monitoring sites, two sites (i.e., site A and C) exceeded the Japan's $PM_{2.5}$ criteria (a daily average of 35 μg m^{-3}). The reason for the relatively high $PM_{2.5}$ in both heavy traffic and industrial areas might be that much of fine particles and their precursors came from the vehicle emissions and fuel combustion and manufacturing processes in factories. The $PM_{2.5}$ data monitored in this study are compared with those measured in other urban and rural areas in Asia during the springtime. Zhang *et al.* [12] reported that $PM_{2.5}$ was 145.3 μg m^{-3} in Beijing during a non-Asian dust period in the springtime. Meanwhile, in Gosan, a typical rural area in Korea, $PM_{2.5}$ during non-Asian dust period was measured at 26.1 μg m^{-3} [13].

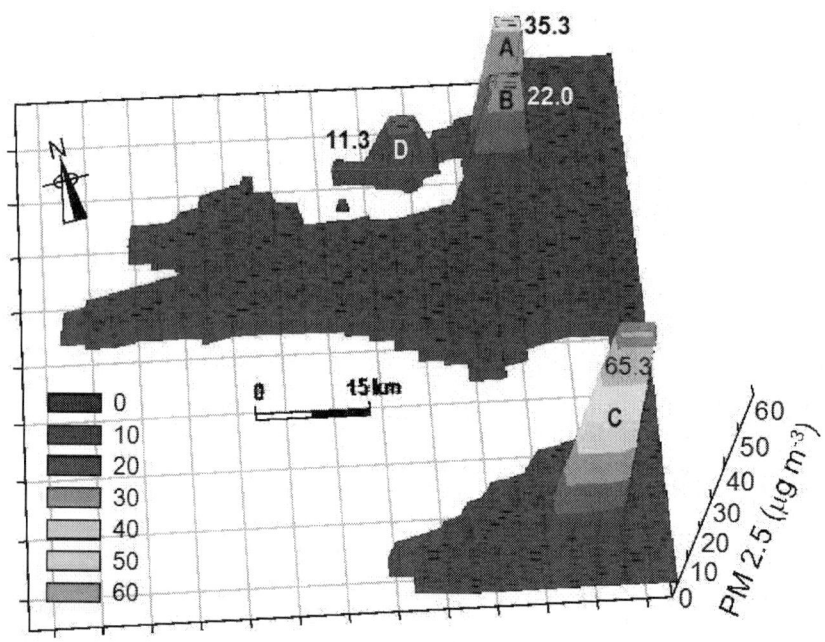

Figure 4. Distribution of PM $_{2.5}$ monitoring data at 4-sampling site

It is a matter of course that a large number of factories contributed to the high $PM_{2.5}$ level and caused the regional worsening air pollution in a residential area (B). However, although the "yellow dust" warning did not issue during our field measurement, there is a possibility of the long-range transport of $PM_{2.5}$ and its precursors from the Asian continent to the local study site in Japan. Thus, it is required to clear the uncertainties regarding the linkage between locally high $PM_{2.5}$ at site C and its long-range transport under the springtime meteological conditions. In this study, in order to determine the source region of aerosols at the site C, the atmospheric backward dispersion model was applied.

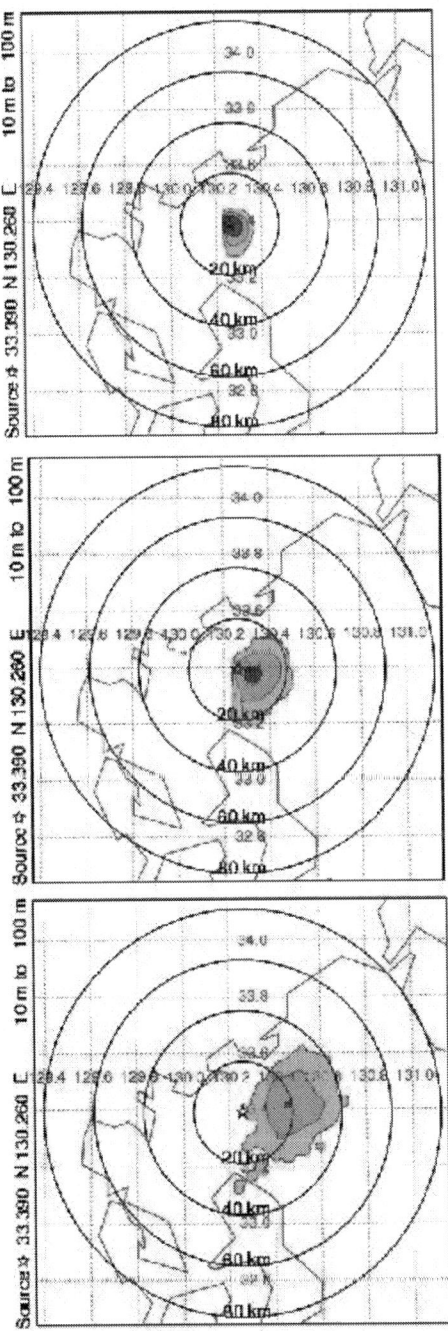

Figure 5. Backward aerosol concentrations simulated by the NOAA ARL HYSPLIT model. The area scales mean the integrated mass concentration [mass m-3] at 10-100 m height of site C (33.39oN; 130.26oE)

Figure 5 displays the backward aerosol dispersion simulated by the NOAA Air Resources Laboratory (ARL) HYSPLIT dispersion-trajectory model started from site C. The area scales mean the integrated mass concentration [mass m^{-3}] at 100 - 1000 m height of site C (33.39°N; 130.26°E).

A detailed model description of HYSPLIT was given in reference [14]. According to the result of the HYSPLIT model, a high value of aerosol concentration at the present receptor (measurement location) was not driven from the Chinese continent but was generated from a local area.

Ambient concentrations of major ionic species (i.e., nitrate, sulfate, and ammonium) associated with $PM_{2.5}$ collected at four-sampling sites were overlapped with a map of Fukuoka Prefecture (Figure 6). As shown in Figure 6, the concentrations of major ionic species turned out to be of considerable variation among four sites. Sulfate was the most abundant species to record the highest concentrations in all urban areas of Fukuoka Prefecture (a heavy traffic area (14.1 µg m^{-3}), an industrial area (30.5 µg m^{-3}), and a residential area (2.8 µg m^{-3}). The sum concentrations of NH_4^+, NO_3^-, and SO_4^{2-} varied in a similar way, as $PM_{2.5}$ (i.e., site C (38.4 µg m^{-3}) > site A (18.3 µg m^{-3}) > site D (5.4 µg m^{-3}) > site B (4.9 µg m^{-3})).

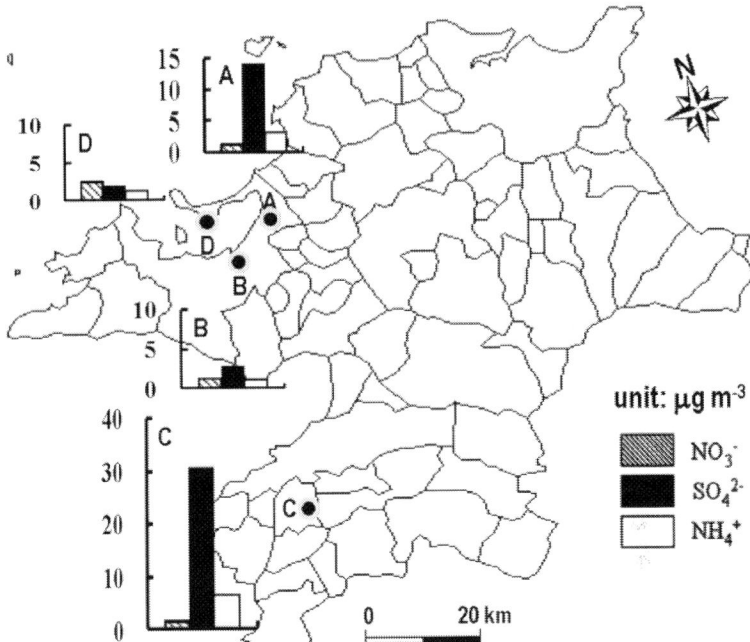

Figure 6. Ambient concentrations of major water-soluble ions associated with PM$_{2.5}$ collected at 4-sampling site in Fukuoka Prefecture

This suggests that the overwhelmingly high level of site C should be associated with the local pollution emissions. In the case of site C and A, the sum of three ionic components correspond to 58.8% and 51.8% of PM$_{2.5}$, respectively. Therefore, it seems reasonable to say that PM$_{2.5}$ in sites C and A was mainly composed by the secondary inorganic aerosol, which was formed by a gas-to-gas reaction in the atmosphere.

The concentrations of three kinds of ionic species in PM$_{2.5}$ in this study are comparable to those of urban areas in Beijing, China, and Durg, India. Gao *et al.* [10] reported that the concentrations of NH$_4^+$, NO$_3^-$, and SO$_4^{2-}$ in PM$_{2.5}$ collected in Beijing were 20.5, 15.2, and 42.3 µg m^{-3}, respectively. On the other hand, those concentrations in Durg, a heavy traffic and industrial area, were marked as 2.1, 3.16, and 6.75 µg m^{-3}, respectively [9]. The occupation ratio of the sum concentration of three ionic components in this area (0.89%) is greatly dissimilar to those of this study.

3.2. Assessment Of Airborne Asbestos Fiber

Even though asbestos is closely regulated in the present, the deposited and accumulated asbestos fibers for the past several tens of years can be resuspended from the ground near the distributing and manufacturing shops of asbestos products.

Asbestos from natural geologic deposits is known as "naturally occurring asbestos" (NOA) [14]. Health risks associated with exposure to NOA are not yet fully understood. As air quality associated with asbestos in ambient outdoor air has seldom been evaluated in Fukuoka Prefecture, we evaluated airborne asbestos in the urban environment there. Six mineral types are defined by the United States Environmental Protection Agency as asbestos [15]. Among them, crocidolite and amosite are commonly known to cause negative health effects such as lung cancer and mesothelioma [4].

Figure 7 shows an example of asbestos image of SEM, the elemental wt% list, and elemental spectrum obtained from EDX analysis.

A part of results of SEM-EDX analysis for asbestos fibers and the processes of classifying asbestos types based on the SEM-EDX elemental wt% data are illustrated in Figure 8.

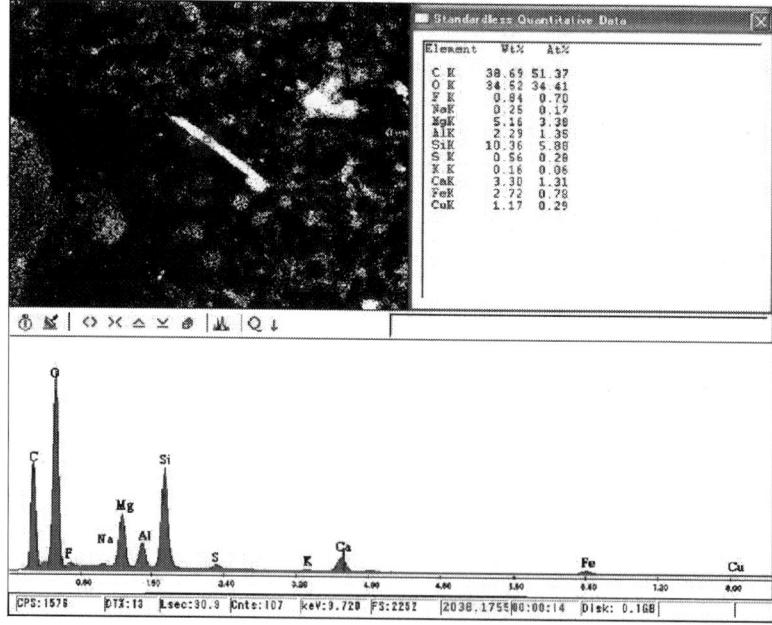

Figure 7. An example of asbestos image of SEM (left upper) and the elemental wt% list (right upper) and spectrum (bottom) obtained from EDX analysis

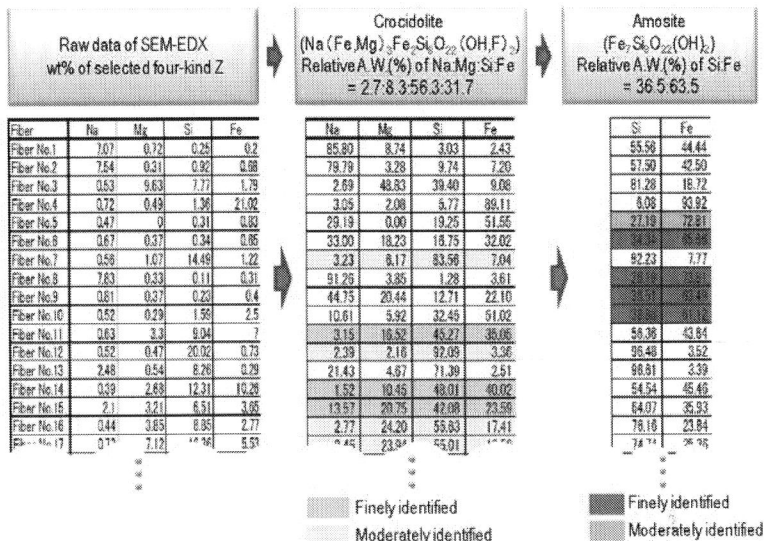

Figure 8. The raw data of SEM-EDX analysis for several asbestos fibers and the processes of classifying asbestos types based on the SEM-EDX elemental wt% data are illustrated in

Figure 9 illustrates the number concentration of asbestos fiber (a sum of crocidolite $(Na(Fe,Mg)_3Fe_2Si_8O_{22}(OH,F)_2)$ and amosite $(Fe_7Si_8O_{22}(OH)_2)$ at four-sampling sites and distribution of point sources of asbestos fiber in Fukuoka Prefecture. The highest airborne asbestos fiber concentration was recorded at site C (14.4 f L^{-1}), followed by site A (5.9 f L^{-1}), site D (3.4 f L^{-1}), and site B (2.5 f L^{-1}). The average concentration of the airborne asbestos fiber at all sites was 6.14 f L^{-1}. This average concentration level is similar to those (7 f L^{-1}) measured at asbestos abatement sites in Korea [16].

In 1989, the Japanese Air Pollution Control Law and related orders were revised to classify asbestos as a "specified dust" and to set up 10 fibers/liter (for including all type of asbestos) as the regulation guideline Concentration at the Boundary of the Asbestos Dusts Generation Facilities (i.e. asbestos products manufacturing facilities) [6]. Asbestos fiber concentration of site C (14.4 f L^{-1}) is considerably higher than the regulated levels of asbestos of the Concentration Standard at the Boundary of the Asbestos Dusts Generation Facilities.

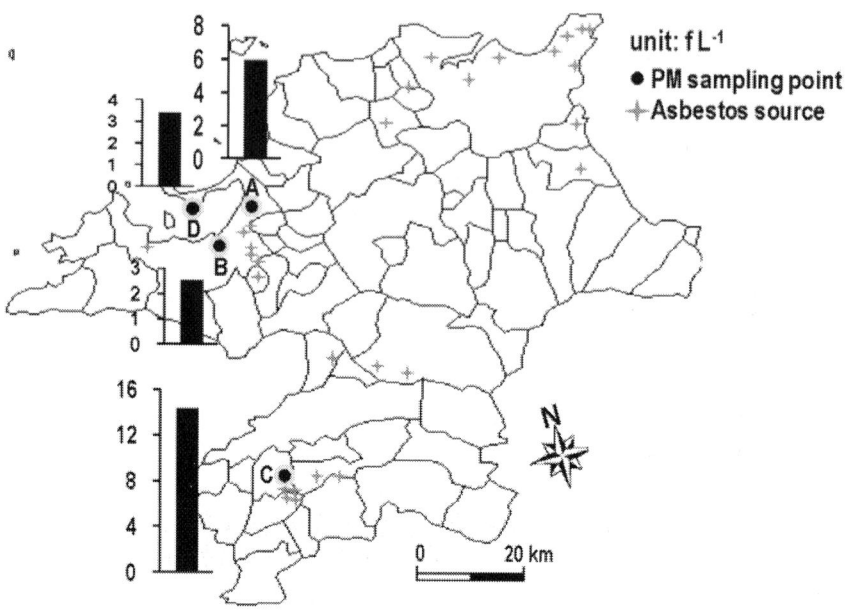

Figure 9. Number Concentration Of Asbestos Fiber (Crocidolite And Amosite) At 4-Sampling Site And Distribution Of Point Sources Of Asbestos Fiber In Fukuoka Prefecture.

To explore the possible reasons for the high asbestos fiber concentration at site C, the forecast atmospheric dispersion was simulated by the NOAA ARL Gaussian model. Figure 10 shows sampling site C in Figure 9 having a dense asbestos point source and the area distribution of the air flume started from intensively distributed companies producing asbestos related products.

According to the result of NOAA's ARL Gaussian model, the C sampling site was in an downwind position. It is therefore suggested that the relative high asbestos concentration at site C was strongly influenced by the clumped distribution of point sources of asbestos fiber. Meanwhile, although the concentration of asbestos measured at site A (heavy traffic area) was lower than the national Concentration Standard at the Boundary of the Asbestos Dusts Generation Facilities, asbestos was detected at levels around 2 times higher than those of sites B and D.

NOAA's ARL Gaussian model shown in Figure 10 indicates that the point sources of asbestos fiber situated densely near site A did not exert a direct influence on the asbestos concentration at site A. For several years, automobile parts that needed insulation from heat and friction were manufactured from dangerous asbestos, due to its excellent heat-resistant qualities. Such parts included brake linings, clutch facings, transmission components, disc brake pads, drum brake linings, and brake blocks [5].

According to both the result of NOAA's ARL Gaussian model (Figure 10) and the information on extensive use of asbestos in automobiles, the asbestos concentration at site A was probably affected by automobiles.

3.3. SPATIAL DISTRIBUTION OF AMBIENT POLLEN

It can be considered that pollen contributes to the organic carbon fraction in totoal suspended particles (TSP). In addition, some plant types can produce pollen in huge quantities. For example, a single ragweed plant can generate a million grains of pollen a day. Therefore, though pollen granules are small and light, they can also attribute to the mass concentration of ambient aerosol particles during the main pollen season [17]. From this point of view, to study the species of pollens, their distribution and concentration can be helpful to understand their ambient behavior and public health, etc.

SEM images of several types of pollens are shown in Figure 11. Classification of pollen grains using SEM was based on morphological characteristics such as shape, size, apertures and ornamentation. The combination of these characteristics makes some pollen grains easily identifiable. However there were also pollen grains sharing several common characteristics that makes identification difficult.

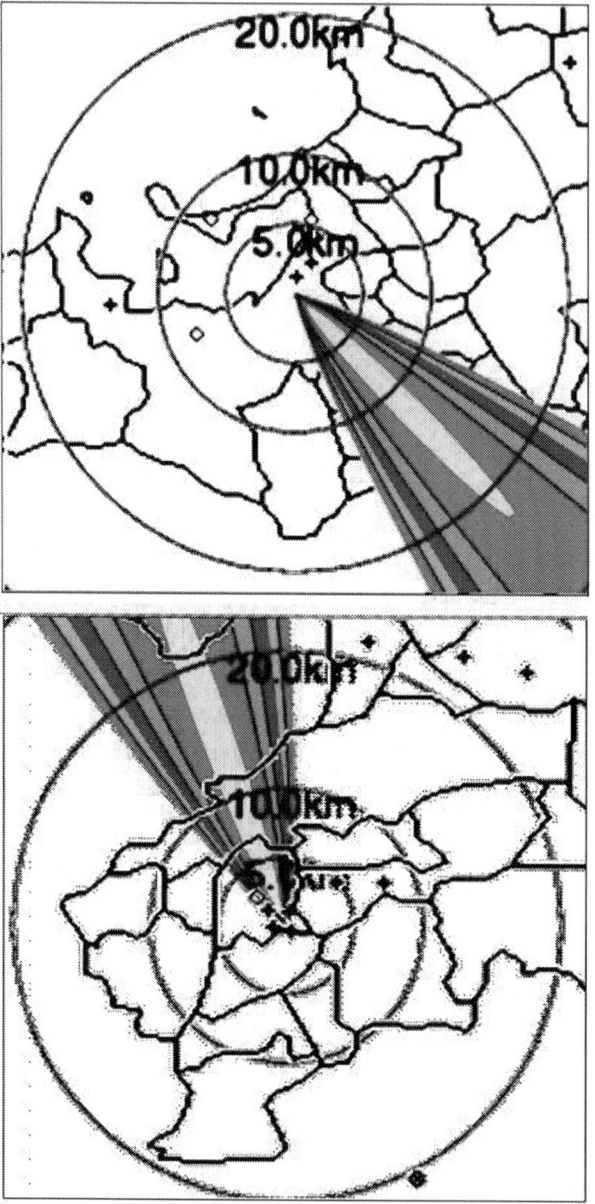

Figure 10. Map Showing Sampling Site C In **Figure 9** Having Dense Asbestos Point Source And The Area Distribution Of Air Flume Started From The Intensively Distributed Companies Producing Asbestos

Figure 12 shows the spatial variation of the number concentration of three-kind of airborne pollen at the four sites in Fukuoka Prefecture. Three-type mainly identified pollen grains are reported here. There was a noticeable spatial difference in the concentrations among three pollen types. Cedar (Cryptomeria, also called the Sugi tree in Japanese) pollen was distributed in advance of other pollens at sites A, B, and C, with the maximum concentration at site A. This cedar pollen is the most common allergen for seasonal allergic rhinitis in Japan. Cedar showed a higher concentration than other types of pollen and it was probably generated from the Sugi tree that is the most important timber tree in Japan.

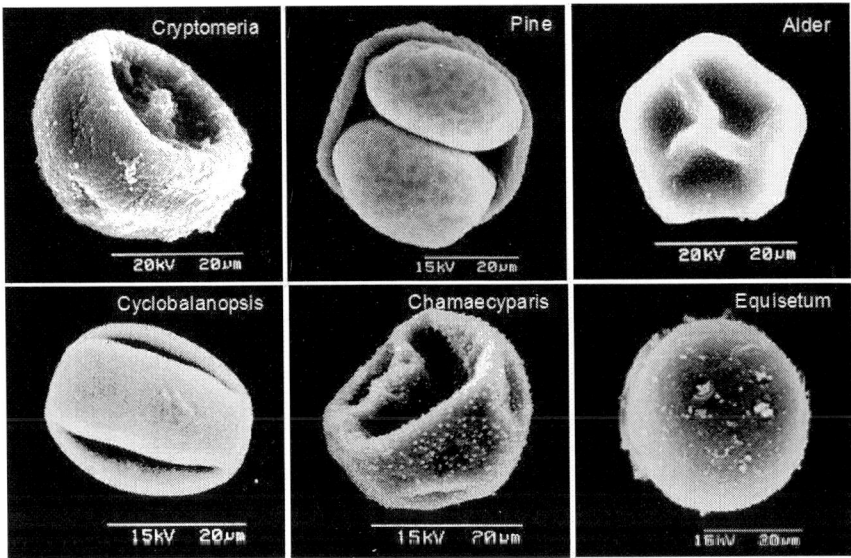

Figure 11. Morphologies Of Several Types Of Pollen Grains (Cryptomeria, Pine, Alder, Cyclobalanopsis, Chamaecyparis, And Equisetum) Identified By Sem Observation

The sum of the number concentration of three pollen types is still high at site A. Although the atmospheric presence of pollen grains is likely to vary depending on the local kind of plant, actual weather situations, and pollinating period, the vegetation of the surface is also important [18]. Cedar pollen easily absorbs water in the atmosphere and then settles down on surface. However, they can be easily resuspended into the atmosphere by the urban surface covered with asphalt and cement. Both hot island and building wind can also promote the resuspension of pollen [19]. Therefore, site A, with its relatively high pollen concentrations, was probably affected by the typical urban surface and local plant.

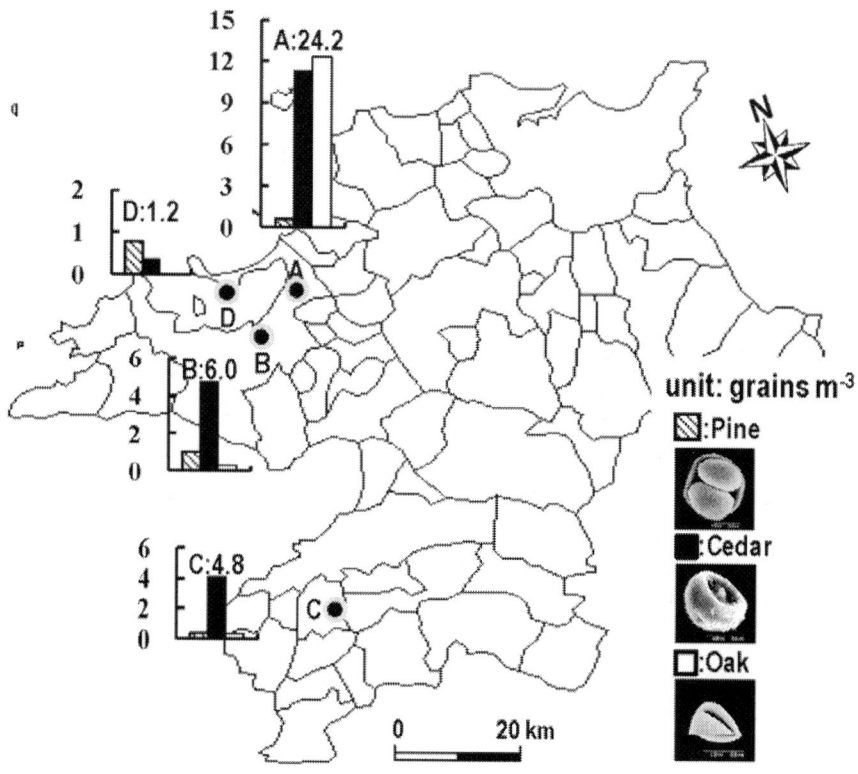

Figure 12. Pollens Number Concentration At Each Sampling Site

4. CONCLUSION

A better knowledge of the impact of both artificial and biological PM on urban atmosphere can help establish improved the management strategies of urban air quality. This study focused on a comprehensive and detailed interpretation for the springtime air quality influenced by both artificial (particulate matter (PM) and asbestos) and biological (pollen) sources in Fukuoka Prefecture, Japan. An intensive measurement of PM was conducted at four characteristic sites (i.e., a heavy traffic area, a residential area, an industrial area, and a desolate area) in the Fukuoka Prefecture during spring of 2007. Analysis of major ionic species in $PM_{2.5}$ was performed by an Ion Chromatography, and asbestos and pollen were identified by Scanning Electron Microscopy with an energy dispersive X-ray spectrometer (EDX). $PM_{2.5}$ concentration (65.3 μg m^{-3}) measured in an industrial area (site C) was extraordinarily high compared to those monitored in other areas; it greatly exceeded the Japan's $PM_{2.5}$ criteria (a daily average of 35 μg m^{-3}). NOAA's HYSPLIT dispersion model suggests that this high level of PM$_{2.5}$ monitored at site C is unlikely to affect the Asian continent. The ambient concentrations of PM$_{2.5}$-related anions (NH$_4^+$, NO$_3^-$, and SO$_4^{2-}$) and their relative contributions to PM$_{2.5}$ were also investigated in four study areas. The concentrations of these major water-soluble ions exhibit not only strong spatial dependence but also different ratios to each other. Asbestos fiber (crocidolite and amosite) concentration values changed in the range of 2.5 to 14.4 f per liter of air. The number of pollen grains showed that Cedar ranked higher in concentration than other types of pollen, with the maximum concentration at site A.

The results of our intensive field measurement suggested that synergic biological effects induced by ambient allergenic pollen and urban fine PM in atmosphere are associated with a peculiar springtime air quality in the Fukuoka Prefecture. Our study also indicates that the specific artificial and natural sources are regionally distributed to influence local air quality and public health. We should be mindful of the fact that in order to improve the understanding of urban air quality regarding the environmental load and repercussions on human health, it will be necessary to continue the monitoring of not only particulate matters but also gaseous materials.

5. ACKNOWLEDGEMENTS

The HYSPLIT aerosol dispersion model (http://www.arl.noaa.gov/ ready.html) developed by the National Oceanic Atmospheric Administration (NOAA) for backward trajectory was very helpful to data interpretation. The authors also express sincere thanks to Prof. G.-U. Kang, Department of Medical Administration, Wonkwang Health Science University for his ion analysis.

REFERENCES

1. Takahashi Y., Tokumam K., Kawashima S. Distribution Chart of Cryptomeria Japonica Forest Through Data Analysis of Landsat-TM. Japanese Journal of Palynology 1992; 38 140-147.

2. Schwartz J., Neas L.M. Fine Particles Are More Strongly Associated than Coarse Particles with Acute Respiratory Health Effects in Schoolchildren. Epidemiology 2000(11) 6-10.

3. Cook P.M., Glass G.E., Tucker J.H. Asbestiform Amphibole Minerals: Detection and Measurement of High Concentrations in Municipal Water Supplies. Science 1974; 185 853-855.

4.   Camus M., Siemiatycki J., Meek M.Sc.B. Nonoccupational Exposure to Chrysotile Asbestos and the Risk of Lung Cancer. New England Journal of Medicine 1998(338) 1565-1571.

5.   Lemen R.A. Asbestos in Brakes Exposure and Risk of Disease. American Journal of Industrial Medicine 2004(45) 229-237.

6.   Monthly Archives. Fukuoka-ken in Japan Encyclopedia. http://www. asbestosremovalsaustralia.com.au/blog/2013/04/Nussbaum (accessed 2 May 2014).

7.   Fukuoka Prefecture. About Fukuoka http. http://www.pref. fukuoka.lg.jp/somu/multilingual/english/about.html (accessed 2 May 2014).

8.   Nussbaum L.F. and Käthe R. Japan encyclopedia. Cambridge: Harvard University; 2005.

9.   Deshmukh D.K., Deb M.K., Tsai Y.I., Mkoma S.L. Water Soluble Ions in PM2.5 and PM1 Aerosols in Durg City, Chhattisgarh, India Aerosol and Air Quality Research 2011(11) 696-708.

10. Gao X., Nie W., Xue L., Wang T., Wang X., Gao R., Wang W., Yuan C., Gao J., Ravi K.P., Wang J., Zhang Q. Highly time-resolved Measurements of Secondary Ions in PM2.5 During the 2008 Beijing Olympics: The Impacts of Control Measures and Regional Transport. Aerosol and Air Quality Research 2013(13) 367-376.

11. Vezey E.L., Skvarla J.J., Vanerpool S.S. Characterizing Pollen Sculpture of Three Elosely-related Capparaceae Species Using Quantitative Image Analysis of Scanning Electron Micrographs, Clarendon: Oxford; 19991.

12. Zhang W.J., Sun Y.L., Zhuang G.S., Xu D.Q. Characteristics and Seasonal Variations of PM2.5, PM10, and TSP Aerosol in Beijing. Biomedical and Environmental Sciences 2006(19) 461-468.

13. Stone E.A., Yoon S.C., Schauer J.J. Chemical Characterization of Fine and Coarse Particles in Gosan, Korea During Springtime Dust events. Aerosol and Air Quality Research 2011(11) 31-43.

14. Interim Guidance. Naturally occurring asbestos (NOA) at school sites http://www.dtsc. ca.gov/Schools/upload/SMBRP_POL_Guidance_Schools_ NOA.pdf. (accessed 3 July 2014).

15. El Dorado County. Naturally occurring asbestos and dust protection. http://www.co.el-dorado.ca.us/emd/apcd/PDF/Naturally_Occuring_Asbestos_June_12.pdf (accessed 3 July 2014).

16. Kim J.Y., Lee S.K., Lee J.H., Lim M.H., Kang S.W., Phee Y.G. A Study on the Factors Affecting Asbestos Exposure Level from Asbestos Abatement in Building Demolition Sites. Korean Industrial Hygiene Journal 2009 (19) 8-15.

17. Farn P.P., Clarence T.N., Patrick J.S. Aerosol characteristics of Arctic haze sampler during AGASP 2. Atmospheric Environment 1990 (24A) 937-939.

18. Mandrioli P., Negrini M.G., Scarani C., Tampieri F., Trombetti F. Mesoscale Transport of Corylus Pollen Grains in Winter Atmosphere. Grana 1980 (19) 227-233.

19. Saito Y., Ide T., Murayama K. Science of Pollenosis. Kagakudouzinn co; 2006. p.14-42.

CHAPTER 10

An Air Quality Management System for Policy Support in Cyprus

Nicolas Moussiopoulos,[1] Ioannis Douros,[1] George Tsegas,[1]
Savvas Kleanthous,[2] and Eleftherios Chourdakis[1]

[1]Laboratory of Heat Transfer and Environmental Engineering, Aristotle
University, University Campus, P.O. Box 483, 54124 Thessaloniki, Greece
[2]Department of Labour Inspection, Ministry of Labour and Social Insurance,
Apelli 12, 1480 Nicosia, Cyprus, Greece

ABSTRACT

The recent air quality directive (2008/50/EC) encourages the introduction of
modelling as a necessary tool for air quality assessment and management.
Towards this aim, an air quality management system (AQMS) has been
developed and installed in the Department of Labour Inspection (DLI) of the
Republic of Cyprus. The AQMS comprises of two operational modules,
providing hourly nowcasting and daily forecasting of the air quality status,
implemented as an integrated model system that performs nested grid
meteorological and photochemical simulations. A third operational module
provides the capability of an interactive configuration of custom emission
scenarios and corresponding model runs covering user-defined domains of
interest. Statistical indicators are calculated at the end of each day for the
measurement locations of DLI's air quality monitoring network. Besides, the
system provides an advanced user interface, which is realised as a web-based
application providing access to model results from any computer with an
internet connection and a web browser.

1. INTRODUCTION

The new air quality directive (2008/50/EC) introduces additional requirements for air quality assessment and management. As a result, along with air quality measurements, modelling tools should also be used by authorities and policy makers in order to assess pollutant concentrations in ambient air. It also requires that up-to-date information on concentrations of all regulated pollutants in ambient air should be readily available to the public.

Towards this aim, an operational web-based system for air quality management developed by the Laboratory of Heat Transfer and Environmental Engineering (LHTEE) has been installed in the Department of Labour Inspection of the Ministry of Labour and Social Insurance in the Republic of Cyprus. The system offers a range of features for nowcasting, forecasting, and performing scenario calculation for air quality in Cyprus. Besides, a number of statistical indices [1] and validation maps are calculated at the end of each day at various points coinciding with the locations of DLI's air quality monitoring stations [2] in order to assess the system's accuracy.

The main goals of the system lie in providing useful up-to-date information to the public on pollutant concentrations in ambient air, as well as supporting local authorities and decision makers in air quality assessment and management.

2. METHODOLOGY

The core of the system handles the compilation of an emissions inventory that includes data from all major activity sectors and functions on the basis of a continuous update of the emissions database. This part of the software dealing with emissions provides functionality for setting up emission scenarios based both on the application of measures as well as targeted modifications. Emission data are prepared as an emissions grid that is subsequently fed into the air quality modelling system [3].

The part of the AQMS dealing with the calculation of ambient concentrations has a functional structure similar to that followed by most AQMSs operating on an around-the-clock basis (see Figure 1). The AQMS comprises of two operation modes, nowcasting and forecasting, which are implemented based on an automated scheme that performs nested grid meteorological and photochemical model simulations. The mesoscale meteorological model MEMO and the chemistry-transport model MARS-aero are used for this purpose, both parts of the EZM (European Zooming Model) system [4]. MEMO [5] is a three-dimensional, non-hydrostatic, prognostic mesoscale model which simulates mesoscale air motion and inert pollutant dispersion at the local-to-regional scale, over complex terrain, allowing multiple nesting. MARS-aero [6] is an Eulerian chemistry-transport model for reactive species operating at similar scales. The modular structure of MARS-aero allows any of four chemical reaction mechanisms for the gaseous phase to be used, while the calculation of secondary aerosols, organic and inorganic, can be enabled at will [7, 8]. Meteorological data such as wind speed, temperature, TKE, surface roughness, Monin-Obukhov length, and friction velocity are required input for the dispersion calculations and are provided by MEMO. Both MEMO and MARS-aero are used in a nested configuration (see Table 1).

Table 1: Description of MEMO and MARS-aero nested grids.

	Depth of nesting	Number of nested grids	Grid size	Grid resolution	Type of nesting
Coarse	2	1	60×60	$25\ km^2$	1-way
Fine		5	50×50	$1\ km^2$	

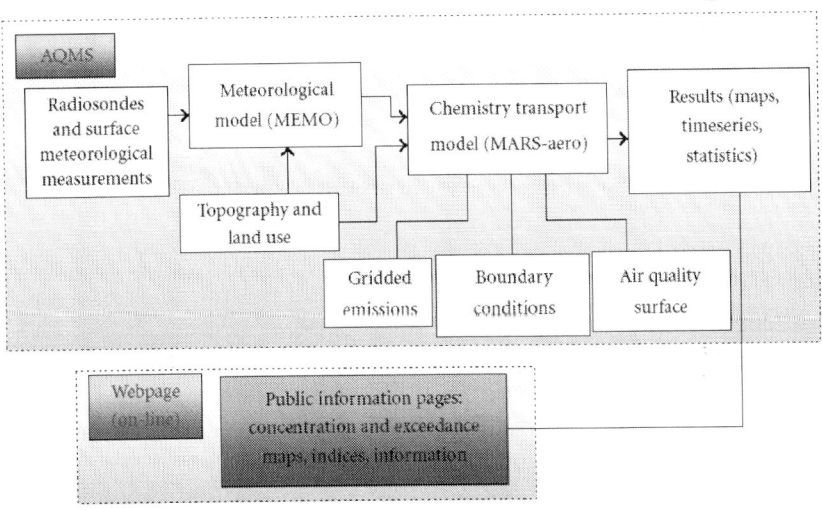

Figure 1: The structure of the air quality management system.

For the initialisation of MEMO, a number of vertical profiles of the key meteorological variables originating from the Global Forecast System (GFS) are used [9]. Such profiles are also assimilated into the model calculations on a 3-hour basis. The downloader module of the automated system undertakes the downloading of the GFS data, processes them, and stores them in a data pool which is kept updated at all times. The scheduler selects only the most recent dataset for input to the MEMO model. Each of these processes keeps a separate event log and diagnostic files accessible to the operator of the system.

The MARS-aero model runs back-to-back with the MEMO model, featuring real time integration of ambient pollutant concentration data that are provided both from valid measurement data, as well as from larger scale models results and are used both as initial and lateral conditions in the model calculations as well as for data assimilation purposes. An upgraded methodology has been developed and incorporated in the system's core for providing increased flexibility in the coupling of the nested domains. Additionally, a dynamic dust concentration component is included in the PM_{10} and $PM_{2.5}$ boundary conditions that are fed in the model core, so as to improve the accuracy of calculated PM concentrations during Saharan dust episodes. These upgrades significantly enhance the operational prognostic skills of the AQMS in the cases of elevated concentration levels that are associated with transboundary transport of air pollutants.

In nowcasting mode the system computes, on an hourly basis, the pollutant concentration fields in Cyprus and, at higher resolution, in the five largest cities. Model results are automatically processed and a range of maps, timeseries graphs, and statistical indices are produced for five major pollutants, namely, PM_{10}, $PM_{2.5}$, NO_2, O_3, and benzene. These results become available to the public via DLI's web page and are also accessible in even greater detail to the staff of the department. In forecasting mode 24-hour photochemical model simulations are driven by corresponding prognostic meteorological simulations of GFS, in order to produce daily air quality forecast maps and graphs for the domains of interest and the prescribed pollutants. Daily forecasts become available to the department and the public through a set of web pages accessible through a similar interface.

The system's user interface is realised as a web-based application which is accessible from any computer having internet access. Through this interface, the authorized DLI users are provided with numerous possibilities for reviewing both the process and the results of the calculations as well as for configuring the information that becomes available in the public web page. The informational public pages represent a very important aspect of the system's structure as, in addition to the concentration levels, the public is also informed on the expected health impacts of the forecasted air quality situation, thus allowing them to manage their activities accordingly.

In order to evaluate the system's performance, suitable statistical indicators are calculated at the end of each day regarding the sites where DLI's air quality measurements take place. An automated procedure operating on an around-the-clock basis undertakes the downloading of the available observation data from DLI's server, processes them, and keeps them in a data base, where historical monitoring data are stored. In addition, model results for the measurement locations of DLI's air quality monitoring network are also kept in appropriate data pool. At the end of each day, a wide range of statistical indices are calculated regarding the stations and pollutants of interest and numerous charts are produced, which demonstrate the system's performance in both nowcasting and forecasting mode. In order to have a better overview of the system's efficiency, validation charts are also produced. These charts present comparisons between the calculated and the observed timeseries concerning air pollutant concentration, wind speed, and wind direction.

The AQMS can also be operated by DLI to support air-quality-related assessment and decision making, by allowing interactive configuration of custom emission scenarios and corresponding model runs. This operational module provides the capability to study emission scenarios and assess their effect on air quality in the five major urban areas of Cyprus or over user-defined domains. The meteorology used for assessment calculations is based on a number of representative meteorological situations which are analysed and duly weighted according to the frequency of their occurrence during a full calendar year.

3. RESULTS

Through the system's interface the public and the authorized DLI users can view the model results, both for nowcasting and forecasting. In the nowcasting section, hourly meteorological and air quality maps are available to any computer with an internet connection through interactive web pages (see Figures 2 and 3). The public is also informed on the expected health impacts of the forecasted air quality situation by messages that are automatically produced and constitute a statement about the expected exposure to air pollution levels on an hourly basis.

Figure 3 demonstrates the gradual dust transport over the Cyprus domain during a typical Saharan dust episode. The system performs reasonably well in predicting the occurrence of Saharan dust episodes. Besides, the model results presented in Figure 3 indicate the high ability of the system in producing the spatial distribution of PM during these episodes.

Figure 2: Snapshots of nowcasting for pollutant concentrations (a) and meteorological parameters (b).

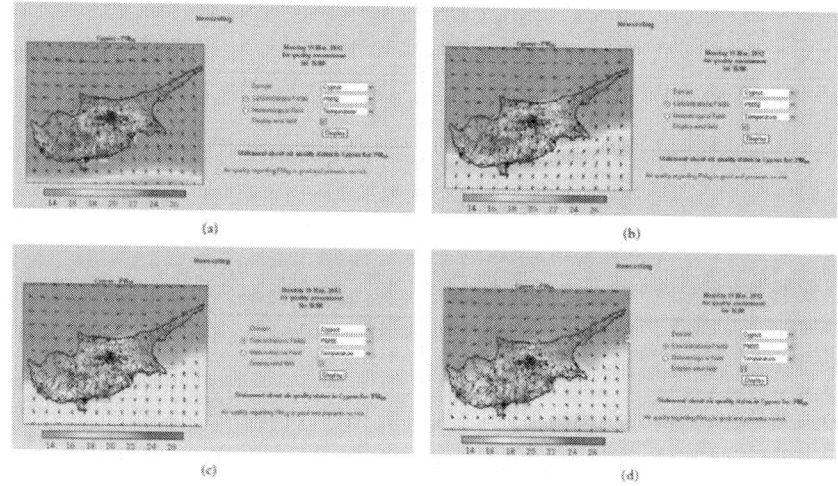

Figure 3: Snapshots of nowcasting results for PM$_{10}$ during a Saharan dust episode.

In the forecasting section of the web interface, a wide range of maps, timeseries graphs, and statistical indices are made accessible to the public, providing information both on forecast pollutant concentrations as well as on meteorological parameters, such as temperature, relative humidity, and wind speed (see Figure 4).

Furthermore, through this interface, the authorized DLI users are provided with geospatial tools for setting up emission scenarios, as well as reviewing the results of the corresponding model runs (see Figure 5). In the scenarios results section, the DLI user can not only access pollutant concentration maps produced by the automatic process of the respective model runs but also to manually produce statistical indices and timeseries graphs for any location inside the calculation domain, using a suitable interactive feature (see Figure 6).

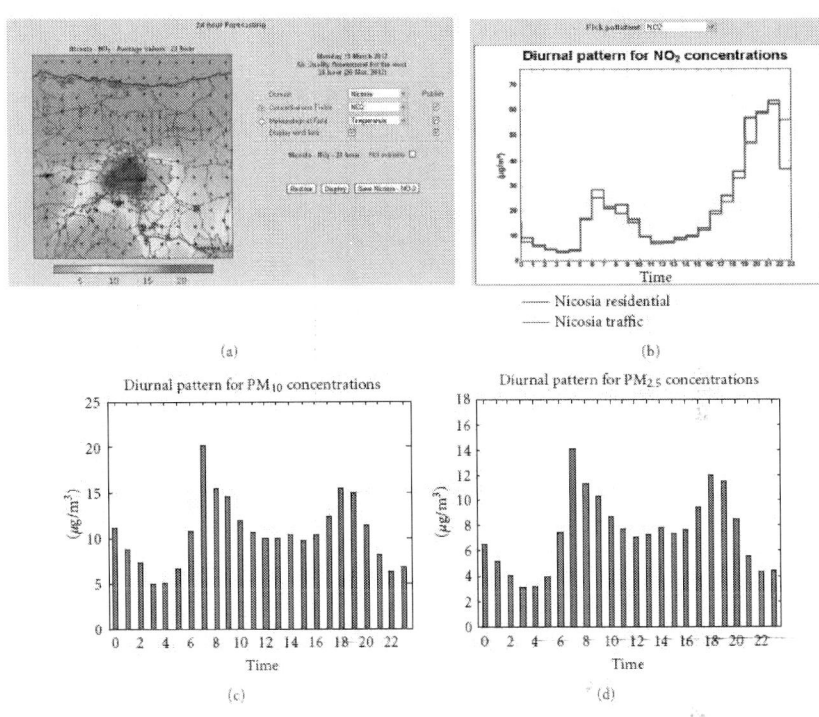

Figure 4: Snapshots of forecasting results for pollutant concentrations.

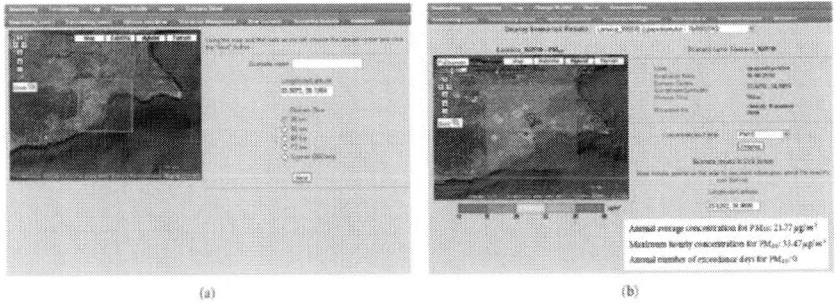

Figure 5: Emission scenario set up (a) and scenarios results (b).

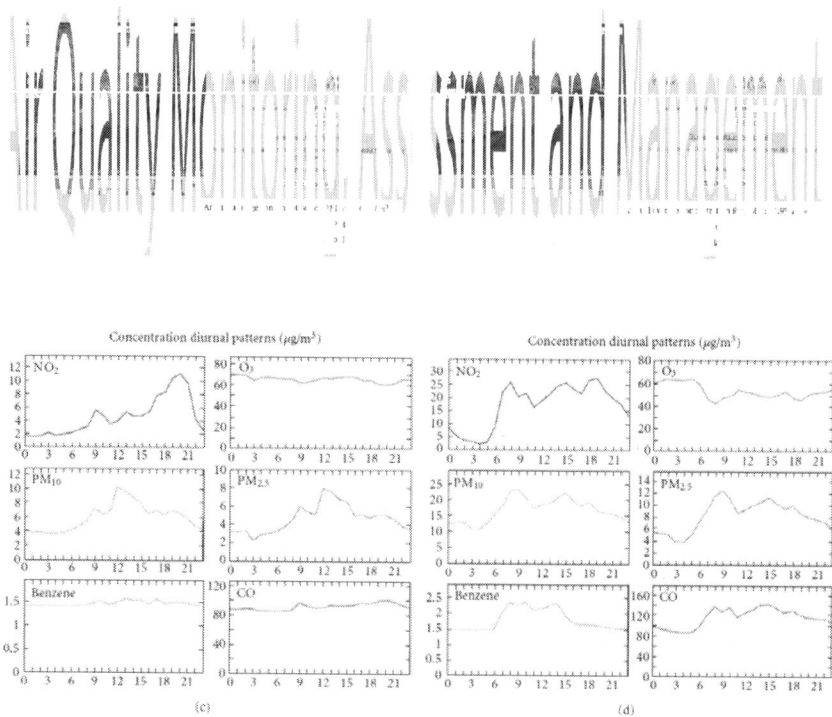

Figure 6: Statistical indices and graphs produced for two different locations inside the computational domain of a particular emission scenario.

In Figure 7 air quality and meteorological validation graphs as they are operationally produced by the system are presented. Each graph shows comparisons between model simulations and the respective observation data regarding air pollutant concentrations, as well as wind speed and direction. Additional information about the model accuracy is also provided in the form of statistical indices, namely, "BIAS", "Average Normalised BIAS" (ANB), "Fractional Bias" (FB), "Root Mean Square Error" (RMSE), "Normalised Mean Square Error" (NMSE), "Correlation Coefficient", and "Index of Agreement". In this field of the graphs, green colour indicates which mode between nowcasting and forecasting operates better according to each statistical index.

As depicted from Figure 7, the system's performance as regards the suburban areas is good in the case of O_3, where concentration levels are determined by the transboundary transport which enters the model as initial and lateral boundary conditions from the results of larger scale models. On the other hand, the supplementary use of air quality measurements in the formation of the boundary conditions leads to increased model efficiency in the cases of air pollutants which are related to human activities, such as NO_2. This fact occurs because most of DLI's measurement stations are located close to areas which contain a wide range of human activities, including transport and industry, and as a result, the obtained concentration values are affected by the aforementioned activities. Thus, the exploitation of observation data as boundary conditions of the model simulations improves the system's computational performance in assessing air pollutant levels mostly influenced by human activities.

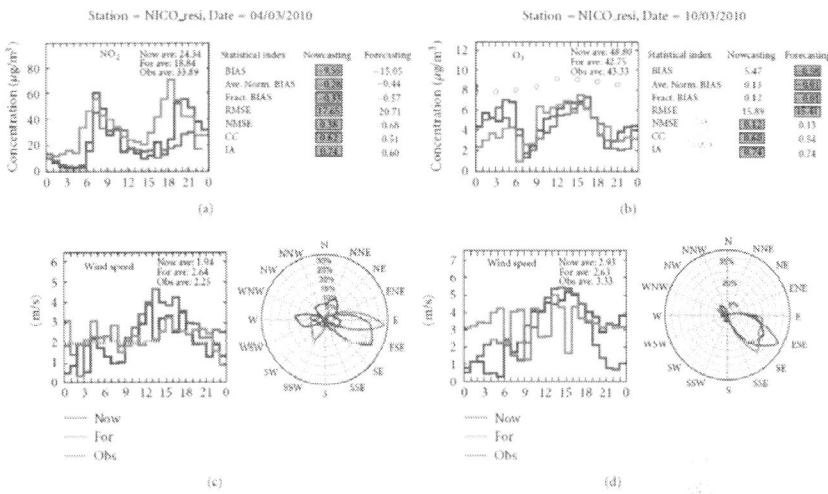

Figure 7: Air quality and meteorological validation graphs as regards the residential station of Nicosia for the 4th and 10th of March 2010 concerning NO_2 and O_3, respectively.

Figure 8 presents a subset of the calculated statistical indicators, namely, "BIAS," "Normalised Mean Square Error," "Correlation Coefficient," and "Index of Agreement," for three typical days regarding nitrogen dioxide (NO_2) and ozone (O_3). As evidenced from the indicators shown in Figure 8, the results of the model are in fairly good agreement with the observed concentrations, the only exceptions being the industrial and traffic stations. This is more or less expected for these hotspots that are highly affected by local and street scale activities.

Figure 9 shows validation maps as regards PM_{10} for a typical winter day and a day in which a slight Saharan dust episode occurs. In these maps the calculated model concentration maps are compared to the concentrations observed at the measurement locations at the same time.

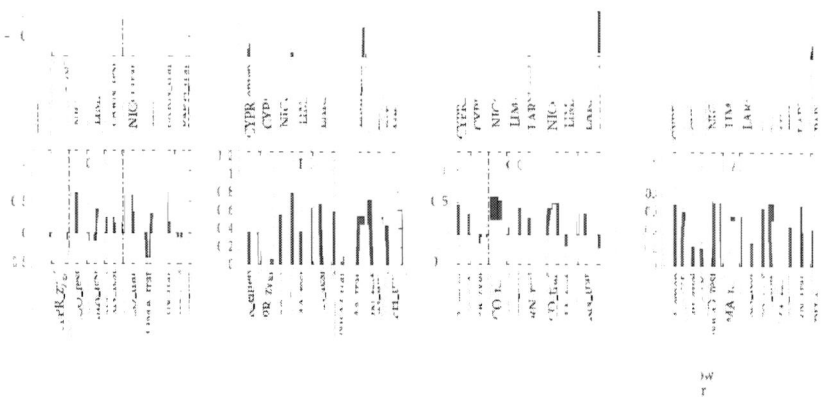

Figure 8: Statistical indices calculated for the 11th of March 2010 as regards NO$_2$ (a) and O$_3$ (b).

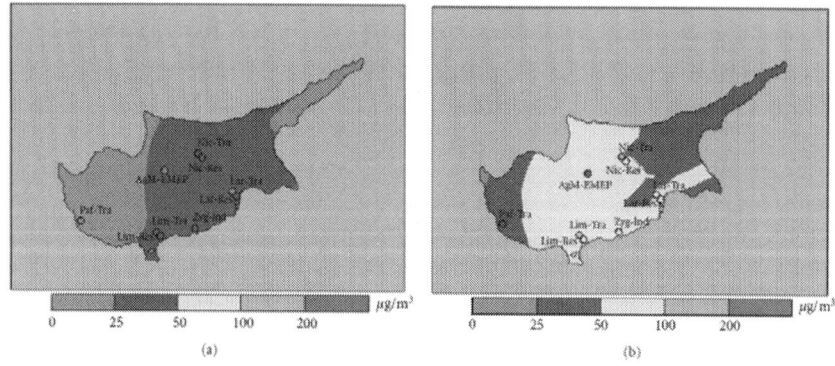

Figure 9: Validation maps as regards PM$_{10}$ for a typical winter day (a) and a day in which a slight Saharan dust episode occurs (b).

4. CONCLUSIONS

The AQMS developed and installed in the Department of Labour Inspection of the Republic of Cyprus is an integrated operational state-of-the-art air quality management system for meteorological and photochemical model simulations which offers a range of features for nowcasting, forecasting, and performing scenario calculations for air quality in areas of interest. Results of the calculations can be reviewed in detail by the DLI, while they can also be accessed in a condensed form by the general public via a user friendly web page. In order to evaluate the system's performance, suitable statistical indicators are calculated at the end of each day for the measurement locations of DLI's air quality monitoring network. Besides, validation charts are also produced, presenting comparisons between the calculated and the observed timeseries concerning air pollutant concentration, wind speed, and wind direction. The AQMS can also be operated by authorised DLI users to support air-quality-related assessment and decision making by performing air quality calculations based on custom emission scenarios. The capabilities offered by this AQMS for producing high-quality assessments of the air quality situation are expected to be a valuable aid to the authorities of Cyprus towards compliance with the relevant EU standards.

ACKNOWLEDGMENTS

This work was cofinanced by the EU in the framework of the transition facility 2005 for Cyprus. The authors would also like to thank the director of DLI Leandros Nicolaides and the DLI team involved in the project.

REFERENCES

1. COST728, Overview of Tools and Methods for Meteorological and Air Pollution Mesoscale Model Evaluation and User Training, 2008.

2. http://www.airquality.dli mlsi.gov.cy/.

3. S. Kleanthous, G. Tsegas, I. Douros, and N. Moussiopoulos, "An air quality management system for Cyprus," in Proceedings of the 7th International Conference on Air Quality-Science and Application (Air Quality '09), R.-M. Hu, Ed., pp. 405–408, CD-ROM edition, Istanbul, Turkey, March 2009.

4. N. Moussiopoulos, "The EUMAC Zooming Model, a tool for local-to-regional air quality studies,"Meteorology and Atmospheric Physics, vol. 57, no. 1–4, pp. 115–133, 1995.

5. N. Moussiopoulos, T. Flassak, P. Sahm, and D. Berlowitz, "Simulations of the wind field in Athens with the nonhydrostatic mesoscale model MEMO," Environmental Software, vol. 8, no. 1, pp. 29–42, 1993.

6. N. Moussiopoulos, P. Sahm, and C. Kessler, "Numerical simulation of photochemical smog formation in Athens, Greece-a case study," Atmospheric Environment, vol. 29, no. 24, pp. 3619–3632, 1995.

7. A. Arvanitis, N. Moussiopoulos, and S. Kephalopoulos, "Development and testing of an aerosol module for regional/urban scales," in Proceedings of

the 2nd Conference on Air Pollution Modelling and Simulation (APMS '01), pp. 277–288, Champs-sur-Marne, April 2001.

8. B. Schell, I. J. Ackermann, H. Hass, F. S. Binkowski, and A. Ebel, "Modeling the formation of secondary organic aerosol within a comprehensive air quality model system," Journal of Geophysical Research D, vol. 106, no. 22, pp. 28275–28293, 2001.

9. http://www.emc.ncep.noaa.gov/gmb/moorthi/gam.html.

CHAPTER 11

Study on an Air Quality Evaluation Model for Beijing City Under Haze-Fog Pollution Based on New Ambient Air Quality Standards

Li Li and Dong-Jun Liu *

Shenzhen Graduate School, Harbin Institute of Technology, Shenzhen 518055, China

ABSTRACT

Since 2012, China has been facing haze-fog weather conditions, and haze-fog pollution and $PM_{2.5}$ have become hot topics. It is very necessary to evaluate and analyze the ecological status of the air environment of China, which is of great significance for environmental protection measures. In this study the current situation of haze-fog pollution in China was analyzed first, and the new Ambient Air Quality Standards were introduced. For the issue of air quality evaluation, a comprehensive evaluation model based on an entropy weighting method and nearest neighbor method was developed. The entropy weighting method was used to determine the weights of indicators, and the nearest neighbor method was utilized to evaluate the air quality levels. Then the comprehensive evaluation model was applied into the practical evaluation problems of air quality in Beijing to analyze the haze-fog pollution. Two simulation experiments were implemented in this study. One experiment included the indicator of $PM_{2.5}$ and was carried out based on the new Ambient Air Quality Standards (GB 3095-2012); the other experiment excluded $PM_{2.5}$ and was carried out based on the old Ambient Air Quality Standards (GB 3095-1996). Their results were compared, and the simulation results showed that $PM_{2.5}$ was an important indicator for air quality and the evaluation results of the new Air Quality Standards were more scientific than the old ones. The haze-fog pollution situation in Beijing City was also analyzed based on these results, and the corresponding management measures were suggested.

KEYWORDS

air quality; comprehensive evaluation; Ambient Air Quality Standards; entropy weighting method; nearest neighbor method

1. INTRODUCTION

With the development of modern industry, industrial equipment and transportation vehicles consume lots of resources and discharge more and more pollutants, and as a result, the atmospheric environment is polluted seriously. Thus the progress of human society is at the expense of the environment [1]. In recent years, areas with polluted air frequently suffer from haze-fog weather conditions in autumn and winter. Especially, in the winter of 2012, a large-scale emergence of haze-fog weather affected the mid-east areas of China, and $PM_{2.5}$, which is one of the key pollution factors in the air environment, became a hot topic in China [2]. As a result the atmospheric environment in China was very harsh, and human health and transportation were seriously affected. The data of the Ministry of Environmental Protection showed that there was large-scale haze-fog pollution in Northeast China, Northwest China, North China, Eastern China and Central China [3,4]. The wide range of haze-fog weather triggered a series of chain reactions, including transportation restrictions, flight delays and increased numbers of patients, *etc.*, and people's life was severely affected [5,6]. In order to protect and improve the living environment and quantitatively analyze the atmospheric environment pollution, new ambient air quality standards were formulated by the Ministry of Environmental Protection [7]. As a result, the "Air Quality Index" (AQI) was introduced to replace the old "Air Pollution Index" (API), and the "Technical Regulations on Ambient Air Quality Index (Trial) (HJ 633-2012)" [8] were published to calculate the AQI.

The severe haze-fog pollution situation has attracted serious concern of academic researchers. Numerous studies have been conducted to investigate the composition, sources, and chemical reactions of the haze-fog pollution in China [9,10,11]. Some scholars claimed the haze-fog pollution was usually accompanied with aerosol concentrations in the atmospheric environment, and had significant impacts on the air quality, human health and visibility [12,13,14].

The formation of haze-fog was very closely linked to atmospheric and meteorological conditions [15,16]. Zhao *et al.* argued that the main source of haze-fog in winter was anthropogenic emissions on a regional scale [17]. Che *et al.* investigated the optical characteristics of the aerosols of haze-fog [18], and Zhang *et al.* analyzed the main chemical components of the aerosols [19]. Wang *et al.* simulated the severe winter of 2013 regional hazes in East Asia and northern China based on simulation models [20].

In order to deal with the haze-fog pollution, the degree of haze-fog pollution should be known first. In this study the ambient air quality was evaluated to investigate the status of haze-fog. Evaluation of ambient air quality is an important part of any atmospheric ecology studies. The assessment of environmental air quality is the process of quantitative description of the ambient air quality by mathematical methods and models. It can help know the present situation and future tendencies of the ambient air quality [21]. Thus, it can provide a scientific basis for the planning and management of ambient air quality. Studies on the evaluation of air quality are an important issue, and have attracted the interest of many scholars. The fuzzy comprehensive evaluation theory was applied in air quality assessment according to the national air quality standard [22]. The level of environmental quality was determined based on the ambient air monitoring data of Dongzhi Country [23]. Artificial neural networks and decision tree models were applied to evaluate the common Air Quality Index in Thessaloniki, Greece [24]. The forest air quality in Yichun Town was evaluated based on BP neural networks [25].

In this study the current situation of haze-fog pollution in China was introduced, and the new ambient air quality standards were analyzed. A comprehensive evaluation model was developed based on an entropy weighting method and nearest neighbor method, and it was applied into the practical problems of air quality in Beijing to analyze the haze-fog pollution status. The entropy weighting method was used to determine the weights of indicators, and the nearest neighbor method was used to evaluate the levels of air quality. Two simulation experiments were implemented. One experiment was with $PM_{2.5}$ and carried out based on the new Ambient Air Quality Standards (GB 3095-2012); the other experiment was without $PM_{2.5}$ and carried out based on the old Ambient Air Quality Standards (GB 3095-1996). We compared their results to investigate the importance of $PM_{2.5}$ and the effects of standards. The situation of haze-fog pollution in Beijing City was then analyzed based on these results.

2. MATERIALS

2.1. Haze-Fog Pollution in China

Haze-fog events have been hot issues in China since 2012. From the year of 2013 onward haze-fog weather has become more and more serious, and this has caused serious harm to human health, traffic safety, and other production and living aspects of human beings. The main indicator affecting the haze-fog was $PM_{2.5}$, and it became a hot topic that concerned people. The average number of haze-fog days in 2013 was 4.7, which was the largest number recorded in the last 52 years. According to the statistics of the China Meteorological Administration, in mid-east China where the haze-fog weather was the most serious, the average of haze-fog days in the year of 2013 was 35.9. This was equivalent to the fact that there was over one month when the mid-east areas of China were in the shadow of haze-fog, and the ratio of haze-fog days in one year was about 10%.

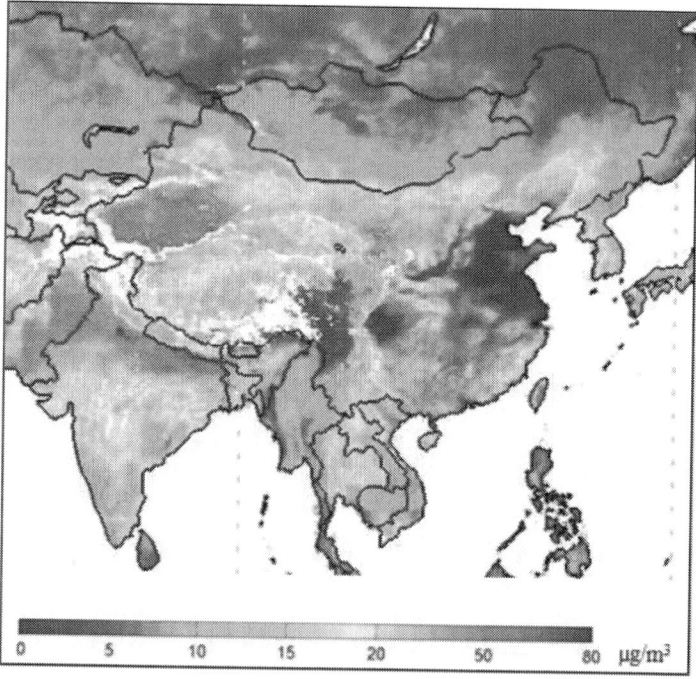

Figure 1. Model map of areas under air pollution based on monitoring data of $PM_{2.5}$.

The $PM_{2.5}$ data in China were monitored through instruments installed on satellites by the National Aeronautics and Space Administration, and the model map of areas under air pollution based on $PM_{2.5}$ monitoring data is shown in Figure 1, where the different colors represent different degrees of $PM_{2.5}$ pollution. The deeper the color of the areas in the map, the heavier the pollution there was. There was continuous haze-fog weather in most parts of China, including Tibet and Xinjiang. The areas with serious haze-fog pollution included the Beijing and Tianjin areas, South Hebei Province, Northeast Henan Province, Western Shandong Province, Jiangsu Province, Anhui Province, Western Zhejiang Province, Northwest Fujian Province, Central Hunan Province, South Jiangxi Province, Central Hubei Province and the Northern Sichuan Basin Area. Southwest China became the only unpolluted land.

The reasons for haze-fog pollution formation are many, and the main reasons can be summarized as follows [26,27]:

(i)   The automobile exhaust is the main source of pollutants. In recent years, there are more and more cars in the cities in China and the components in automobile exhaust are the main components of the haze-fog;

(ii) Secondary pollution from factories is also an important reason. There is much benzene and aldehydes in chemical pollution emissions, and they are important components of haze-fog;

(iii) The relative humidity near the ground in the haze-fog areas is relatively high, and the ground has lots of dust, so particulate matter can easily form;

(iv) Burning garbage and burning coal in winter for heating can also generate pollutants.

2.2. New Air Quality Standards

In order to protect and improve the living environment and analyze the atmospheric environment pollution quantitatively, new ambient air quality standards were formulated. The standards formulated the function division of air quality, grading standards, pollutant indicators, time to acquire pollutants, concentration limit values, sampling and analysis methods, and effectiveness of data statistics. The air quality standard in effect before 2012 in China was "Ambient Air Quality Standards (GB 3095-1996)" [28]. They were published and implemented in 1996 and modified in 2000. However, because the haze-fog pollution in China is more and more serious, the regulation of the measurement of PM_{10} and $PM_{2.5}$ in ambient air, was published by the Ministry of Environmental Protection, and started to be implemented from 1 January 2011. The measurement of $PM_{2.5}$ was standardized for the first time in the regulation, but it was not included in the mandatory monitoring indicators at that time. In

February 2012, the State Council of China issued the new air quality standards: "Ambient Air Quality Standards (GB 3095-2012)" [29]. This regulation will be implemented in 1 January 2016. In this new standard the $PM_{2.5}$ values were mandatorily included. Three methods of $PM_{2.5}$ monitoring were published in the regulation of "$PM_{2.5}$ automatic monitoring instrument specifications and requirements (trial)" published in May 2012.

The threshold values of some main indicators in the Ambient Air Quality Standards (GB 3095-1996) and (GB 3095-2012) are shown in Table 1 and Table 2. The main differences between the new Ambient Air Quality Standards (GB 3095-2012) and the old ones (GB 3095-1996) were as follows:

a) The three levels of air environment function classification were adjusted to two levels, and the threshold values were adjusted accordingly;

b) (PM$_{2.5}$ was included in the standards, and 8-hour average threshold values for O_3 were also included;

c) (The threshold values of PM_{10} (annual mean) and NO_2 (1-hour average) were adjusted, and they were more severe than before;

d) The regulations of the validity of data statistics were adjusted.

Table 1. Threshold values of ambient air pollutant indicators in GB 3095-1996.

No.	Pollutant Indicator	Average Time	Level I	Level II	Level III	Unit
1	SO_2	Annual mean	20	60	100	$\mu g/m^3$
		24-hour average	50	150	250	
		1-hour average	150	500	700	
2	NO_2	Annual mean	40	40	80	$\mu g/m^3$
		24-hour average	80	80	120	
		1-hour average	120	120	240	
3	CO	24-hour average	4	4	6	mg/m^3
		1-hour average	10	10	20	
4	O_3	1-hour average	120	160	200	$\mu g/m^3$
5	PM_{10}	Annual mean	40	100	150	$\mu g/m^3$
		24-hour average	50	150	250	

Table 2. Threshold values of ambient air pollutant indicators in GB 3095-2012.

| No. | Pollutant Indicator | Average Time | Threshold Values | | Unit |
			Level I	Level II	
1	SO_2	Annual mean	20	60	$\mu g/m^3$
		24-hour average	50	150	
		1-hour average	150	500	
2	NO_2	Annual mean	40	40	$\mu g/m^3$
		24-hour average	80	80	
		1-hour average	200	200	
3	CO	24-hour average	4	4	mg/m^3
		1-hour average	10	10	
4	O_3	8-hour average	100	160	$\mu g/m^3$
		1-hour average	160	200	
5	PM_{10}	Annual mean	40	70	$\mu g/m^3$
		24-hour average	50	150	
6	$PM_{2.5}$	Annual mean	15	35	$\mu g/m^3$
		24-hour average	35	75	

3. METHODS

3.1. Entropy Weighting Method

In the evaluation system, determination of weights for all the indicators is an important process that can measure the degree of impact of indicators. When the weight of an indicator is high, it has great impact on the capability; otherwise, the impact is lower. In information theory, information entropy is an important concept, and it can measure the amount of useful information that the data provides in a system [30]. The basic criteria of entropy weighting method are as follows: when the data of the multiple evaluated objects on one indicator show great differences, the entropy value of this indicator must be low according to information theory. This shows that this indicator can contribute much useful information and thus the weight of this indicator should be set high; otherwise, when the entropy value of an indicator was high, it may contribute little useful information according to the information theory. Its weight should be correspondingly low [31]. The procedures for weighting the indicators are as follows:

i.   The original data of all indicators should be normalized, and this can eliminate the impact of dimension. For one benefit indicator, the higher its value is, the better its effect on the air quality was. The normalization equation is:

$$r_{ij} = \frac{x_{ij} - \min_{i}\{x_{ij}\}}{\max_{i}\{x_{ij}\} - \min_{i}\{x_{ij}\}} \tag{1}$$

ii.   For a cost indicator, the lower its value is, the better its effect on the air quality is. The normalization equation is:

$$r_{ij} = \frac{\max_{i}\{x_{ij}\} - x_{ij}}{\max_{i}\{x_{ij}\} - \min_{i}\{x_{ij}\}} \tag{2}$$

iii.   where, x_{ij} (i=1, 2, ..., m, and j =1, 2, ..., n) is the monitoring value of the j-th object on the i-th indicator, and r_{ij} is the dimensionless value normalized.

iv.   For the evaluation problem with m indicators and n evaluated objects, the entropy value p_i of the i-th indicator can be defined as:

$$p_i = -k \sum_{j=1}^{n} f_{ij} \ln f_{ij} \tag{3}$$

where $f_{ij} = r_{ij} / \sum_{j=1}^{n} r_{ij}$, k=1/ln n, i=1, 2, ..., m. When f_{ij}= 0, we set f_{ij} ln f_{ij} =0.

The weight of the i-th indicator λ_i can be calculated according to the information entropy theory:

$$\lambda_i = \frac{1 - p_i}{m - \sum_{j=1}^{m} p_i} \tag{4}$$

where $0 \le \lambda_i \le 1$, and $\sum_{i=1}^{m} \lambda_i = 1$.

The entropy weighting method is a good objective weighting method, and it can reflect the degree of effective amount of information provided. In this study the entropy weighting method is used to weight the indicators of the air quality.

3.2. Nearest Neighbor Method

The nearest neighbor method is one of the clustering analysis methods. The purpose of the clustering analysis method is to classify the samples with close distances to the clustering centers as the same class based on some criterion. The basic ideas of the nearest neighbor method are as follows: for an evaluation problem with multiple objects, it is supposed that there are multiple samples x_i (i=1, 2, ..., N). For one sample x_i to be classified, we investigate the distances between x_i and x^* which is the known clustering center. The cluster of sample x_i is defined as the cluster with the nearest distance from the known cluster [32]. According to this classification ideology, the distance of nearest neighbor method can be set as:

$$d_i = \min_i \left\| x_i - x^* \right\|$$ (5)

The Euclidean distance function is employed in this study, and the equation is:

$$d_i(x_i - x^*) = \sqrt{\sum_i (x_i - x^*)^2}$$ (6)

In the evaluation system, different indicators are often with different dimensions, and the data of indicators should be normalized first. The nearest neighbor method is a practical evaluation method. It is suitable for processing multi-indicator evaluation problems. It has a simple principle, easy calculation and high practicability. When this method was applied to the evaluation work, good comparability assessment results could be obtained.

4. EVALUATION MODEL AND ALGORITHM

For the issue of air quality evaluation, a comprehensive evaluation model based on the entropy weighting method and nearest neighbor method was developed. The entropy weighting method was used to weight the various air quality indicators, and the importance of each indicator was analyzed. The nearest neighbor method was utilized to evaluate the air quality according to the Ambient Air Quality Standards. The algorithm of the evaluation model is shown in Figure 2.

Then the comprehensive evaluation model was applied to the practical problem of evaluating the air quality in Beijing to analyze the haze-fog pollution status. Because the most important changes of the Ambient Air Quality Standards were the introduction of $PM_{2.5}$, we set two simulation experiments, one is with the indicator of $PM_{2.5}$, and the other not. The simulation experiment with $PM_{2.5}$ was carried out based on the new Ambient Air Quality Standards (GB 3095-2012), and the other experiment without $PM_{2.5}$ was based on the old Ambient Air Quality Standards (GB 3095-1996). We compared their results, and analyzed the air quality of Beijing City at the same time. Thus the ambient air quality of Beijing City was analyzed comprehensively, and it is hoped that the results will provide helpful suggestions for the health and life of the people. The study procedures of this paper were as follows:

- Step 1: Initialize, and collect the original data of all indicators of air quality;

- Step 2: Define the indicators and their weights. The entropy weighting method was used to determine the weights of all indicators, and their importance was analyzed;

- Step 3: Construct the evaluation model based on the nearest neighbor method, and it was used to evaluate the air quality of Beijing City;

- Step 4: Use the model to evaluate the air quality in two simulation experiments. One experiment was with $PM_{2.5}$ and carried out based on the new Ambient Air Quality Standards (GB 3095-2012); the other experiment was without PM2.5 and carried out based on the old Ambient Air Quality Standards (GB 3095-1996);

- Step 5: Compare the results of the two simulation experiments, and analyze the air quality of Beijing City according to the evaluation results of the model;

- Step 6: Draw conclusions, and provide reasonable suggests to decision making according to the research results.

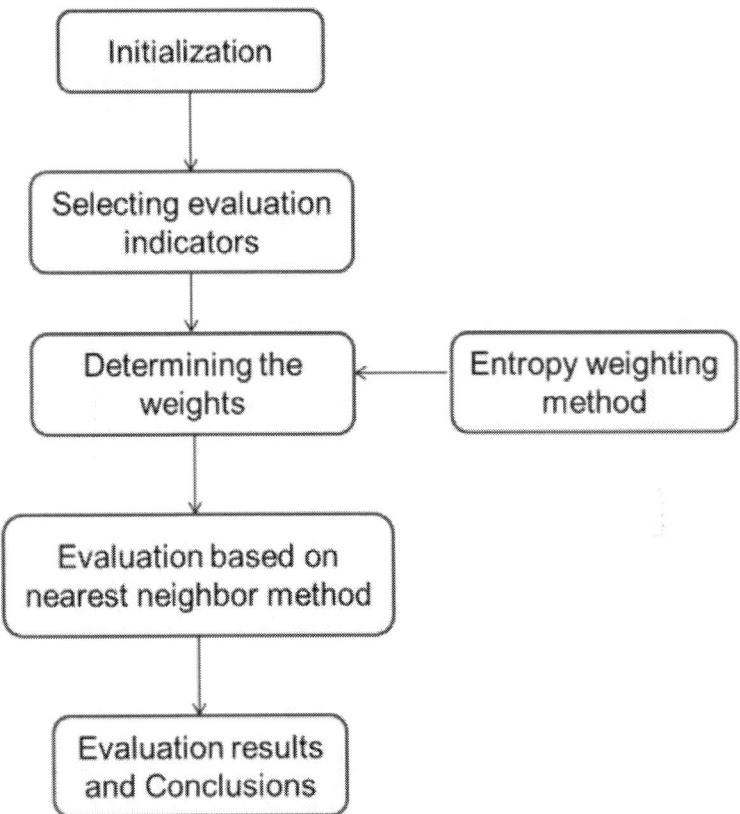

Figure 2. Evaluation model algorithm.

5. RESULTS AND ANALYSIS

5.1. Simulation Experiments

The two simulation experiments are described in this section. One experiment was with $PM_{2.5}$ and carried out based on the new Ambient Air Quality Standards (GB 3095-2012); the other experiment was without $PM_{2.5}$ and carried out based on the old Ambient Air Quality Standards (GB 3095-1996). The original data were the values of some important indicators in February 2014, which was one of the most polluted months in Beijing. The original data, as shown in Table 3, were from the China National Environmental Monitoring Center, and they were 24-hour averages of all pollutants. The indicators which reflected the air quality were $PM_{2.5}$, PM_{10}, CO, NO_2, and SO_2.

Table 3. Air quality data of Beijing city in February 2014.

Date	$PM_{2.5}$ ($\mu g/m^3$)	PM_{10} ($\mu g/m^3$)	CO (mg/m^3)	NO_2 ($\mu g/m^3$)	SO_2 ($\mu g/m^3$)
02–01	135	146	1.99	57	51
02–02	63	93	1.19	30	18
02–03	5	31	0.30	8	4
02–04	26	39	0.62	21	19
02–05	87	101	1.58	48	52
02–06	119	131	2.08	65	67
02–07	94	60	1.45	51	34
02–08	72	43	1.27	38	23
02–09	8	15	0.45	17	15
02–10	23	27	0.74	37	25
02–11	100	114	2.09	78	58
02–12	111	115	2.15	77	53
02–13	190	176	2.85	84	74
02–14	265	295	3.11	94	73
02–15	387	443	3.75	107	102
02–16	296	204	3.18	87	79
02–17	104	50	2.03	71	65
02–18	70	84	1.13	61	37
02–19	66	74	1.23	61	34
02–20	161	171	2.16	78	43
02–21	257	284	2.67	103	77
02–22	262	299	3.32	101	99
02–23	212	248	3.61	92	128
02–24	259	327	4.74	119	133
02–25	353	390	5.29	121	75
02–26	315	250	3.70	105	80
02–27	15	22	0.43	22	12
02–28	77	102	1.50	71	32

Using the formulas in Section 3.1, the weights of all indicators were calculated based on the entropy weighting method according to the data in Table 3 and the results are shown in Table 4.

Table 4. Entropy values and weights of indicators.

	$PM_{2.5}$	PM_{10}	CO	NO_2	SO_2
Entropy value	1.3887	1.3999	1.4002	1.3595	1.3944
Weight	0.2001	0.2059	0.2060	0.1850	0.2030

From Table 4 we could see that the weights of CO and PM_{10} were high, while that of NO_2 was low. The differences between all indicators were not much, and this indicated that all indicators were very important to the evaluation system.

In next step the nearest neighbor method was used to evaluate the air quality of Beijing City in February. The air quality levels were set as the clustering centers, and we investigated the distances between each sample and the clustering centers. Using the formulas in Section 3.2, we calculated the distances, and the level of air quality was determined according to the principle of minimum distance. Two simulation experiments would be implemented. One experiment was with $PM_{2.5}$ and carried out based on the new Ambient Air Quality Standards (GB 3095-2012); the other experiment was without PM2.5 and carried out based on the old Ambient Air Quality Standards (GB 3095-1996). We compared their results to investigate the importance of $PM_{2.5}$ and the effects of the standards.

In the first simulation experiment based on the new Ambient Air Quality Standards (GB 3095-2012), five indicators were included ($PM_{2.5}$, PM_{10}, CO, NO_2, and SO_2). The air quality was evaluated and the results, shown in Table 5, were calculated according to the formulas in Section 3.2. In Table 5, d_1 stands for the distances between the daily values of all indicators and the threshold values of Level I, while d_2 stands for the distances from the daily values of all indicators to the threshold values of Level II in the standards (GB 3095-2012). In addition, from Table 2 we could see that the NO_2 threshold values at Level I and Level II were the same, and so were those of CO, therefore, we could ignore these two indicators, which may reduce the evaluation workload.

Table 5. Evaluation results based on Ambient Air Quality Standards (GB 3095-2012).

Date	d_1	d_2	Rank	Date	d_1	d_2	Rank
02–01	0.2949	0.2357	II	02–15	0.7511	0.6146	II
02–02	0.3725	0.3740	I	02–16	0.4535	0.3520	II
02–03	0.4313	0.4695	I	02–17	0.1865	0.2107	I
02–04	0.3616	0.3981	I	02–18	0.2890	0.2945	I
02–05	0.2374	0.2242	II	02–19	0.2993	0.3118	I
02–06	0.2222	0.1624	II	02–20	0.3565	0.2860	II
02–07	0.3072	0.3195	I	02–21	0.4715	0.3435	II
02–08	0.3477	0.3726	I	02–22	0.4801	0.3460	II
02–09	0.3840	0.4328	I	02–23	0.3987	0.2759	II
02–10	0.3361	0.3816	I	02–24	0.5262	0.3954	II
02–11	0.2277	0.1964	II	02–25	0.6732	0.5408	II
02–12	0.2538	0.2209	II	02–26	0.5080	0.3948	II
02–13	0.3108	0.2103	II	02–27	0.3953	0.4385	I
02–14	0.4941	0.3665	II	02–28	0.3173	0.3098	II

In the second simulation experiment, four indicators were included (PM_{10}, CO, NO_2, and SO_2). $PM_{2.5}$ was excluded. The old Ambient Air Quality Standards (GB 3095-1996) was adopted to evaluate the air quality. The evaluation results are shown in Table 6. In Table 6, d_1 and d_2 also stand for the distances between the daily values of all indicators and the threshold values of Level I and Level II, while d_3 stands for the distances between the daily values of all indicators and the threshold values of Level III in the standards (GB 3095-1996).

Table 6. Evaluation results based on Ambient Air Quality Standards (GB 3095-1996).

Date	d_1	d_2	d_3	Rank	Date	d_1	d_2	d_3	Rank
02–01	0.2352	0.2185	0.5498	II	02–15	0.5602	0.3759	0.3075	III
02–02	0.3264	0.4082	0.7463	I	02–16	0.2739	0.0797	0.3365	II
02–03	0.4602	0.5583	0.8978	I	02–17	0.2139	0.2068	0.5293	II
02–04	0.3907	0.4757	0.8148	I	02–18	0.2634	0.3188	0.6491	I
02–05	0.2657	0.2721	0.6063	I	02–19	0.2510	0.3207	0.6531	I
02–06	0.2416	0.1715	0.4977	II	02–20	0.2090	0.2046	0.5154	II
02–07	0.2471	0.3238	0.6625	I	02–21	0.3616	0.2069	0.3545	II
02–08	0.2927	0.3896	0.7306	I	02–22	0.4150	0.2118	0.2602	II
02–09	0.4183	0.5116	0.8511	I	02–23	0.4581	0.2247	0.2093	III
02–10	0.3397	0.4276	0.766	I	02–24	0.5639	0.3584	0.1071	III
02–11	0.2029	0.1771	0.4996	II	02–25	0.5080	0.3789	0.2808	III
02–12	0.1884	0.1828	0.5086	II	02–26	0.3299	0.1601	0.2703	II
02–13	0.2441	0.0825	0.3819	II	02–27	0.4074	0.5056	0.8454	I
02–14	0.3471	0.1894	0.3452	II	02–28	0.2202	0.2893	0.6152	I

5.2. Analysis and Discussion

As shown in Table 5, the air qualities of less than half a month in February were at Level I, and those of the other dates were at Level II. The results of Experiment 2 in Table 6 were basically consistent with those of Experiment 1. The days with Level II air quality in Table 5 were at Level II or III in Table 6, while the days with Level I air quality in Table 6 were still at Level I in Table 5.

However, there were some differences between the results of our two simulation experiments. The air qualities on 5 and 28 February were at Level II in Experiment 1 in Table 5, while those were at Level I in Experiment 2 in Table 6. This illustrated the air quality became worse due to the introduction of $PM_{2.5}$. If the air quality were evaluated based on the old Ambient Air Quality Standards (GB 3095-1996), the results tended to be not scientific or deviate from the actual situation. $PM_{2.5}$, as a new indicator that reflects air quality, plays an important role in the evaluation process, and it is receiving more and more attention in

China. Thus the new Ambient Air Quality Standards (GB 3095-2012) are more scientific than the old ones.

There were two continuous periods when the air qualities were severe polluted (at Level II in Table 5, or Level II and III inTable 6), that is, from the 11th to the 16th and the 20th to the 26th. The data of all indicators in those days were high, especially the $PM_{2.5}$ indicator, which was one of the most important factors influencing the air quality of Beijing City. The $PM_{2.5}$ concentration on 11 to 16 February in Beijing increased obviously in Table 3, because it was affected by the fireworks during the Spring Festival and adverse weather conditions. On 20 to 26 February, due to the adverse weather conditions, Beijing was experiencing serious haze-fog pollution, and this haze-fog pollution event was wide ranging, of severe degree, and long lasting.

6. CONCLUSIONS

Haze-fog pollution in China was very severe during autumn and winter in recent years. In order to protect and improve the living environment and analyze the atmospheric environment pollution quantitatively, new ambient air quality standards were issued. A comprehensive evaluation model based on an entropy weighting method and nearest neighbor method was developed according to the new Ambient Air Quality Standards (GB 3095-2012), compared with the old standards (GB 3095-1996). The model was applied into assess the air quality in Beijing in February 2014. The simulation results showed that $PM_{2.5}$ played an important role in the evaluation process, and could affect the air quality to a large extent. The results based on the new Ambient Air Quality Standards (GB 3095-2012) were more scientific than the old standards (GB 3095-1996). In February 2014 in Beijing there were two continuous periods when the air qualities was severe polluted, that is, the 11th to the 16th and the 20th to the 26th. The air quality in Beijing during less than half a month in February was at Level I, and those of the other dates were at Level II according to the new standards. This was affected by the fireworks during the Spring Festival and adverse weather conditions, and the haze-fog pollution situation in Beijing was still severe.

The haze-fog pollution is heavy in China, and the management and control of haze-fog is very important. The prevention of haze-fog pollution is a systematic project. This requires the environmental protection departments to take the lead, and it also needs social consensus to urge the relevant work. Measures from two aspects are suggested to reduce the haze-fog pollution. One is to reduce the pollution emissions from the sources. We must make efforts to adhere to plans to reduce pollutant emissions. The other is to establish emergency plans for heavy pollution weather. The measures can be taken in advance according to the weather forecast in order to reduce or avoid the occurrence of weather-driven heavy pollution.

ACKNOWLEDGMENTS

Research works in this paper are financially supported by Research Planning Foundation in Humanities and Social Sciences of the Ministry of Education of China (Grant No. 13YJAZH044) and National Science Foundation of China(Grant No. 61173052).

AUTHOR CONTRIBUTIONS

Li Li conceived the study idea, developed the evaluation model, and provided a lot of instructive suggestions contributed to the manuscript revision. Dong-Jun Liu contributed to the study design, collected the data, gathered and measured information on the variables of interest, developed the evaluation model with Li Li, developed the programs and performed data analysis and discussion and wrote the initial manuscript draft.

REFERENCES

1. Xie, Y.S.; Li, Z.Q.; Li, L.; Wang, L.; Li, D.H.; Chen, C.; Li, K.T.; Xu, H. Study on influence of different mixing rules on the aerosol components retrieval from ground-based remote sensing measurements. *Atmos. Res.* **2014**, *145–146*, 267–278.

2. Li, W.J.; Shao, L.Y.; Buseck, P.R. Haze types in Beijing and the influence of agricultural biomass burning. *Atmos. Chem. Phys.* **2010**, *10*, 8119–8130.

3. Wang, Z.F.; Li, J.; Wang, Z.; Yang, W.Y.; Tang, X.; Ge, B.Z.; Yan, P.Z.; Zhu, L.L.; Chen, X.S.; Chen, H.S. Modeling study of regional severe hazes over Mid-eastern China in January 2013 and its implications on pollution prevention and control. *Sci. China Earth Sci.* **2014**, *57*, 3–13.

4. Shen, G.; Xue, M.; Yuan, S.Y.; Zhang, J.; Zhao, Q.Y.; Li, B.; Wu, H.S.; Ding, A.J. Chemical compositions and reconstructed light extinction coefficients of particulate matter in a mega-city in the western Yangtze River Delta, China. *Atmos. Environ.* **2014**, *8*, 14–20.

5. Tie, X.; Wu, D.; Brasseur, G. Lung cancer mortality and exposure to atmospheric aerosol particles in Guangzhou, China. *Atmos. Environ.* **2009**, *43*, 2375–2377.

6. Hanafy, M.E.; Roggemann, M.C.; Guney, D.O. Detailed effects of scattering and absorption by haze and aerosols in the atmosphere on the average point spread function of an imaging system. *J. Opt. Soc. Am. A Opt. Image Sci. Vis.* **2014**,*31*, 1312–1319.

7. Tao, J.; Zhang, L.; Ho, K.; Zhang, R.; Lin, Z.; Zhang, Z.; Zhang, Z.S.; Lin, M.; Cao, J.J.; Liu, S.X.; Wang, G.H. Impact of $PM_{2.5}$ chemical compositions on aerosol light scattering in Guangzhou—The largest megacity in South China. *Atmos. Res.* **2014**, *135*, 48–58.

8. Technical Regulations on Ambient Air Quality Index (Trial) (HJ 633–2012). 2012. Available online: http://kjs.mep.gov.cn/hjbhbz/bzwb/ dqhjbh/ jcgfffbz/201203/t20120302_224166.htm (accessed on 24 July 2014).

9. Guo, S.; Hu, M.; Wang, Z.B.; Slanina, J.; Zhao, Y.L. Size-resolved aerosol water soluble ionic compositions in the summer of Beijing: Implication of regional secondary formation. *Atmos. Chem. Phys.* **2010**, *10*, 947–959.

10. Li, W.J.; Zhou, S.Z.; Wang, X.F.; Xu, Z.; Yuan, C.; Yu, Y.C.; Zhang, Q.Z.; Wang, W.X. Integrated evaluation of aerosols from regional brown hazes over northern China in winter: Concentrations, sources, transformation, and mixing states. *J. Geophys. Res. Atmos.* **2011**, *116*.

11. Sun, Y.L.; Zhuang, G.S.; Tang, A.; Wang, Y.; An, Z. Chemical characteristics of $PM_{2.5}$ and PM_{10} in Haze-Fog episodes in Beijing. *Environ. Sci. Technol.* **2006**, *40*, 3148–3155.

12. Pope, C.A.; Burnett, R.T.; Thun, M.J.; Call, E.E.; Krewski, D.; Ito, K.; Thurston, G.D. Lung cancer, cardiopulmonary mortality, and long-term exposure to fine particulate air pollution. *J. Am. Med. Assoc.* **2002**, *287*, 1132–1141.

13. Deng, X.Y.; Tie, X.X.; Wu, D.; Zhou, X.J.; Bi, X.Y.; Tan, H.B.; Li, F.; Jiang, C.L. Long-term trend of visibility and its characterizations in the Pearl River Delta (PRD) region, China. *Atmos. Environ.* **2008**, *42*, 1424–1435.

14. Gao, J.; Wang, T.; Zhou, X.H; Wu, W.S.; Wang, W.X. Measurement of aerosol number size distributions in the Yangtze River delta in China: Formation and growth of particles under polluted conditions. *Atmos. Environ.* **2009**, *43*, 829–836.

15. Lee, K.H.; Kim, Y.J.; Kim, M.J. Characteristics of aerosol observed during two severe haze events over Korea in June and October 2004. *Atmos. Environ.* **2006**, *40*, 5146–5155.

16. Li, Z.Q.; Gu, X.; Wang, L.; Li, D.; Li, K.; Dubovik, O.; Schuster, G.; Goloub, P.; Zhang, Y.; Li, L.; Ma, Y.; Xu, H. Aerosol physical and chemical properties retrieved from ground-based remote sensing measurements during heavy haze days in Beijing winter. *Atmos. Chem. Phys.* **2013**, *13*, 5091–5122.

17. Zhao, X.J.; Zhao, P.S.; Xu, J.; Meng, W.; Pu, W.W.; Dong, F.; He, D.; Shi, Q.F. Analysis of a winter regional haze event and its formation mechanism in the North China Plain. *Atmos. Chem. Phys.* **2013**, *13*, 5685–5696.

18. Che, H.; Xia, X.; Zhu, J.; Li, Z.; Dubovic, O.; Holben, B.; Goloub, P.; Chen, H.; Estelles, V.; Cuevas-Agulló, E.; *et al.* Column aerosol optical properties and aerosol radiative forcing during a serious haze-fog month over North China Plain in 2013 based on ground-based sunphotometer measurements. *Atmos. Chem. Phys.* **2013**, *13*, 29685–29720.

19. Zhang, J.K.; Sun, Y.; Liu, Z.R.; Ji, D.S.; Hu, B.; Liu, Q.; Wang, Y.S. Characterization of submicron aerosols during a serious pollution month in Beijing (2013) using an aerodyne high-resolution aerosol mass spectrometer. *Atmos. Chem. Phys.* **2013**, *13*, 19009–19049.

20. Wang, L.T.; Wei, Z.; Yang, J.; Zhang, Y.; Zhang, F.F.; Su, J.; Meng, C.C.; Zhang, Q. The 2013 severe haze over the southern Hebei, China: Model evaluation, source apportionment, and policy implications. *Atmos. Chem. Phys.* **2013**, *13*, 28395–28451.

21. Guleda, O.E.; Ibrahim, D.; Halil, H. Assessment of urban air quality in Istanbul using fuzzy synthetic evaluation. *Atmos. Environ.* **2004**, *38*, 3809–3815.

22. Yan, Y.; Zhao, Y.N.; Zhou, G.C.; Bi, M.T.; Feng, S. Application of fuzzy comprehensive evaluation theory in air quality assessment. *J. Chem. Pharm. Res.* **2014**, *6*, 13–21.

23. Zhang, H.; Qing, C.S.; Yu, Y.P.; Li, S.J. Air quality assessment of Dongzhi Country based on fuzzy comprehensive evaluation. In Proceedings of the 3rd International Conference on Environmental Technology and Knowledge Transfer, Heifei, China, 13–14 May 2010; pp. 678–681.

24. Kyriakidis, L.; Karatzas, K.; Kukkonen, J.; Papadourakis, G.; Ware, A. Evaluation and analysis of artificial neural networks and decision trees in forecasting of common air quality index in Thessaloniki, Greece. *Eng. Intell. Syst.* **2013**, *21*, 111–124.

25. Wang, K.; Wang, W.S.; Zhang, X.; Sun, L.X. Evaluation of forest air quality based on BP neural network. *J. Harbin Inst. Technol.* **2010**, *42*, 1278–1281.

26. Shen, G.F.; Yuan, S.Y.; Xie, Y.N.; Xia, S.J.; Li, L.; Yao, Y.K.; Qiao, Y.Z.; Zhang, J.; Zhao, Q.Y.; Ding, A.J.; *et al.* Ambient levels and temporal variations of $PM_{2.5}$ and PM_{10} at a residential site in the mega-city, Nanjing, in the western Yangtze River Delta, China. *J. Environ. Sci. Health* **2014**, *49*, 171–178.

27. Wang , H.L.; Zhu, B.; Shen, L.J.; Zhang, Z.F.; Liu, X.H. The mass concentration and chemical compositions of the atmospheric aerosol during the Spring Festival in Nanjing. *China Environ. Sci.* **2014**, *34*, 30–39.

28. Ministry of Environmental Protection. Ambient Air Quality Standards (GB 3095-1996). 2012. Available online: http://kjs.mep.gov.cn/hjbhbz/bzwb/dqhjbh/dqhjzlbz/201203/W020120410330232398521.pdf (accessed on 29 February 2012).

29. Ministry of Environmental Protection. Ambient Air Quality Standards (GB 3095-2012). 2012. Available online: http://www.zzemc.cn/em_aw/Content/GB3095-2012.pdf (accessed on 29 February 2012).

30. Liu, L.; Zhou, J.Z.; An, X.L.; Zhang, Y.C.; Yang, L. Using fuzzy theory and information entropy for water quality assessment in Three Gorges region, China. *Expert Syst. Appl.* **2010**, *37*, 2517–2521.

31. Zou, Z.H.; Yun, Y.; Sun, J.N. Entropy method for determination of weight of evaluating indicators in fuzzy synthetic evaluation for water quality assessment. *J. Environ. Sci.* **2006**, *18*, 1020–1023.

32. Kazuo, H.; Yasunobu, T. Effective algorithms for the nearest neighbor method in the clustering problem. *Pattern Recognit.* **1993**, *26*, 741–746.

CHAPTER 12

A Survey of Wireless Sensor Network Based Air Pollution Monitoring Systems

Wei Ying Yi [1,2], Kin Ming Lo [1], Terrence Mak [3], Kwong Sak Leung [1,2,*],
Yee Leung [2,4] and Mei Ling Meng [5,6]

[1] Department of Computer Science and Engineering, The Chinese University of Hong Kong, Shatin NT, Hong Kong, China
[2] Institute of Future Cities, The Chinese University of Hong Kong, Shatin NT, Hong Kong, China
[3] Department of Electronics and Computer Science, University of Southampton, University Road, Southampton S017 1BJ, UK
[4] Department of Geography and Resource Management, The Chinese University of Hong Kong, Shatin NT, Hong Kong, China
[5] Department of Systems Engineering and Engineering Management, The Chinese University of Hong Kong, Shatin NT, Hong Kong, China
[6] Stanley Ho Big Data Decision Analytics Research Center, The Chinese University of Hong Kong, Shatin NT, Hong Kong, China

ABSTRACT

The air quality in urban areas is a major concern in modern cities due to significant impacts of air pollution on public health, global environment, and worldwide economy. Recent studies reveal the importance of micro-level pollution information, including human personal exposure and acute exposure to air pollutants. A real-time system with high spatio-temporal resolution is essential because of the limited data availability and non-scalability of conventional air pollution monitoring systems. Currently, researchers focus on the concept of The Next Generation Air Pollution Monitoring System

(TNGAPMS) and have achieved significant breakthroughs by utilizing the advance sensing technologies, MicroElectroMechanical Systems (MEMS) and Wireless Sensor Network (WSN). However, there exist potential problems of these newly proposed systems, namely the lack of 3D data acquisition ability and the flexibility of the sensor network. In this paper, we classify the existing works into three categories as Static Sensor Network (SSN), Community Sensor Network (CSN) and Vehicle Sensor Network (VSN) based on the carriers of the sensors. Comprehensive reviews and comparisons among these three types of sensor networks were also performed. Last but not least, we discuss the limitations of the existing works and conclude the objectives that we want to achieve in future systems.

KEYWORDS

air pollution monitoring; Wireless Sensor Network (WSN); real-time monitoring; high spatio-temporal resolution; low-cost ambient sensor

1. INTRODUCTION

Over the past few years, air pollution has drawn a lot of interest in terms of research and everyday life. According to data from Google Search, about 46 million results are related to "2014 Air Pollution", while the number of results related to "2014 Nobel Prize" is only about 27 million (accessed on 2014-8-20). The public concern on air pollution increases significantly due to the serious hazards to the public health, as described in [1]. Heart disease, Chronic Obstructive Pulmonary Disease (COPD), stroke and lung cancer are highly related to air pollution. People breathing in air of poor quality could suffer from difficulty in breathing, coughing, wheezing and asthma. In addition to the human health, air pollution also has a major effect on the global environment and the worldwide economy. It is well known that acid rain, haze and global climate change are caused by air pollution. In 2010, the European Commission threatened the UK with legal actions for the breaching of PM_{10} (PM_X stands for particulate matter with diameter of less than or equal to X μm) limit values. The UK could pay £300 million per year for this [2].

In order to mitigate the impacts of air pollution on human health, global environment and worldwide economy, governments have put tremendous efforts on air pollution monitoring. With detailed information of the air pollution situation, scientists, policy makers and planners are able to make informed decisions on managing and improving the living environment [3].

Countries adopting proper policies on air pollution can reduce the public health expenses as described above [4].

Traditionally, air pollution situation is monitored by conventional air pollution monitoring systems with stationary monitors. These monitoring stations are highly reliable, accurate and able to measure a wide range of pollutants by using the conventional analytical instruments, such as gas chromatograph-mass spectrometers [5].

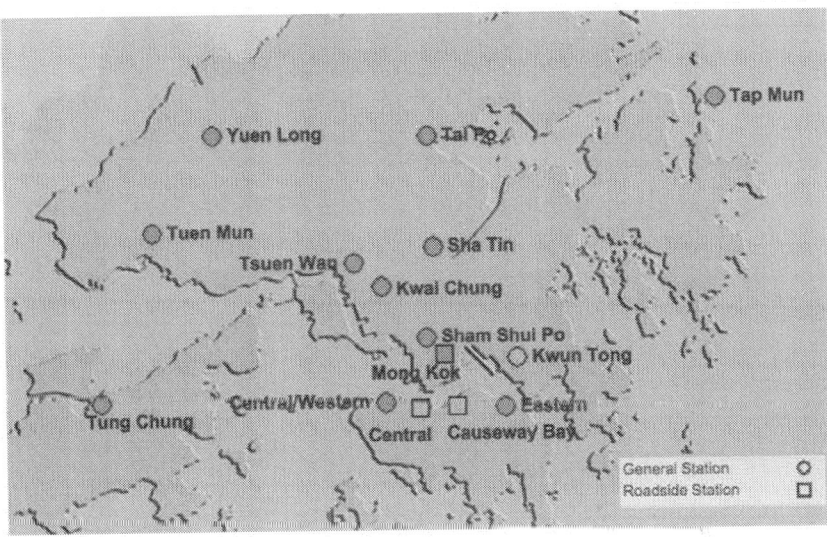

Figure 1. Deployment of stationary monitors in Hong Kong [6].

Table 1. The number of stationary monitors in selected cities.

City	Number of Stationary Monitors	Coverage Area	Coverage Per Monitor (Number of Football Fields)
Beijing, China	35 [7]	16,000 km^2	64,025
Hong Kong, China	15 [6]	2700 km^2	25,210
New York, USA	44 [8]	1200 km^2	3820
London, UK	123 [9]	1600 km^2	1822

The drawbacks of the conventional monitoring instruments are their large size, heavy weight and extraordinary expensiveness. These lead to sparse deployment of the monitoring stations (see Figure 1 and Table 1). In order to be effective, the locations of the monitoring stations need careful placement because the air pollution situation in urban areas is highly related to human activities (e.g., construction activities) and location-dependent (e.g., the traffic choke-points have much worse air quality than average) [10,11,12]. Changes in urban arrangement, activities or regulation may affect both the species and the concentrations of air pollutants, which require relocating stations or adding new stations. These requirements are typically hard or even impossible to fulfill due to the cost inefficiency in acquisition and maintenance of the monitoring stations. Moreover, the conventional monitoring instruments involve long-term time-consuming average models. The air pollution situation is updated hourly or even daily. Hence, the air pollution maps built by the conventional air pollution monitoring systems are with extremely low spatial and temporal resolutions.

Such low spatio-temporal resolution is sufficient for ambient background monitoring but extremely inadequate for the public to be aware of their personal exposure to air pollution and cannot reflect their personal health risks. In [13], researchers noted that the pollutant concentrations within a street may vary over a space with magnitude of few meters and over time with magnitude of few seconds. The conventional monitoring systems cannot detect this phenomenon because of their limited data availability and non-scalability characteristics. Furthermore, when road traffic is the major pollution source, which is always the case in urban areas, acute exposure to the public is prevalent [14]. Evidences show that acute exposure to or short-term change of pollutants may trigger or worsen some health events or diseases [15,16,17].

In order to increase the spatio-temporal resolution of the air pollution information, researchers are pushing the air pollution monitoring systems to the limit by combining the low-cost portable ambient sensors and the Wireless Sensor Network (WSN) into one system which is known as The Next Generation Air Pollution Monitoring System (TNGAPMS) [18]. By utilizing the low-cost portable ambient sensors and the WSN, the air pollution information can be updated in minutes or even seconds [19]. Also, the low-cost portable sensors enable the mobility and the feasibility in large-scale deployment of the sensor nodes. The spatial and temporal resolutions of the pollution information are significantly increased in TNGAPMS. TNGAPMS fills the gap between the conventional monitoring systems and the air quality models because the air pollution information at locations without monitoring stations is accomplished by air quality models or estimations [20]. TNGAPMS also helps researchers understand the distribution of the air pollutants more efficiently and accurately to improve the air quality models. The public users can even measure their personal exposures to pollutants using wearable sensor nodes [21].

Researchers anticipate that the real-time, high spatio-temporal (The spatial resolution of the air pollution information is in scale of tens to hundreds of square meters while the air pollution information of a specific location has reporting interval less than few minutes and is available to the users, including researchers, public users, and policy makers, with minimal or no delay.) air pollution information can help advise the public to take proper actions according to their individual health needs (e.g., asthmatics could choose an alternative healthier route to minimize the personal air pollution exposure), and raise public awareness about the air pollution that further leads to change of public "unclean" activities (e.g., driver with better driving habits can reduce pollutants' emission).

The remainder of this paper is organized as follows. In Section 2, the air quality standards defined by different agencies all over the world are introduced. In Section 3, we discuss the limitations of the measurement equipment used in conventional air pollution monitoring systems and the opportunities provided by the low-cost portable ambient sensors. In Section 4, 20 state-of-the-art systems of TNGAPMS are presented and classified into three categories. The advantages and disadvantages of each category are described. In Section 5, we focus on the comparisons of the three categories of existing works classified in Section 4. Finally, we point out the limitations of the existing works and conclude the objectives we want to achieve when building a future TNGAPMS in Section 6.

2. AIR QUALITY STANDARDS

Pollutants are emitted by human activities and natural sources. Hundreds of hazardous pollutants in our living environment have been identified [22]. However, six of these pollutants are well studied and ubiquitous in our daily lives, including carbon monoxide (CO), nitrogen dioxide (NO_2), ground level ozone (O_3), sulfur dioxide (SO_2), particulate matter (PM) and lead (Pb) [23]. The health effects (see Table 2) and environmental effects caused by these pollutants can be found in [24,25,26,27,28,29].

Governments and organizations have put regulation limits on these pollutants to reduce the risks. The United States Environmental Protection Agency (EPA), the World Health Organization (WHO), the European Commission (EC), the Chinese Ministry of Environmental Protection (MEP) and the Environmental Protecting Department (EPD) of Hong Kong have declared different standard limits for these pollutants (see Table 3).

In order to help the public understand the current air quality easily, the government and organization agencies introduced an indicator called Air Quality Index (AQI). AQI measures the "condition or state of each relative to the requirements of one or more biotic species and/or to any human need or purpose" [30]. In a word, it tells the public how "good" the current air quality is or the forecast air quality will be. Different agencies may use different air quality indices [31,32,33,34].

Table 2. The six common pollutants and their health effects.

Pollutant	Health Effects
Carbon Monoxide (CO)	Reducing oxygen capacity of the blood cells leads to reducing oxygen delivery to the body's organs and tissues. Extremely high level can cause death.
Nitrogen Dioxide (NO_2)	High risk factor of emphysema, asthma and bronchitis diseases. Aggravate existing heart disease and increase premature death.
Ozone (O_3)	Trigger chest pain, coughing, throat irritation and congestion. Worsen bronchitis, emphysema and asthma.
Sulfur Dioxide (SO_2)	High risk factor of bronchoconstriction and increase asthma symptoms.
Particulate Matter ($PM_{2.5}$ & PM_{10})	Cause premature death in people with heart and lung diseases. Aggravate asthma, decrease lung function and increase respiratory symptoms like coughing and difficulty breathing.
Lead (Pb)	Accumulate in bones and affect nervous system, kidney function, immune system, reproductive systems, developmental systems and cardiovascular system. Affect oxygen capacity of blood cells.

To illustrate the concept of AQI, an AQI example introduced by the Environmental Protection Department (EPD) of Hong Kong [35] called Air Quality Health Index (AQHI) system is given (see Table 4). The AQHI system provides a better understanding on health risks to the public and suggests detail precautionary actions with respect to each AHQI level [36].

3. AIR POLLUTION MONITORING EQUIPMENT

Conventional air pollution monitoring systems are mainly based on sophisticated and well-established instruments. In order to guarantee the data accuracy and quality, these instruments use complex measurement methods [44] and a lot of assisting tools including temperature controller (cooler and heater), relative humidity controller, air filter (for PM), and build-in calibrator [45]. As consequences, these instruments are typically with high cost, high power consumption, large volume, and heavy weight. Thanks to technology advance, ambient sensors with low cost, small size and fast response time (in the order of seconds or minutes) is available recently. However, no low-cost portable ambient sensor can achieve the same data accuracy and quality as conventional monitoring instruments [46] (see Table 5 and Table 6).

Table 3. Different standards of the six common pollutants.

Pollutant		EPA [37]	WHO [38–40]	EC [41]	MEP [42]	EPD [43]
Carbon Monoxide (CO)		9 ppm (8 h) 35 ppm (1 h)	100 mg/m³ (15 min) 15 mg/m³ (1 h) 10 mg/m³ (8 h) 7 mg/m³ (24 h)	10 mg/m³ (8 h)	10 mg/m³ (1 h) 4 mg/m³ (24 h)	30 mg/m³ (1 h) 10 mg/m³ (8 h)
Nitrogen Dioxide (NO₂)		100 ppb (1 h) 53 ppb (1 year)	200 µg/m³ (1 h) 40 µg/m³ (1 year)	200 µg/m³ (1 h) 40 µg/m³ (1 year)	200 µg/m³ (1 h) 80 µg/m³ (24 h) 40 µg/m³ (1 year)	200 µg/m³ (1 h) 40 µg/m³ (1 year)
Ozone (O₃)		75 ppb (8 h)	100 µg/m³ (8 h)	120 µg/m³ (8 h)	200 µg/m³ (1 h) 160 µg/m³ (8 h)	160 µg/m³ (8 h)
Sulfur Dioxide (SO₂)		75 ppb (1 h) 0.5 ppm (3 h)	500 µg/m³ (10 min) 20 µg/m³ (24 h)	350 µg/m³ (1 h) 125 µg/m³ (24 h)	500 µg/m³ (1 h) 150 µg/m³ (24 h) 60 µg/m³ (1 year)	500 µg/m³ (10 min) 125 µg/m³ (24 h)
Particulate Matter	PM₂.₅	35 µg/m³ (24 h) 12 µg/m³ (1 year)	25 µg/m³ (24 h) 10 µg/m³ (1 year)	25 µg/m³ (1 year)	75 µg/m³ (24 h) 35 µg/m³ (1 year)	75 µg/m³ (24 h) 35 µg/m³ (1 year)
	PM₁₀	150 µg/m³ (24 h)	50 µg/m³ (24 h) 20 µg/m³ (1 year)	50 µg/m³ (24 h) 40 µg/m³ (1 year)	150 µg/m³ (24 h) 70 µg/m³ (1 year)	100 µg/m³ (24 h) 50 µg/m³ (1 year)
Lead (Pb)		0.15 µg/m³ (3 month)	0.5 µg/m³ (1 year)	0.5 µg/m³ (1 year)	1 µg/m³ (3 month) 0.5 µg/m³ (1 year)	1 µg/m³ (3 month) 0.5 µg/m³ (1 year)

Table 4. Air Quality Health Index (AQHI) of Hong Kong Environmental Protection Department.

Health Risk Category	AQHI
Low (Green)	1
	2
	3
Moderate (Orange)	4
	5
	6
High (Red)	7
Very High (Brown)	8
	9
	10
Serious (Black)	10+

Currently, the air pollution data at locations without monitoring stations are obtained by air quality models or estimations [20]. However, the data from the air quality models lack of cross-validation and verification. The low-cost portable ambient sensors provide a huge opportunity in increasing the spatio-temporal resolution of the air pollution information and are even able to verify, fine-tune or improve the existing ambient air quality models.

In the following subsections, the working mechanisms of the low-cost portable ambient sensors that are widely used in TNGAPMS are introduced. As a matter of fact, except the air pollution detecting technologies mentioned in Section 3.1 and Section 3.2, there are other detecting technologies such as Surface Acoustic Wave (SAW) [47,48,49], Quartz Tuning Fork (QTF) [50,51], Raman Lidar [52,53] and Differential Ultra Violet Absorption Spectroscopy (DUVAS) [54,55] that we will not discuss for unpopularity reason.

Table 5. Instruments used in air quality monitoring systems (Part A).

Pollutant	Example Product	Measurement Method	Resolution	Accuracy	Range	Price (USD)
PM$_{2.5}$	Met One Instrument BAM-1020 Beta Attenuation Monitor [56]	Beta Attenuation	1 µg/m^3	±1 µg/m^3	0–1000 µg/m^3	About $25,000
	Met One Instrument Aerocet 831 Aerosol Mass Monitor [57]	Light Scatting	0.1 µg/m^3	±10% of reading	0–1000 µg/m^3	About $2000
	Alphasense OPC-N2 Particle Monitor [58]	Light Scatting	Not Provided	Not Provided	Not Provided	About $500
	Sharp Microelectronics DN7C3CA006 PM2.5 Module [59]	Light Obscuration (Nephelometer)	Not Provided	Not Provided	25–300 µg/m^3	About $20
PM$_{10}$	Teledyne Model 602 BetaPLUS Particle Measurement System [60]	Beta Attenuation	0.1 µg/m^3	±1 µg/m^3	0–1500 µg/m^3	About $30,000
	Met One Instrument Aerocet 831 Aerosol Mass Monitor [57]	Light Scatting	0.1 µg/m^3	±10% of reading	0–1000 µg/m^3	About $2000
	Alphasense OPC-N2 Particle Monitor [58]	Light Scatting	Not Provided	Not Provided	Not Provided	About $500
	Sharp GP2Y1010AU Air Quality Sensor [61]	Light Obscuration (Nephelometer)	Not Provided	Not Provided	0–500 µg/m^3	About $20
Lead (Pb)	Operation in Lab	-	-	-	-	-

3.1. Gas Sensor

Nowadays, many different technologies for gas detection are available, each with certain advantages and disadvantages. To date, there are five types most suitable and widely used low-cost portable gas sensors, namely electrochemical sensors, catalytic sensors, solid-state (semiconductor) sensors, non-dispersive infrared radiation absorption (NDIR) and photo-ionization detector (PID) sensors (see Table 7). All of these sensors are low cost, light weight (less than one hundred grams) and with fast response time (in tenths seconds or few minutes). However, no single type of sensors is able to measure all the hazard gases (hundreds of hazard gases have been identified). Each type of sensors is sensitive to specific kinds of hazard gases.

Table 6. Instruments used in air quality monitoring systems (Part B).

Pollutant	Example Product	Measurement Method	Resolution	Accuracy	Range	Price (USD)
Carbon Monoxide (CO)	Teledyne Model T300U Gas Filter Correlation Carbon Monoxide Analyzer [62]	IR Absorption with Gas Filter Correlation Wheel	0.1 ppb	±0.5% of reading	0-100 ppb or 0-100 ppm	About $30,000
	Aeroqual Series 500 with CO Sensor Head [63]	Electrochemical Sensor	10 ppb	±0.5 ppm at 0-5 ppm or ±10% at 5-25 ppm	0-25 ppm	About $2000
	Alphasense B4 Series CO Sensor [64]	Electrochemical Sensor	4 ppb	Not Provided	0-1000 ppm	About $200
	Hanwei MQ-7 CO Sensor [65]	Solid-State Sensor	Not Provided	Not Provided	20-2000 ppm	About $10
Nitrogen Dioxide (NO$_2$)	Teledyne Model T500U Nitrogen Dioxide Analyzer [66]	Cavity Attenuated Phase Shift Spectroscopy	0.1 ppb	±0.5% of reading	0-5 ppb or 0-1 ppm	About $30,000
	Aeroqual Series 500 with NO$_2$ Sensor Head [67]	Electrochemical Sensor	1 ppb	±0.02 ppm at 0-0.2 or ±10% at 0.2-1 ppm	0-1 ppm	About $2000
	Alphasense B4 Series NO$_2$ Sensor [68]	Electrochemical Sensor	12 ppb	Not Provided	0-20 ppm	About $200
	SGXSensorTech MICS-2714 NO$_2$ Sensor [69]	Solid-State Sensor	Not Provided	Not Provided	0.05-10 ppm	About $10
Ozone (O$_3$)	Teledyne Model 265E Chemiluminescence Ozone Analyzer [70]	Chemiluminescence Detection	0.1 ppb	±0.5% of reading	0-100 ppb or 0-2 ppm	About $25,000
	Aeroqual Series 500 with O$_3$ Sensor Head [71]	Solid-State Sensor	1 ppb	±5 ppb	0-150 ppb	About $2000
	Alphasense B4 Series O$_3$ Sensor [72]	Electrochemical Sensor	4 ppb	Not Provided	0-5 ppm	About $200
	Hanwei MQ-131 O$_3$ Sensor [73]	Solid-State Sensor	Not Provided	Not Provided	10-1000 ppm	About $10
Sulfur Dioxide (SO$_2$)	Teledyne Model 6400T/6400E Sulfur Dioxide Analyzer [74]	UV Fluorescence	0.1 ppb	±0.5% of reading	0-50 ppb or 0-20 ppm	About $30,000
	Aeroqual Series 500 with SO$_2$ Sensor Head [75]	Electrochemical Sensor	10 ppb	±0.15 ppm at 0-0.5 ppm or ±10% at 0.5-10 ppm	0-10 ppm	About $2000
	Alphasense B4 Series SO$_2$ Sensor [76]	Electrochemical Sensor	5 ppb	Not Provided	0-100 ppm	About $200
	Hanwei MQ-136 SO$_2$ Sensor [77]	Solid-State Sensor	Not Provided	Not Provided	0-200 ppm	About $50

Table 7. Comparison of the five types of gas sensors.

Sensor Type	Detectable Gases	Linearity	Cross Sensitivity	Power Consumption	Maintenance	Response Time (T90)	Life Expectancy
Electro-chemical [78]	Gases which are electrochemically active, about 20 gases	Linear at room temperature	Can be eliminated by using chemical filter	Lowest, very little power consumption	Low	<50 s	1-2 years
Catalytic [79]	Combustible gases	Linear at 400 °C to 600 °C	No meaning when measuring mixed gases	Large, need to heat up to 400 °C to 600 °C	Low sensitivity with time due to poisoning and burning out	<15 s	Up to 3 years
Solid-state [80]	About 150 different gases	Linear at operational temperature	Can be minimized by using appropriate filter	Large, need heating element to regulate temperature	Low	20 s to 60 s	10+ years
Non-dispersive Infrared [81]	Hydrocarbon gases and carbon dioxide	Nonlinear, need linearize procedure	All hydrocarbons share a similar absorption band, make them all cross sensitive	Small, mainly consume by the infrared source	The least	<20 s	3-5 years
Photo-ionization [82]	Volatile organic compounds (VOCs)	Relatively linear	Any VOCs with ionization potentials less than the ionizing potential of the lamp used will be measured	Medium, mainly consume by the ultraviolet source	The lamp requires frequent cleaning	<3 s	Depend on the Ultraviolet lamp, normally 6000 h

Although, till now there is no low-cost portable gas sensor can achieve the same data accuracy and quality as conventional monitoring instruments. The low-cost portable gas sensors provide a fair enough accuracy and detection range [46]. What is more, all sensors need to be calibrated (When calibrating a sensor, the sensor is exposed to a specific kind of pollution gas with predefined concentration, the parameters of the sensor are adjusted such that the difference between the predefined gas concentration and the sensor output is minimized.) before operation and after a specific operational time. The necessity of calibration and the calibration procedures can be found in [83].

As described in Section 2, there are four types of hazard gases that we want to monitor most. They are carbon monoxide (CO), nitrogen dioxide (NO_2), ground level ozone (O_3) and sulfur dioxide (SO_2). Combining the descriptions in [46,84] and the comparisons (with respect to sensor detectable gases, linearity, cross sensitivity, power consumption, maintenance, response time and life expectancy) in Table 7, two best types of sensors for these four types of hazard gases are determined.

- CO: Can be well detected by solid-state and electrochemical sensors.

- NO_2: Can be well detected by solid-state and electrochemical sensors. Need to consider the interference gas O_3. Proper methods can be applied to reduce the interference.

- O_3: Can be well detected by solid-state and electrochemical sensors. Need to consider the interference gas NO_2. Proper methods can be applied to reduce the interference.

- SO_2: Can only be well detected by solid-state and electrochemical sensors. It poisons the catalytic sensors. The sensitivity of NDIR sensors is not high enough.

In a word, the solid-state and electrochemical sensors are the most suitable types of sensors to monitor these four types of hazard gases in building the TNGAPMS scenario. In fact, these two types of sensors are the basic elements in most of the existing works presented in Section 4. The operational principles of these two types of sensors are introduced as follows.

3.1.1. Solid-state Gas Sensor [80]

The working principle of the solid-state ambient gas sensors was discovered when researchers were dealing with the semiconductor p-n junctions, which are sensitive to environmental gases.

A solid-state sensor consists of one or several metal oxides like tin oxide or aluminum oxide (the type of metal oxide being used depends on the target ambient gas the sensor aims for), and a heating element. The metal oxides can be processed into a paste, which is called bead-type sensor (see Figure 2). The metal oxides can also be deposited onto a silica chip similar to making semiconductors, which is called chip-type sensor (see Figure 3). When the metal oxides are exposed to the ambient gases, the gases will dissociate into charged ions or complexes that make the electrons accumulate on the surface of the metal oxides. The accumulation of electrons changes the conductivity of the metal oxides. By measuring the conductivity change, researchers are able to deduce the concentration of a specific kind of ambient gas.

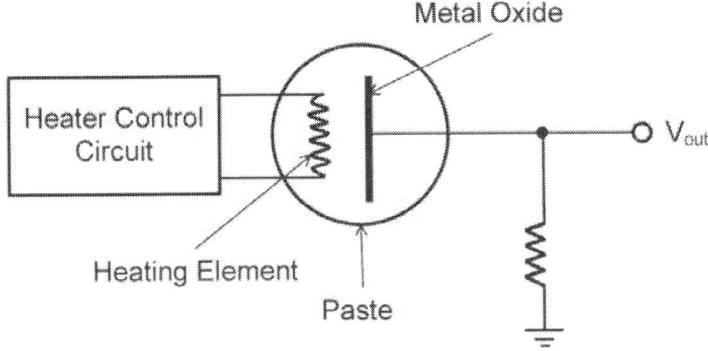

Figure 2. Bead-type sensor.

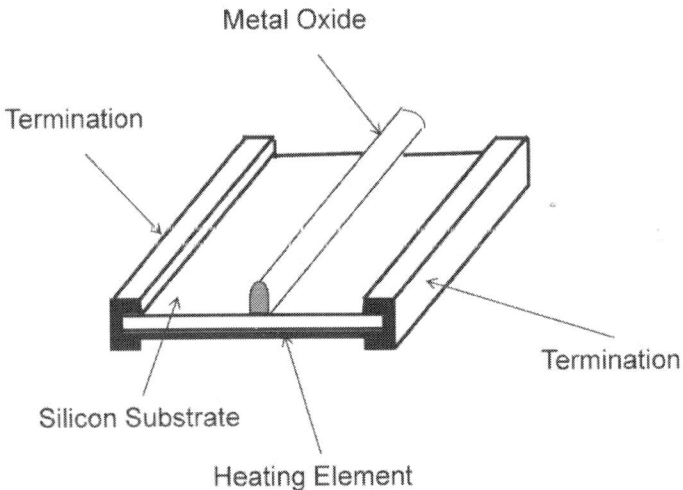

Figure 3. Chip-type Sensor.

In order to increase the reaction rate that results in a strong electrical signal, a heating element is used inside the solid-state ambient gas sensor. The heating element is also used to regular the temperature because the response (conductivity change) of a specific kind of ambient gas is different in different temperature ranges.

3.1.2. Electrochemical Gas Sensor [78]

The working mechanisms of the electrochemical ambient gas sensors are electrochemical reactions (oxidation-reduction reactions, to be specific) within the sensors. The reaction between the sensor and the ambient gas molecules produces an electrical signal (current) proportional to the concentration of the ambient gas.

An electrochemical sensor consists of a Working Electrode (WE) and a Counter Electrode (CE). For sensors requiring an external driving voltage, a Reference Electrode (RE) is needed. These two or three electrodes are separately deployed into the electrolyte within the sensor (see Figure 4).

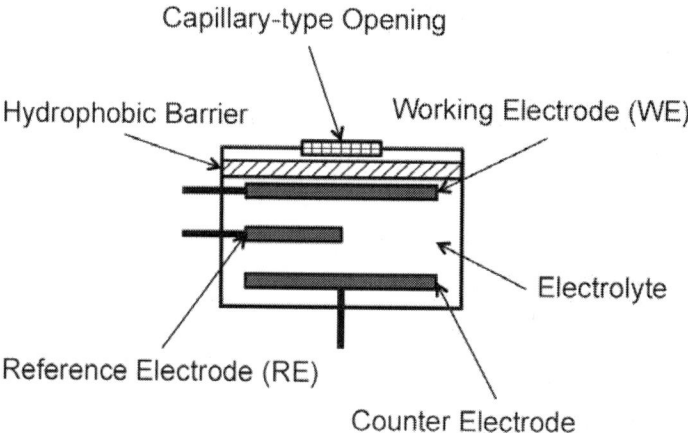

Figure 4. Basic Electrochemical Sensor.

Different sensors may use different types of selective membranes, electrolyte and working electrodes in order to improve the sensor's selectivity to a specific kind of ambient gas. To allow enough amount of ambient gas to react with the sensor while preventing electrolyte leakage, the ambient gas first goes through a capillary-type opening and a hydrophobic barrier. When the ambient gas reaches the working electrode, the oxidation-reduction reaction occurs. The specifically developed electrode for an interested ambient gas catalyzes these reactions. By measuring the current between the Working Electrode (WE) and the Counter Electrode (CE), researchers are able to deduce the concentration of the target ambient gas. For sensor with Reference Electrode (RE), the reference electrode is used to control the oxidation and reduction reactions and reduce the potential drift on working electrode due to deterioration (may not work when the electrodes are fouled).

Note that, most of the electrochemical ambient gas sensors require a small amount of oxygen and humidity to function properly. Also, wind velocity influences the chemical equilibrium on the sensor's surface and further influences the sensor's readings [85].

3.2. Particulate Matter Sensor

The measurement of particulate matter (PM) is not straightforward and there are many techniques (used in conventional monitoring systems and TNGAPMSs) available for measuring the mass concentrations of PM. Due to the complex nature of PM, different measurement techniques may give different results [86]. Some conventional monitoring instruments use a heating element to eliminate the effect of changing humidity and temperature. However, the heating element evaporates the semi-volatile species and influences the measurement results. Therefore, some instruments use a special dryer instead of a heating element (e.g., the Nafion dryer [87]).

The available techniques for measuring the concentration of PM can be grouped into two categories. One is direct reading instrument which provides continuous measurements (sampling interval is in seconds or minutes) on the concentration of PM in ambient air (see Table 8). The other one is filters-based gravimetric sampler, which collects the PM onto a filter that needs to be weighted periodically in lab. The weighting procedure is a time and human resources consuming task, which leads to a large delay (in days) between collection and reporting. However, the filters-based gravimetric technique is usually used as the reference method in government agencies. One should note that the reference methods are not the absolute methods but subject to many artifacts (temperature and humidity change and semi-volatile compounds).

The commonly used continuous measurement techniques of PM in ambient air are listed as follows.

3.2.1. Tapered Element Oscillating Micro-Balance (TEOM) Analyzers [88]

The TEOM analyzers are widely used in the conventional air pollution monitoring systems. The operation principle of TEOM is that the oscillation frequency of the tapered glass tube is proportional to the mass of the tube. The

PMs deposited onto the tube will change the mass and oscillation frequency of the tube. By measuring the oscillation frequency change of the tube and the volume of air sampled, researchers are able to deduce the mass concentration ($\mu g/m^3$) of PM in ambient air.

Note that the air is sampled through a size selective inlet. For example, a PM_{10} size selective inlet rejects 50% (no design can reject 100%) of the particles with diameter more than 10 μm and let through particles with diameter of 10 μm and less. In order to eliminate the effect of humidity change, a heating element or a dryer is used.

3.2.2. β-Attenuation Analyzers [89]

The β-Attenuation Analyzers or β-Attenuation Monitors (BAM) are the most widely used PM measurement instruments in the conventional air pollution monitoring systems. The air is first sampled through a size selective inlet (PM_{10} or $PM_{2.5}$) with or without heater/dryer that minimizes the water contained in the air. Then the air goes through a paper filter, which catches the PM. The paper filter with PM is exposed to β-attenuation source. After the measurement interval, researchers are able to deduce the mass of the PM on the filter by measuring the radiation intensity of the filter.

3.2.3. Black Smoke Method [90]

The black smoke technique collects the PM on a paper filter over 24 h period through a size selective inlet. The darkness of the paper filter is then measured by a reflectometer and converted to the PM's mass concentration. This kind of monitoring equipment is relative simple, robust and cost-efficient. However, the mass concentration is derived by measuring the darkness of the filter and the darkness of PM varies in different locations. This makes the darkness-to-mass coefficient change from time to time and location to location.

3.2.4. Optical Analyzers [91]

The optical analyzers utilize the interaction between the ambient PM and the imaging, laser or infrared light. These analyzers are small, lightweight and battery operated. Base on the optical principle, the optical analyzers can be classified into three categories, namely direct imaging, light scatting and light obscuration (nephalometer) analyzers.

- Light Scatting:

  This category of optical analyzers uses a high-energy laser as the light source. When a particle passes through the detection chamber that only allows single particle sampling, the laser light is scattered by the particle. A photo detector detects the scatting light. By analyzing the intensity of the scatting light, researchers can deduce the size of the particle. Also, the number of particle counts can be deduced by counting the number of detecting light on the photo detector (see Figure 5). The advantage of this approach is that a single analyzer can detect particles with different diameters simultaneously (i.e., $PM_{2.5}$, PM_5 and PM_{10}). However, the particle counts need to be converted to mass concentration by calculation (depends on the particle counts, particle types and particle shapes) and this will introduce errors that further affect the precision and accuracy of the analyzers.

- Direct Imaging:

  In a direct imaging particle analyzer, a beam of halogen light illuminates the particles and the shadow of each particle is projected to a high definition, high magnification and high resolution camera. The camera records the passing particles. The video is then analyzed by computer software to measure the PM's attributes. Both size and counts of the PMs in the ambient air can be obtained. What's more, the color and the shape of the particles can also be detected.

- Light Obscuration (Nephelometer):

  This category of optical analyzers uses the fastest particle concentration ($\mu g/m^3$) measurement method with high precision and low detection limited. A nephelometer is an instrument that measures the size and mass concentration of PM in the ambient air. In a nephelometer, a near infrared LED is used as the light source and a silicon detector is used to measure the total light scattered (which is majorly responsible for the total light extinction) by the PMs (see Figure 6). By analyzing the intensities (in magnitude) of the scattered light and the shape of the scattering pattern, both the size distribution and the mass concentration can be determined right away [92].

The comparisons of these four types of PM measuring techniques are shown in Table 8. Because of the high data resolution and accuracy, large size, heavy weight and high cost, the TEOMs and BAMs are typically used in the conventional air pollution monitoring systems. Although the readings from the light scattering and the light obscuration optical analyzers are with relative low resolution and accuracy, and the particle-count-to-mass-concentration coefficient is different from time to time and location to location, these two types of PM sensors are widely used in hand-held monitoring devices and TNGAPMSs because of their small size, light weight, low cost and simultaneously measuring ability.

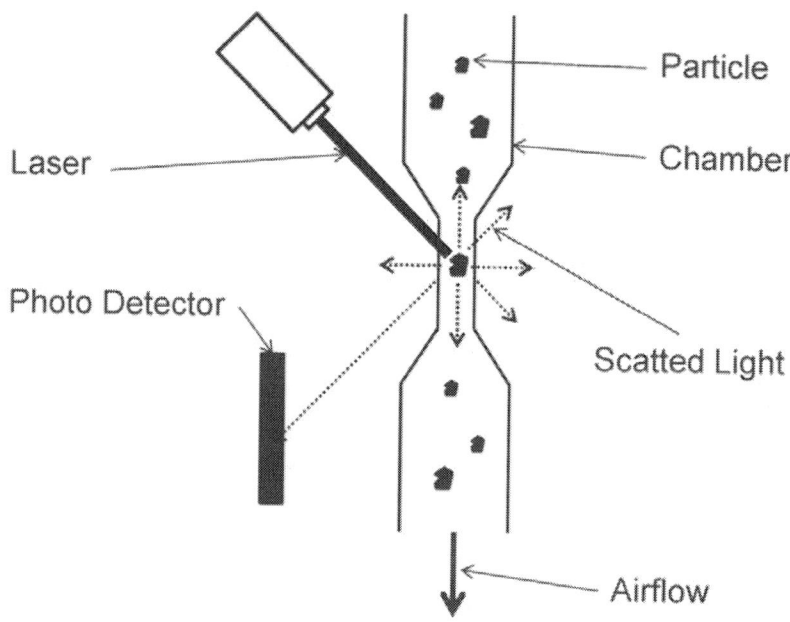

Figure 5. Basic Light Scatting Particle Counter.

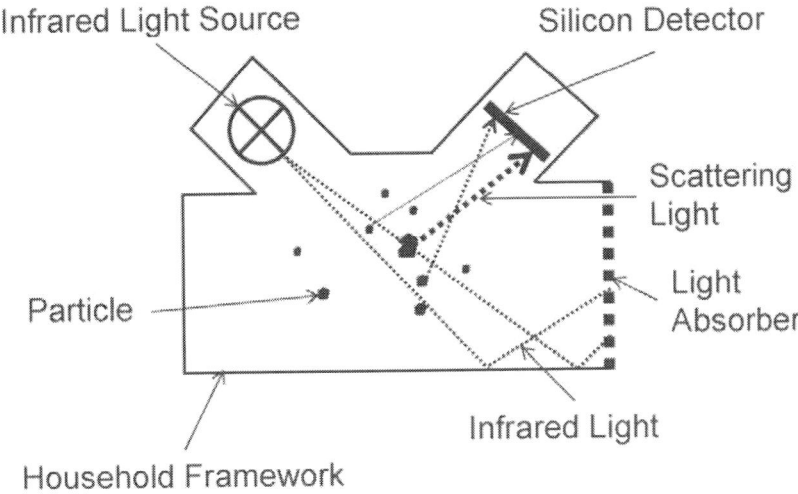

Figure 6. Basic Nephelometer.

Table 8. Comparison of four types of particulate matter (PM) measurement methods.

Measurement Method	Advantages	Disadvantages	Accuracy
Tapered Element Oscillating Microbalance (TEOM) analyzers	Provide real time (<1 h) data with high precision.	A heater must be used which leads to lose of semi-volatile material. Usually with large size, heavy weight and high cost.	±0.5 µg/m³
β attenuation analyzers (BAM)	Provide real time (<1 h) data with high precision.	A radioactive source is used. If heater is used some semi-volatile material may be lost. Need to replace the paper filter periodically. Usually with large size, heavy weight and high cost.	±1.0 µg/m³
Black smoke method	Simple, robust and inexpensive. Easy to maintain. Short sample time (in minutes).	Measure the darkness rather than the mass concentration of the particulate matters. Darkness-mass factor may change from time to time and location to location.	±2.0 µg/m³, or higher
Optical analyzers	Small, light weight and usually battery operated. Short sample time (in seconds or minutes). Can measure different sizes of particles simultaneously.	Depends on some assumptions of particle characteristics (e.g. each particle is perfect bean-like shape). These assumptions may be different from time to time and location to location.	Depends on the analyzer type and usually not specifically declared by the manufacture.

4. STATE-OF-THE-ART WSN BASED AIR POLLUTION MONITORING SYSTEMS

Twenty state-of-the-art TNGAPMSs that significantly improve the spatio-temporal resolution of the air pollution information and the quality of services provided are presented in this section. The existing works are classified into three categories based on the carriers of the sensor nodes, and the advantages and disadvantages of each category are discussed.

Air pollution in urban areas with ubiquitous emission sources attracts extensive attentions worldwide due to the tremendous impacts on human lives at anytime and anywhere. Networks of monitoring stations using traditional measurement instruments have been deployed to mitigate these impacts. Data acquired by these stations can be utilized for building pollution maps and models that provide authorized environmental situation information and prediction. However, limitations in spatio-temporal resolution and Quality of Services (QoS) are prevalent in these systems [93,94,95]. These limitations result in issues and problems of the conventional air pollution monitoring systems, like non-scalability of system, limited data availability on personal exposure, and out-of-the-fact warnings on acute exposure.

In order to address these prevalent problems, researchers have put lots of efforts into the concept of TNGAPMS by utilizing the advance sensing techniques, MicroElectroMechanical Systems (MEMS), and Wireless Sensor Networks (WSN).

According to the definition of participatory sensing [96,97] and vehicular wireless sensor networks [98,99], and our insights while reviewing the related works, the existing works are classified into three categories based on the carriers of the sensor nodes, namely Static Sensor Network (SSN. Sensor nodes are usually mounted on the streetlight or traffic light poles, or carefully selected locations.), Community Sensor Network (CSN. Sensor nodes are carried by the public communities, usually by volunteers or people who are keen on air quality.), and Vehicle Sensor Network (VSN. Sensor nodes are carried by the public transportations or specially equipped cars.).

These existing works greatly improve the spatio-temporal resolution and QoS of the air pollution information compared with that of conventional monitoring systems. However, in TNGAPMSs, it is impossible to use the same high-end measurement instruments as the ones utilized in stationary monitors of Conventional Stationary Monitoring Network (CSMN). Hence, whenever we deal with the TNGAPMSs, we face the same interesting trade-off as shown in Figure 7. In the following subsections, the three types of sensor networks (SSN, CSN and VSN) are discussed in detail.

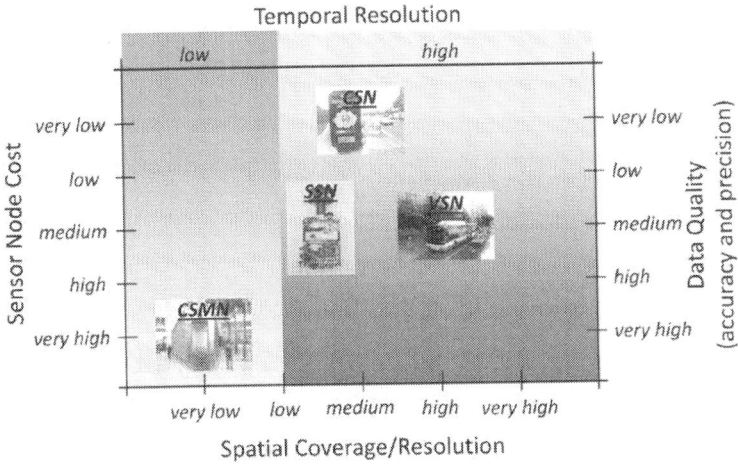

Figure 7. Trade-off between tolerable sensor node cost, obtainable measurement coverage/resolution, expected data quality and achievable measurement temporal resolution for Conventional Stationary Monitoring Network (CSMN), Static Sensor Network (SSN), Community Sensor Network (CSN) and Vehicle Sensor Network (VSN) [100].

4.1. Static Sensor Network (SSN)

In SSN systems, the sensor nodes are typically mounted on the streetlight or traffic light poles, or walls (see Figure 8) By utilizing the low-cost ambient sensors, the number of sensor nodes in SSN systems is much larger than that in the conventional monitoring systems. Air pollution information with high spatio-temporal resolution is achievable in SSN systems. Authorized air pollution information is available to the public through web pages, Web Apps, mobile Apps, etc.

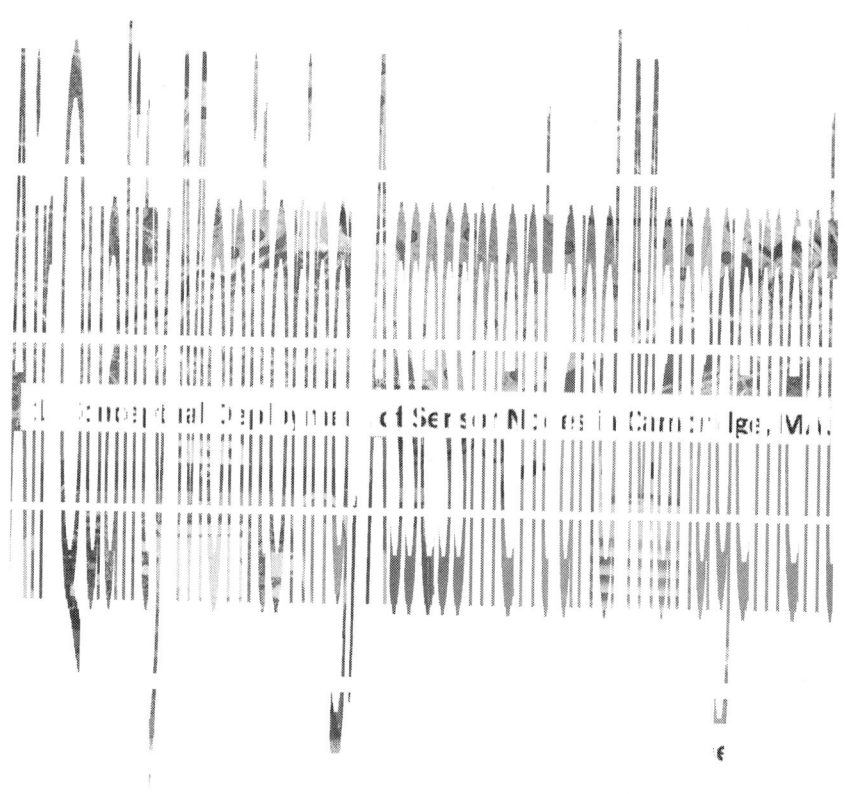

Figure 8. Example of the SSN system architecture and prototype. Red dots are the sensor nodes. Green dots are the gateways that forwarding the acquired data to the Contaminant Source. Figures are adapted from [101].

Carrier: The sensor nodes are usually mounted on the streetlight or traffic light poles, or carefully selected locations.

Related Works

In [101], the project CitySense was presented. This paper claimed that most research groups of WSN evaluate their ideas by simulations, small-scale test deployments or large-scale test deployments with narrow range of target environment, which may have potential issues and problems in real-life large-scale applications. The motivations of the CitySense project are to provide an urban-scale wireless networking testbed, which is able to support a wide range of applications including outdoor air pollution monitoring. Each sensor node consists of a Linux PC, dual 802.11 a/b/g radios and a wide range of sensors. These sensors nodes are mounted on and powered by the streetlights. Sensing data are uploaded to the server through Wi-Fi and authorized air pollution information is available to the public through a customized Web App.

In [102], a WSN based urban air quality monitoring system was proposed. This system consists of a set of sensor nodes, a gateway and a centralize control system provided by the LabVIEW program. Each sensor node integrates with a ZigBee communication link, a CO sensor and a battery. And the gateway is consisted of a Global System for Mobile (GSM) communication link and a wind speed and direction sensor. Data from the sensor nodes are uploaded to the gateway and further forwarded to the central system. This system was deployed to the main roads in Taipei city and the experiment results illustrated that the system can provide micro-scale air pollution information in real-time.

In [103], an outdoor ambient real-time air quality monitoring system was proposed, implemented and tested. In this system, the concentration of O_3, NO_2, CO and H_2S are sensed and transmitted back to the backend server through the GPRS wireless communication link every minute. Authorized air pollution information is available to the public through the customized Web and mobile Apps. A solar panel was utilized to solve the power constraint issue of the sensor nodes (stationary).

In [104], an innovative system named Wireless Sensor Network Air Pollution Monitoring System (WAPMS) was proposed and simulated to monitor the outdoor air pollution situations in Mauritius. This system comprises of an array of sensor nodes and a communication system that gathers the air pollution data to the server. The air pollution data are acquired and passed to the cluster heads by the sensor nodes autonomously. The cluster heads then forward the data to the server. In order to minimize the power consumption in the WSN, a novel data aggregation algorithm named Recursive Converging Quartiles (RCQ) was proposed and implemented. Moreover, a hierarchical routing protocol was utilized to maximize the sensor nodes' energy efficiency.

In [105], an outdoor WSN based air quality monitoring system (WSNAQMS) for industrial and urban areas was proposed. The sensor node consists of a set of gas sensors (O_3, CO and NO_2) and a ZigBee wireless communication link based on the Libelium's [106] gas sensing capable mote. Data are uploaded to the central server through the ZigBee communication link. Authorized air pollution information is available to the public through Email, SMS and customized Web App. This framework is claimed to be simple and reusable in other applications.

Also the failure sensor nodes can be identified efficiently and the energy consumption of each sensor node is minimized. Moreover, a simple Clustering Protocol of Air Sensor (CPAS) network was proposed, which proved to be efficient (in simulation) in terms of network energy consumption, network lifetime, and the data communication rate. The QoSs of the network such as delay, accuracy and reliability (fault tolerance) were also considered.

In [107], a WSN based indoor air pollution monitoring system was presented. The focuses were the power consumption on sensor, sensor node and network levels. Several methodologies that greatly improved the lifetime (up to 3 years) of the monitoring system have been proposed and simulated. The sensor node equips with several sensors (accelerometer, temperature and relative humidity sensors, CO, VOCs and motion sensors), a ZigBee communication link and a battery. In the simulation, 36 sensor nodes were place in the first floor of a 4-story building. Data acquired by the sensor nodes were available to the researchers only.

In [108], an indoor and an outdoor air pollution monitoring architectures based on Wi-Fi were proposed. In this paper, only the indoor one was implemented and tested. Each sensor node consists of several sensors (temperature and relative humidity sensors, CO, methane and solvent vapors sensors) and a Wi-Fi communication link. In order to mitigate the influence factors (temperature and relative humidity) of the gas sensors, a neural network was implemented to obtain the temperature and relative humidity correcting values for the pollutants' concentrations. Sensed data were processed by a PC and published to a customized web page.

Adavantages:

- Loose constraint on energy consumption (The sensor nodes are typically powered by batteries with large capacity or energy harvest devices or power line.)

- No locating device (The location of a sensor node is known once it was deployed since the sensor node is stationary.)

- Loose limitations on weight and size (The carrier of the sensor node is able to carry sufficient enough weight.)

- Multiple sensors per node (One sensor node can equip with several types of sensors because of the loose limitations on weight and size.)

- Accurate and reliable data (Sensor node can integrate with assisting tools because of the loose limitations on weight and size.)

- Guaranteed network connectivity (Once the stationary sensor node joined the network, the topology is fixed and the connectivity is guaranteed.)

- Well calibrated and maintained sensors (The sensor nodes can be well calibrated and maintained by the professionals periodically.)

Disadvantages or Challenges:

- Careful placement of sensor nodes requirement (This is because of the location dependence of air pollutants.)

- Large number of sensor nodes requirement (Data with sufficient geographic coverage and spatial resolution are only achievable by increasing the number of the stationary sensor nodes.)

- Resource wasting in certain level (The stationary sensor nodes are in sleep mode most of the time because continuously updating data at one location is pointless [13].)

- Inconveniences of calibration and maintenance (The professionals need to visit all stationary sensor nodes, which is a time and manpower consuming task, to perform operations.)

- 2-Dimensional data acquisition (Only the air quality of urban surface is monitored.)

- Customized network requirement (A customized wireless or wired network is required when the cellular network is not utilized.)

4.2. Community Sensor Network (CSN)

In CSN (or Participatory Sensing) systems, the sensor nodes are typically carried by the users (see Figure 9). By utilizing the low-cost portable ambient sensors and the ubiquitous smart phones, users are able to acquire, analyze and share the local air pollution information [96]. Authorized air pollution information is available to the public through web pages, Web Apps, mobile Apps, etc.

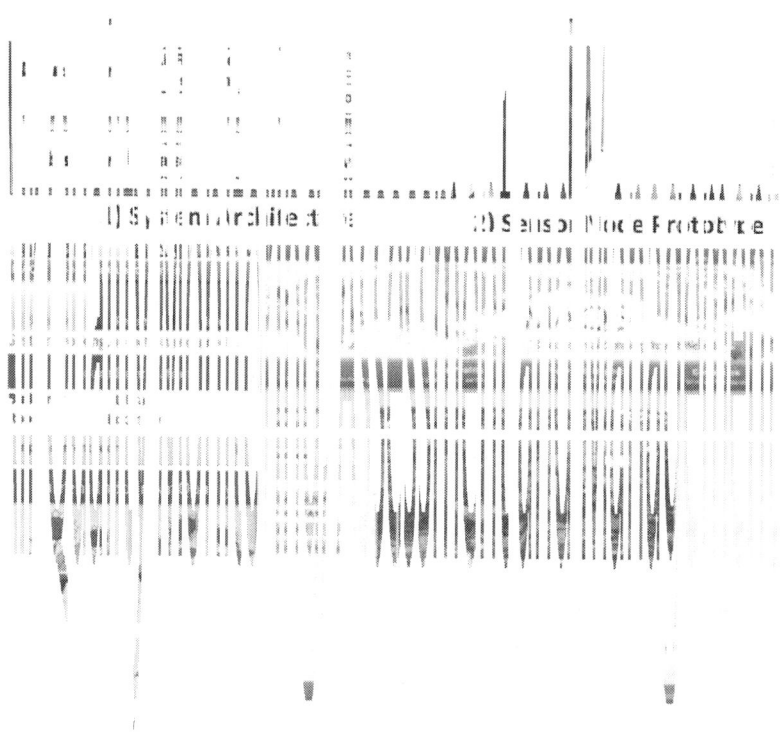

Figure 9. Example of the Community Sensor Network (CSN) system architecture and prototype. Figures are adapted from [109].

Carrier: The sensor nodes are carried by the public or professional users, usually by volunteers or people who are keen on air quality.

Related Works

In [10], a low-power and low-cost mobile sensing system for outdoor participatory air pollution monitoring called GasMobile was introduced. The sensor node composes of a small-size, low-cost O_3 sensor and an off-the-shelf smart phone. The sensor communicates with the smart phone through the USB port. Data (tagged with location information from the build-in GPS module) are uploaded to the server through the cellular network. Authorized information is available to the public through the customized Web and mobile Apps. Two methods were proposed and implemented to improve the data quality of the sensor. This paper claimed that air pollution information with high spatial resolution can be achieved by the community-driven sensing infrastructure like OpenSense [110].

In [97], an outdoor air quality sensing system (P-Sense) based on the participatory sensing technology was presented. Each sensor node consists of a set of sensors (CO_2, CO, VOCs, H_2, temperature and relative humidity) and a Bluetooth link. Data are acquired by the sensors and transmitted to the smart phone through the Bluetooth link. The smart phone then uploaded the data to the server through the cellular network. Authorized air pollution information is available to the public through the customized Web and mobile Apps. Several research issues that need to be addressed before practical deployment of the P-sense system were also pointed out.

In [109], a personalized mobile indoor air quality sensing system called MAQS was presented. Each sensor node consists of several sensors (CO_2, CO, O_3, temperature and relative humidity sensors) and a Bluetooth link communicating with the smart phone. The smart phone further forwards the data to the server using a build-in Wi-Fi module, which was also utilized for localization. Authorized air pollution information is available to the public through the customized Web and mobile Apps. Three novel techniques were proposed and implemented to improve the data accuracy and energy efficiency of the system.

In [111], a hardware and software platform for outdoor participatory air quality monitoring, called N-SMART was introduced. By attaching sensors (CO, NO_X, temperature and Bluetooth) to a GPS-enabled cellphone, the raw air pollution data, which help understand the impacts of air pollution on both individuals and communities, are gathered. The sensor node communicates with the cellphone through the Bluetooth wireless link. Note that, this paper didn't focus on the implementation but the design of the sensing platform. Several research challenges like unpredictable user behaviors and movements, and user privacy problems were discussed in this paper.

In [112], an outdoor urban noise pollution monitoring system called NoiseTube was proposed and implemented. Although it is not an urban air pollution monitoring system, the system architecture and implementation are very similar. Each sensor node is the smart phone itself. The noise data (tagged with location information from the build-in GPS module) are collected by the build-in microphone. Collected data are uploaded to the server through the cellular network. Authorized noise pollution information is available to the public through the customized web page and mobile App.

In [113], a Volatile Organic Compounds (VOCs) sensor node with high selectivity and sensitivity was developed. The authors focused more on the development of the novel tuning fork sensors than the implementation of the air quality monitoring system. Each sensor node consists of several tuning fork sensors (detecting VOCs, temperature and relative humidity) and a Bluetooth device communicating with the smart phone. A customized mobile App for visualizing the sensing data was implemented.

Adavantages:

- Cost efficiency (The sensor node utilizes the cellphone's GPS module and the cellular network, or even the cellphone's computational power.)

- Coupled data generators and consumers (Local or personal air pollution information is available.)

- Public-driven property (The cost of the sensor nodes and the data transmission can be apportioned by the users. It is costly and infeasible for a single agency to acquire all the sensor nodes.)

- Automatic gathering property (The sensor nodes are densely distributed at locations with gathering people automatically. Data with higher spatial resolution and accuracy are achievable in such case.)

- Mobility of sensor nodes (The mobility of the cellphones or users enlarges the sensor node's geographic coverage.)

- Public behaviors acquisition ability (Information such as the public movement patterns, and interaction between air quality and public behaviors, is achievable.)

Disadvantages or Challenges:

- Low data accuracy and reliability (The sensor nodes are typically put in pockets or handbags. Also, the users spend significant amount of time indoor or inside cars [114]).

- Privacy issues (The users may not want to make their location information public for privacy issues).

- Badly calibrated and maintained sensors (Professional calibrations of sensors performed by the public users are very unlikely. Frequent collections and calibrations of sensors by the professionals are infeasible).

- Serious constraint on energy consumption (The sensor nodes is typically powered by cellphone's battery or battery with small capacity).

- Uncontrolled or semi-controlled mobility (The routes of the sensor nodes or users are pre-determined. The sensor nodes may squeeze into a small place with crowded people and cause redundant sampling. Some locations may never be visited).

- 2-Dimensional data acquisition (Only the air quality of urban surface is monitored).

- Serious limitations on weight and size (The sensor node should be portable, which affects the accuracy, reliability and number of sensors equipped, because it is carried by user).

-

4.3. Vehicle Sensor Network (VSN)

In VSN systems, the sensor nodes are typically carried by the public transportations like buses or taxis (see Figure 10). By utilizing the low-cost portable ambient sensors and the mobility of vehicles, one sensor node is able to achieve sufficient large geographic coverage [99,115]. Authorized air pollution information is available to the public through web pages, Web Apps, mobile Apps, etc.

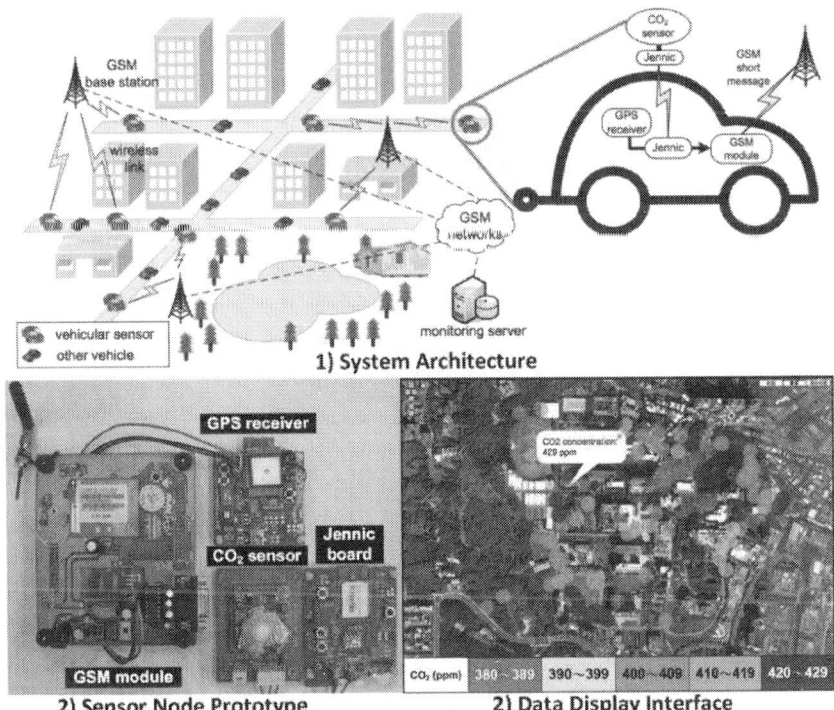

Figure 10. Example of the Vehicle Sensor Network (VSN) system architecture and prototype. Figures are adapted from [98].

Carrier: The sensor nodes are carried by the public transportations (buses, trains and taxis) or specially equipped cars.

Related Works

In [11], the Mobile Air Quality Monitoring Network (MAQUMON) was presented. This system is composed by a number of car-mounted sensor nodes measuring the concentrations of O_3, CO and NO_2. Each sensor node utilizes a GPS module for acquiring time and location information and a Bluetooth link for communicating with the laptop inside car. Collected data are then uploaded to the server through the laptop's Wi-Fi link. Authorized air pollution information is accessible through the sensor node's LCD display or the SensorMap Web App.

In [19], a distributed infrastructure based on the WSN and Grid computing for real-time comprehensive air pollution monitoring and mining was presented. In this system, two types of sensor nodes are utilized, namely the Mobile Sensor Node (MS node) and the Static Sensor Node (SS node). The sensor node consists of a Generic Ultra Violet Sensor Technologies and Observation (GUSTO) sensor (able to detect SO_2, NO_X, O_3 and VOCs) and a wireless link (ZigBee or Wi-Fi or others). The MS nodes are mounted on the public transportations and transmitting data to the SS nodes. The SS nodes are able to perform data acquisition and fusion, and further forward the data (from MS nodes and SS nodes) to the central server. Currently, the air pollution information is only available to the researchers. A distributed data mining-algorithm for identifying the relationships between the urban transport and the environment was also proposed.

In [98,116], a vehicular wireless sensor network architecture was proposed and implemented to achieve the micro-climate monitoring. A CO_2 sensor is mounted outside the car to monitor the concentration of CO_2. A ZigBee intra-vehicle wireless network is utilized to communicate with the inside-car processing unit, which equipped with a micro-controller, a GSM short message module and a GPS module. Data are then sent to the GSM base stations and further forwarded to the monitoring server. Authorized air pollution information is available to the public through a Web App. In order to balance the accuracy of sensed data and the cost of communication, an on-demand approach that adjusts vehicles' reporting rates was proposed.

In [99], a mobile sensor node prototype that can be mounted on vehicle was introduced and tested. Each sensor node consists of a set of sensors (CO, PM, NO, NO_2 and VOCs) for detecting the pollutants' concentrations, a GPS module for collecting the location information, and a GPRS or Wi-Fi module for communicating with the server. Analyzed data are available to the public through a Web App. This paper claimed that the proposed system demonstrated higher spatial coverage at the expense of lower temporal resolution compared with the SSN systems.

In [115], a low-cost air pollution monitoring system using vehicular sensor network was proposed to complement the conventional air pollution monitoring networks. Each sensor node consists of a set of sensors (temperature, relative humidity, NO_2, CO_2, CO and O_3), a GPS module, and a ZigBee wireless link. Data acquired by the sensor nodes are transmitted back to the central computer for further analysis through the static ZigBee accessing points. These sensor nodes are mounted on the public transportations, like buses. By utilizing the mobility of the public transportations, even with a few sensor nodes, the urban air pollution information with fine-grained (high spatial resolution) level was achieved. In this paper, the air pollution information is only available to the researchers.

In [117], a fine-grained vehicular-based mobile air pollution measuring approach was presented. The proposed schema can utilize multiple types of mobile sensor nodes including the proposed Mobile Sensing Box (MSB) and other personal sensing devices. The MSB consists of two ambient sensors (CO and PM) for data collection, a GPS module for location and time information acquisition and a cellular module for data transmission. The car mounted with a MSB travels around the city. Real-time data are received and analyzed by the Cloud Server. Authorized air pollution information is available to the public through the customized Web and mobile Apps.

In [118], a GPRS Sensor Array for outdoor air pollution monitoring was proposed, implemented and tested. The system consists of a mobile sensing unit, which was mounted on the public transportation, and an Internet enabled server. Each sensing unit integrated with a set of sensors (CO, NO_2 and SO_2), a GPS and GPRS modules. Data with location information are sent to the server through the cellular network (GPRS) for further processing and analysis. Authorized air pollution information is available to the public through the customized Web App.

Adavantages:

- Loose constraint on energy consumption (The sensor nodes are powered by the vehicles' batteries.)

- Loose limitations on weight and size (The carrier of the sensor node is able to carry sufficient enough weight.)

- Multiple sensors per node (One sensor node can equip with several types of sensors because of the loose limitations on weight and size.)

- Accurate and reliable data (Sensor node can integrate with assisting tools because of the loose limitations on weight and size.)

- High mobility of sensor nodes (The highly mobile vehicles significantly enlarge the sensor node's geographic coverage.)

- Feasibility in maintenance (The vehicles mounted with sensor nodes can be driven to a specific location. Professionals can perform maintenance on large amount of sensor nodes simultaneously.)

- Well calibrated and maintained sensors (This is because of the feasibility in maintenance of the VSN systems.)

- Automatic gathering property (The sensor nodes are densely distributed at locations with gathering public transportations automatically. Data with higher spatial resolution and accuracy are achievable in such case.)

Disadvantages or Challenges:

- Uncontrolled or semi-controlled mobility (The routes of the sensor nodes or public transportations are pre-determined. The sensor nodes may squeeze into a small place with crowded transportations and cause redundant sampling. Some locations may never be visited.)

- Redundant sampling issues (The vehicles may be trapped into traffic jams or parked in parking lots that cause redundant sampling. This issue compromises the spatial and temporal resolutions.)

- Cost inefficiency on carriers (The specially equipped cars may cost a huge amount of money.)

- Locating and communication devices requirement (The system requires GPS modules, and wireless modules or cellular modules.)

- Customized network requirement (A customized wireless network is required when the cellular network is not utilized. The network connectivity may not be guaranteed due to the mobility of vehicles.)

- 2-Dimensional data acquisition (Only the air quality of urban surface is monitored.)

- Spatial-to-Temporal resolution trade-off (Higher spatial coverage at the expense of lower temporal resolution [99].)

In this section, 20 state-of-the-art TNGAPMSs are discussed and classified into three categories, namely the SSN, CSN and VSN. Summary information (with respect to the Carrier, WSN Type, Sensor Type, Power Source, Locating Device, Computational Power, Operation Environment, Sensing Periodic, Number of

Sensor Nodes in System, Geographic Coverage and Data Availability) of these systems is shown in Table 9 and Table 10. Although these systems greatly improve the pollution information's spatio-temporal resolution compared with the conventional monitoring systems, there exist some issues or challenges in these TNGAPMSs that we will discuss in Section 6.

Table 9. Summary information of the 20 systems in literature works (Part A) (* means unknown).

Sensor Network Type	System	Carrier	WSN Type	Sensor Type	Power Source	Locating Device	Computational Power of Sensor Node (Clock Speed/SRAM/Storage)
SSN	In [105]	Not mentioned	ZigBee	Electrochemical (O_2, CO, NO_x)	Not mentioned	None	Arduino (14 MHz/512 KB/2 GB)
	In [101]	Streetlight pole	Wi-Fi (802.11 a/b/g)	Solid-state (CO, NO, O_3)	Power line	None	Linux based embedded PC (200 MHz/128 MB/1 GB)
	In [104]	Not mentioned	Not mentioned	Not mentioned	Not mentioned	Not mentioned	Not mentioned
	In [102]	Streetlight pole	Zigbee + Cellular network (GSM)	Solid-state (CO)	Battery	None	Octopus II (1 MHz/10 KB/1 MB)
	In [108]	Wall	Wi-Fi (802.11 b/g)	Solid-state (CO, $VOCs$)	Not mentioned	None	TPu4830 (*/*/512KB)
	In [107]	Wall	ZigBee	Solid-state (CO, $VOCs$)	Battery	None	JN5148 (32MHz/128KB/*)
	In [103]	Station	Cellular network (GPRS)	Solid-state (CO, NO_2, O_3, H_2S)	Battery, Solar panel	None	Arduino (16 MHz/8 KB/2 GB)
CSN	In [10]	Public user	Cellular network	Solid-state (O_3)	Battery	Cellphone GPS module	HTC HERO smartphone
	In [111]	Public user	Not mentioned	Not mentioned	Cellphone Battery	Cellphone GPS module	LG VX9700 smart phone
	In [106]	Public user	Wi-Fi	NDIR (CO_2), Solid-state (CO, O_3)	Battery	Cellphone Wi-Fi module	Arduino (16 MHz/2 KB/32 KB)
	In [112]	Public user	Cellular network	Microphone	Cellphone battery	Cellphone GPS module	NOKIA N95 cellphone
	In [97]	Public user	Cellular network	Solid-state (CO, $VOCs$), Catalytic (H_2), Electrochemical (CO)	Battery	Cellphone GPS module	PRC200 Sanyo cellphone
	In [113]	Public user	Cellular network	QTF ($VOCs$)	Battery	Cellphone GPS module	Motorola Q phone
VSN	In [115]	Public transportation	ZigBee	Solid-state (CO, NO_2, O_3, CO_2)	Bus battery	GPS module	Arduino (16 MHz/8 KB/*)
	In [13]	Car	Wi-Fi	Solid-state (CO, NO_2, O_3)	Battery	GPS module	8051 uC (*/4KB/2MB)
	In [98]	Car	Cellular network (GSM)	NDIR (CO_2)	Car battery	GPS module	JN5139 (16 MHz/96 KB/192 KB)
	In [117]	Car	Cellular network	Solid-state (CO), Optical analyzer (PM)	Bus battery	GPS module	Arduino (16 MHz/8 KB/128 KB)
	In [118]	Bus	Cellular network (GPRS)	Electrochemical (CO, SO_2, NO_2)	Not mentioned	GPS module	HCS12/9S12 (25 MHz/12 KB/512 KB)
	In [19]	Public transportation	Wi-Fi or ZigBee or Others	DUVAS (O_3, NO, NO_2, SO_2, $VOCs$)	Not mentioned	Not mentioned	Not mentioned
	In [99]	Car	Wi-Fi or Cellular network (GPRS)	Optical analyzer (PM), Solid-state (CO, NO_2, NO, $VOCs$)	Not mentioned	GPS module	Renesas H8S (*/*/*)

Table 10. Summary information of the 20 systems in literature works (Part B).

Sensor Network Type	System	Operation Environment	Sensing Periodic	Number of Sensor Node in System	Geographic Coverage	Data Availability
SSN	In [105]	Outdoor roadside	200 to 300 s	60 to 200	300 m × 500 m	Email, SMS, Web App
	In [101]	Outdoor	Not mentioned	about 100	Harvard campus	Web App
	In [104]	Outdoor	Not mentioned	300 to 1200	Port Louis	Not mentioned
	In [102]	Outdoor roadside	10 min	9	Intersection circle of Keelung Road and Roosevelt Road	Researcher only
	In [108]	Indoor	5 to 60 s	Not mentioned	One floor of a building	Web page
	In [107]	Indoor	Adaptive	36	One floor of a building	None
	In [103]	Outdoor	1 min	4	1 Km²	Web App, mobile App
CSN	In [10]	Outdoor roadside	5 s	Not mentioned	Citywide	Web App, mobile App
	In [111]	Outdoor	Not mentioned	Not mentioned	Not mentioned	Not mentioned
	In [109]	Indoor	6 s	Not mentioned	One floor of a building	Web App, mobile App
	In [112]	Outdoor	1 s	Not mentioned	Citywide	Web page, mobile App
	In [97]	Outdoor	Not mentioned	Not mentioned	Not mentioned	Web App, mobile App
	In [113]	Outdoor	Not mentioned	Not mentioned	Not mentioned	Web App, mobile App
VSN	In [115]	Outdoor roadside	Not mentioned	1	Not mentioned	None
	In [13]	Outdoor roadside	1 min or few times per hour	Not mentioned	Citywide	Web App
	In [98]	Outdoor roadside	3 s	16	National Chiao-Tung University campus	Web App
	In [117]	Outdoor roadside	5 s	2	Citywide	Web App
	In [118]	Outdoor roadside	Not mentioned	1	American University of Sharjah campus	Web App
	In [19]	Outdoor roadside	1 min	18	Not mentioned	None
	In [99]	Outdoor roadside	Not mentioned	1	Nanyang Technological University and neighboring industrial estate	Web App

5. COMPARISON OF THE THREE TYPES OF SENSOR NETWORKS

In this section, the comparisons between SSN, CSN and VSN are presented. The six properties for comparison are listed as follows. Each property is described in detail with respect to the **Ranking** (the ranking of SSN, CSN and VSN of specific property, the higher the better), **Reasons** (reasons for why we choose this property for comparison) and **Explanation** (detail explanation of the ranking).

5.1. Mobility/Geographic-Coverage

Ranking: VSN > CSN > SSN

Reasons

The mobility of the carrier enables a sensor node to cover sufficient large geographic areas within a short period of time. Higher spatial resolution of the sensed data can be achieved and fewer number of sensor nodes are required compared with systems using stationary carriers.

Explanation

The sensor nodes carried by the public transportations in VSN systems are with the highest mobility among the three types of sensor networks. Following is the sensor nodes carried by the public users in CSN systems because the users travel much slower than the vehicles and the users spend most of time indoor or inside cars [114]. The stationary sensor nodes in SSN systems are with the lowest or zero mobility. Intuitively, the geographic coverage of a sensor node is proportional to the mobility of the carrier.

5.2. Temporal Resolution

Ranking: SSN > VSN > CSN

Reasons

One of the objectives of TNGAPMS is to increase the temporal resolution of the acquired air pollution information. And the air pollution information from all TNGAPMSs has much higher temporal resolution than that from the conventional monitoring systems. However, the temporal resolutions of the acquired pollution information in SSN, CSN and VSN are slightly different due to several reasons.

Explanation

In terms of building a pollution map, the pollution information from SSN systems has the highest temporal resolution. Then comes the pollution information from VSN systems, followed by that from CSN systems. The ranking is based on the assumptions that the sensor nodes' sensing rates are identical in different systems and the sensors have a limited effective coverage [99]. In a single sensor node case, the pollution information's temporal resolution of a specific location (a circular area with a specific radius) in SSN systems is the sensor node's sensing rate itself. However, in VSN and CSN systems, the pollution information's temporal resolutions at a specific location depend on how frequent the location is visited and how often the pollution data are sensed at that location. Intuitively, the mobility of VSN and CSN systems lowers the temporal resolutions of the acquired pollution information. Moreover, the pollution information's temporal resolution is further reduced by redundant sampling issues like traffic jams, parked vehicles and indoor stay of users (In this case, the average temporal resolutions of SSN, CSN and VSN systems are compared. In SSN systems, only one location is monitored. In CSN and VSN systems, one sensor node typically covers several locations and this results in lower average temporal resolutions when redundant sampling issues happened).

In terms of monitoring personal exposure, the pollution information's temporal resolution for people wearing the sensor nodes in CSN systems is the highest. For people without carrying the sensor nodes, the temporal resolution of the pollution data on personal exposure depends on the pollution map.

5.3. Cost Efficiency

Ranking: CSN > VSN > SSN

Reasons

The air pollution situation in rapid industrializing countries is much more critical than that in industrialized countries [38]. Several pollution sources (over-polluting industry, poorly tuned diesel engines and burning of trash) in developing countries contribute to the air pollution much more significantly than that in developed countries [111]. Moreover, the governments in developing countries spend less fraction of their GDPs on environmental protection than developed countries [119]. In a word, the environmental protection agencies in developing countries are dealing with serious air pollution situation with little amount of money. Hence, the cost efficiency of the air pollution monitoring system becomes a non-negligible property for comparison.

Explanation

In sensor node level, the CSN systems have the highest cost efficiency, followed by the SSN systems and the VSN systems. In CSN systems, the users' cellphones are fully utilized, including the build-in GPS and wireless communication modules, and the computational powers. The sensor nodes in CSN systems typically require no locating, communicating and computing devices and hence the cost efficiency is enhanced. In SSN systems, the stationary sensor nodes require no locating device but the communicating and computing devices because the location of a specific sensor node is known once it is deployed. For the sensor nodes in VSN systems, the GPS modules are essential due to the mobility of the carriers. Also, the communicating and computing devices are needed. Hence, the cost efficiency of VSN systems is the lowest in sensor node level.

In system level, the vast majority of the system cost is contributed by the acquisition of sensor nodes. Moreover, if the number of sensor nodes in system enlarges, a larger database for data storage and management, a faster wireless sensor network for data transmission and a more powerful computing center for data processing and decision making in real-time are required. As described in Subsection 5.1, the SSN systems require the largest amount of sensor nodes to cover a specific area, followed by the CSN systems and the VSN systems. Hence, in system level, the VSN systems have the highest cost efficiency followed by the CSN systems and the SSN systems.

The final ranking is achieved by averaging the rankings in sensor node level and system level.

5.4. Endurance

Ranking: SSN > VSN > CSN

Reasons

The endurance of the sensor nodes is a major property for comparison because it will further influence the Maintenance property and the Data Quality property. A sensor node with energy constraint (e.g., powered by a small capacity battery) requires replacing battery frequently, which increases the burden of maintenance. Moreover, the energy constraint of sensor nodes limits the use of conditioning appliances (e.g., temperature controllers, humidity controllers, gas pumps, etc.) that help improve the data quality.

Explanation

The sensor nodes in CSN systems are with the lowest duration compared to the sensor nodes in VSN systems and SSN systems because they are powered by cellphone or portable batteries. In VSN systems, the sensor nodes are powered by vehicles' batteries. The power supply is guaranteed once the vehicle started. In SSN systems, the sensor nodes are powered by large capacity batteries, energy harvest devices or even power lines. The duration of the sensor nodes can be counted as infinity if they are powered by power lines. Hence, the sensor nodes in SSN systems are with the highest duration compared to the sensor nodes in CSN systems and VSN systems.

5.5. Maintenance

Ranking: VSN > SSN > CSN

Reasons

In order to guarantee the data quality, maintenance on the sensor nodes like changing dead batteries, replacing malfunction components or calibrating sensors are indispensable. As a matter of fact, all TNGAPMSs require massive deployment of the sensor nodes and the sensors used in these systems need frequent calibration to be efficient [83]. We expect that, in real-life large-scale deployment, the maintenance on the sensor nodes will occupy the vast majority of efforts of the maintenance on the whole system. The feasibility of maintenance on the sensor nodes is critical in this case.

Explanation

In CSN systems, the sensor nodes are carried out by the public users who are lack of special knowledge and equipment or even unlikely to explicitly maintain the sensor nodes. Moreover, it is infeasible for the professionals to collect and maintain all the sensor nodes frequently. Hence the sensor nodes in CSN systems are typically not well maintained. In SSN systems, the sensor nodes are well maintained by the professionals. However, the professionals need to visit all locations with sensor node deployment to conduct the maintenance periodically. Tremendous amount of manpower and time are required in this case and the flexibility of maintenance is reduced compared to VSN systems. In VSN systems, the sensor nodes carried by the public transportations can be driven to a specific location on demand for maintenance by the professionals. Manpower and time are saved because the professionals are able to maintain large amount of sensor nodes simultaneously. As a result, the flexibility of maintenance of VSN systems is the highest among these three types of systems.

5.6. Data Quality

Ranking: SSN > VSN > CSN

Reasons

Good data quality is essential for developing TNGAPMSs. The data quality of the low-cost portable ambient sensors used in TNGAPMSs is poorer than that of the high-end instruments used in the conventional air pollution monitoring systems. However, the low-cost portable ambient sensors still provide a fair enough accuracy and detection range [46], and flexibility in massive deployment.

Explanation

In CSN systems, the constraints on weight, size and power consumption (usually powered by small capacity batteries) of the sensor nodes are extraordinary serious. These sensor nodes are typically light weight and small size, and impossible to equip with assisting instruments like temperature and humidity controllers. Moreover, the sensor nodes are not well maintained and may be put into bags or pockets that further lower the data quality. In VSN systems, these constraints are not as critical as that in CSN systems. Adding assisting tools to the sensor nodes is possible and the sensor nodes are well maintained by the professionals. However, the high mobility of the sensor nodes becomes a major factor affecting the accuracy of the sensor readings due to the varying air flow around the sensor head [10]. In SSN systems, limitations on the weight, size and power consumption (powered by power line or renewable energy source) of the sensor nodes are relaxed. The sensor nodes are able to equip with assisting equipment to ensure the data quality. The network connectivity and the sensor node's power supply are guaranteed and the reliability of the sensed data is improved due to stationary characteristic. The data quality of SSN systems is the highest among these three types of systems, followed by the data quality of VSN systems and CSN Systems.

After we have an in-deep understanding of these six comparison properties described above, we graded these properties of SSN, CSN and VSN systems using the grading code defined as: '0' means 'None'; '1' means 'Low/Short/Inconvenient'; '2' means 'Medium' and '3' means 'High/Long/Convenient'. The final grade of each comparison property of SSN, CSN and VSN systems following the descriptions above are shown in Figure 11.

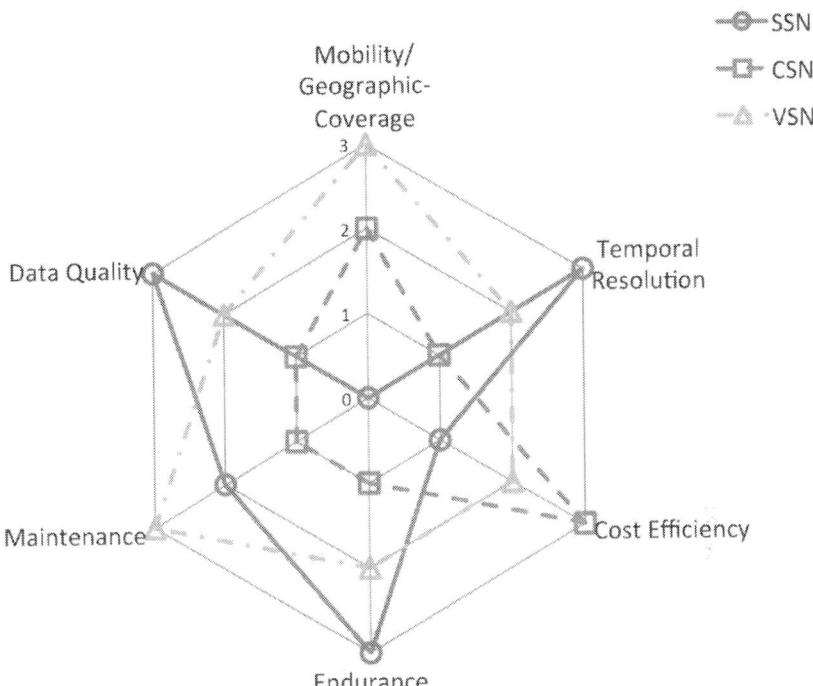

Figure 11. Grading result of the six major comparison properties in Static Sensor Network (SSN), Community Sensor Network (CSN) and Vehicle Sensor Network (VSN) ('0' means 'None'; '1' means 'Low/Short/Inconvenient'; '2' means 'Medium' and '3' means 'High/Long/Convenient').

6. DISCUSSION AND CONCLUSIONS

Air pollution is an essential environmental issue due to the tremendous impacts on public health, global environment, and worldwide economy. Urban air pollution with non-uniform distribution trend arises the necessity for pollution monitoring with high spatio-temporal resolution, which the conventional air pollution monitoring systems cannot provide because of the limited data availability and non-scalability of the systems. By utilizing the advance sensing technologies, MicroElectroMechanical Systems (MEMS) and Wireless Sensor Network (WSN), researchers are pushing the concept of The Next Generation Air Pollution Monitoring System (TNGAPMS) to the limit and have achieved great progresses. Many of state-of-the-art air pollution monitoring systems have been implemented and tested. All of these systems evidence that an air pollution monitoring system with high spatio-temporal resolution, cost and energy efficiency, deployment and maintenance feasibility, convenient accessing ability for the public or professional users are achievable. However, from Section 4 and

Section 5, we can conclude that there are still some issues or challenges of these existing systems that need to be addressed. Also there are some abilities or characteristics of these existing systems that we want to carry forward or enhance when building the future systems.

6.1. Issues and Challenges Need to Be Addressed

Lack of 3-Dimensional Data Acquisition Ability: All the systems presented in Section 4 are only able to monitor the air pollution situation of urban surface or roadside while the necessities and importance of the 3-Dimensional air pollution information are highlighted [120,121,122]. Current LIDARs or satellites based 3-Dimensional monitoring systems face the same issues as the conventional monitoring systems. We anticipate that 3-Dimensional air pollution information with high spatio-temporal resolution can be acquired in real-time by mounting the portable sensor nodes on the multi-rotors Unmanned Aerial Vehicles (UAVs).

Infeasibility of Active Monitoring: The sensor nodes in SSN, CSN and VSN systems presented are all passive monitoring sensor nodes (sensor nodes periodically update data). We believe that active monitoring (users can fully control the sensor network including the formation and routes of the sensor nodes) provides higher flexibility and QoS.

Uncontrolled or Semi-Controlled Carriers: The carriers in SSN, CSN and VSN systems are uncontrolled or semi-controlled because they are either stationary or with pre-determined routes. We anticipate that fully controlled carriers have higher mobility and make active monitoring possible. By utilizing the fully controlled carriers (i.e., the multi-rotors UAVs), feasibility in deployment, cost efficiency of systems and convenience in maintenance can be achieved.

6.2. Abilities and Characteristics Need to Be Carried Forward

In fact, all the abilities or characteristics of these state-of-the-art TNGAPMSs need to be carried forward and some of them can be improved in future systems.

Mobility of Carriers: The mobility of the carriers enables one sensor node to cover a sufficient large geographic area within a short period of time. The number of sensor nodes required is reduced and the system cost and maintenance are relaxed. In fact, the multi-rotors (There are regulations for drones in some areas and we put aside this issue for a moment in this paper.)

will not be trapped into traffic-jams or stop by unreachable areas as the carriers in VSN systems. The multi-rotors have much higher mobility than the carriers in VSN systems.

Feasibility of Maintenance: The system's feasibility on maintenance need to be carried forward and enhanced because it affects the data quality and cost efficiency of the system. If the fully controlled UAVs were utilized in the system, professional maintenance on large number of sensor nodes can be performed simultaneously by driving all UAVs to a specific location. In this case, the quality of the sensed data is guaranteed while the time and manpower for maintenance are saved.

Add-on Sensor Ability: We note that all the sensor nodes in the existing TNGAPMSs are with no add-on ability. Reconfigurations on the hardware and software of the sensor nodes are needed whenever the sensing species are modified. In real-life large-scale applications, there could be hundreds or even thousands of sensor nodes in the system. Sensor nodes with add-on (the sensor node is able to identify the type of sensor mounted and chooses the suitable program to handle the sensing data) ability are essential in this case. Properties like modifiable sensing and transmitting intervals, remote programmable ability, cost and energy efficiencies and failure check feature are also essential.

Last but not least, all the existing state-of-the-art TNGAPMSs claim that they have a better spatio-temporal resolution than the conventional air pollution monitoring systems (which is obvious). However, none of them has ever considered how good they are, not to mention the comparisons among the SSN, CSN and VSN systems, with respect to real-time performance, spatio-temporal resolution and QoS. And this will be a major direction of our future works.

REFERENCES

1. World Health Organization. 7 Million Premature Deaths Annually Linked to Air Pollution. Available online: http://www.who.int/mediacentre/news/releases/2014/air-pollution/en/ (accessed on 20 August 2015).

2. European Commission. Air quality: Commission Sends Final Warning to UK Over Levels of Fine Particle Pollution. Available online: http://europa.eu/rapid/press-release_IP-10-687_en.htm?locale=en (accessed on 20 August 2015).

3. World Health Organization. Monitoring Ambient Air Quality for Health Impact Assessment. 1999. Available online: http://www.euro.who.int/__data./pdf_file/0010/119674/E67902.pdf (accessed on 20 August 2015).

4.   World Health Organization. Ambient (Outdoor) Air Quality and Health. Available online: http://www.who.int/mediacentre/factsheets/fs313/en/ (accessed on 20 August 2015).

5.   Amorim, L.C.A.; Carneiro, J.P.; Cardeal, Z.L. An optimized method for determination of benzene in exhaled air by gas chromatography-mass spectrometry using solid phase microextraction as a sampling technique. J. Chromatogr. B **2008**, 865, 141–146.

6.   Environmental Protection Department of Hong Kong. Air Quality Health Index. Available online: http://www.aqhi.gov.hk/en.html (accessed on 22 August 2015).

7.   Beijing Municipal Environmental Protection Bureau. Beijing Environmental Statement 2013. Available online: http://www.bjepb.gov.cn/ bjepb/resource/cms/2014/06/2014061911140819230.pdf (accessed on 22 August 2015).

8.   New York State Department of Environmental Conservation. New York State Air Quality Monitoring Center Home. Available online: http://www.dec.ny.gov/airmon/index.php (accessed on 22 August 2015).

9.   King's College London. London Air Quality Network: Monitoring Sites. Available online: http://www.londonair.org.uk/london/ asp/ Public Episodes.asp (accessed on 22 August 2015).

10. Hasenfratz, D.; Saukh, O.; Sturzenegger, S.; Thiele, L. Participatory Air Pollution Monitoring Using Smartphones. In Mobile Sensing: From Smartphones and Wearables to Big Data; ACM: Beijing, China, 2012.

11. Völgyesi, P.; Nádas, A.; Koutsoukos, X.; Lédeczi, A. Air Quality Monitoring with SensorMap. In Proceedings of the 7th International Conference on Information Processing in Sensor Networks (IPSN '08), St. Louis, MO, USA, 22–24 April 2008; pp. 529–530.

12. Richards, M.; Ghanem, M.; Osmond, M.; Guo, Y.; Hassard, J. Grid-based analysis of air pollution data. Ecol. Model. **2006**, 194, 274–286.

13. Dobre, A.; Arnold, S.J.; Smalley, R.J.; Boddy, J.W.D.; Barlow, J.F.; Tomlin, A.S.; Belcher, S.E. Flow field measurements in the proximity of an urban intersection in London, UK. Atmos. Environ. **2005**, 39, 4647–4657.

14. Air Quality Expert Group. Nitrogen Dioxide in the United Kingdom; Technical Report; Department for the Environment, Food and Rural Affairs: London, UK, 2004.

15. Yu, O.; Sheppard, L.; Lumley, T.; Koenig, J.Q.; Shapiro, G.G. Effects of Ambient Air Pollution on Symptoms of Asthma in Seattle-Area Children Enrolled in the CAMP Study. Environ. Health Perspect. **2000**, 108, 1209–1214.

16. PopeIII, C.A.; Verrier, R.L.; Lovett, E.G.; Larson, A.C.; Raizenne, M.E.; Kanner, R.E.; Schwartz, J.; Villegas, G.; Gold, D.R.; Dockery, D.W. Heart rate variability associated with particulate air pollution. Am. Heart J. **1999**, 138, 890–899.

17. Peters, A.; Dockery, D.W.; Muller, J.E.; Mittleman, M.A. Increased Particulate Air Pollution and the Triggering of Myocardial Infarction. Circulation **2001**, 103, 2810–2815.

18. United States Environmental Protection Agency. Next Generation Air Measuring Research. Available online: http://www2.epa.gov/air-research/next-generation-air-measuring-researchr/ (accessed on 12 July 2015).

19. Ma, Y.; Richards, M.; Ghanem, M.; Guo, Y.; Hassard, J. Air Pollution Monitoring and Mining Based on Sensor Grid in London. Sensors **2008**, 8, 3601–3623.

20. Bravo, M.A.; Fuentes, M.; Zhang, Y.; Burr, M.J.; Bell, M.L. Comparison of exposure estimation methods for air pollutants: Ambient monitoring data and regional air quality simulation. Environ. Res. **2012**, 116, 1–10.

21. Budde, M.; El Masri, R.; Riedel, T.; Beigl, M. Enabling Low-cost Particulate Matter Measurement for Participatory Sensing Scenarios. In Proceedings of the 12th International Conference on Mobile and Ubiquitous Multimedia (MUM '13), Luleå, Sweden, 2–5 December 2013; ACM: New York, NY, USA, 2013; pp. 19:1–19:10.

22. International Sensor Technology. Hazardous Gas Data. 1997. Available online: http://www.intlsensor.com/pdf/hazgasdata.pdf (accessed on 25 August 2015).

23. United States Environmental Protection Agency. What are the Six Common Air Pollutants? Available online: http://www.epa.gov/airquality/urbanair/ (accessed on 25 August 2015).

24. United States Environmental Protection Agency. Carbon Monoxide Home. Available online: http://www.epa.gov/airquality/carbonmonoxide/ (accessed on 27 August 2015).

25. United States Environmental Protection Agency. Nitrogen Dioxide Home. Available online: http://www.epa.gov/airquality/nitrogenoxides/ (accessed on 27 August 2015).

26. United States Environmental Protection Agency. Ground Level Ozone. Available online: http://www.epa.gov/airquality/ozonepollution/ (accessed on 27 August 2015).

27. United States Environmental Protection Agency. Sulfur Dioxide Home. Available online: http://www.epa.gov/airquality/sulfurdioxide/ (accessed on 27 August 2015).

28. United States Environmental Protection Agency. Particulate Matter Home. Available online: http://www.epa.gov/airquality/particlepollution/ (accessed on 27 August 2015).

29. United States Environmental Protection Agency. Lead Home. Available online: http://www.epa.gov/airquality/lead/ (accessed on 27 August 2015).

30. Johnson, D.L.; Ambrose, S.H.; Bassett, T.J.; Bowen, M.L.; Crummey, D.E.; Isaacson, J.S.; Johnson, D.N.; Lamb, P.; Saul, M.; Winter-Nelson, A.E. Meanings of Environmental Terms. J. Environ. Qual. **1997**, 26, 581–589.

31. United States Environmental Protection Agency. Air Quality Index (AQI)— A Guide to Air Quality and Your Health. Available online: http://www.airnow.gov/index.cfm?action=aqibasics.aqi (accessed on 1 September 2015).

32. European Commission. Indices Definition. Available online: http://www.airqualitynow.eu/about_home.php (accessed on 1 September 2015).

33. Ministry of Environmental Protection of the People's Republic of China. PRC National Environmental Protection Standard: Technical Regulation on Ambient Air Quality Index. 2012. Available online: http://kjs.mep.gov.cn/hjbhbz/bzwb/dqhjbh/jcgfffbz/201203/W020120410332725219541.pdf (accessed on 1 September 2014).

34. Environmental Protection Department of Hong Kong. About AQHI. Available online: http://www.aqhi.gov.hk/en/what-is-aqhi/about-aqhi.html (accessed on 1 September 2015).

35. Wong, T.W.; Tam, W.W.S.; Yu, I.T.S.; Wong, A.H.S.; Lau, A.K.H.; Ng, S.K.W.; Yeung, D.; Wong, C.M. A Study of the Air Pollution Index Reporting System; Technical Report. Environmental Protection Department of Hong Kong: Hong Kong, China, 2012.

36. Environmental Protection Department of Hong Kong. Health Advice. Available online: http://www.aqhi.gov.hk/en/health-advice/sub-health-advice.html (accessed on 1 September 2015).

37. United States Environmental Protection Agency. National Ambient Air Quality Standards. Available online: http://www.epa.gov/air/criteria.html (accessed on 27 August 2015).

38. World Health Organization. Air Quality Guidelines. 2005. Available online: http://www.who.int/phe/health_topics/outdoorair/outdoorair_aqg/en/ (accessed on 27 August 2015).

39. World Health Organization. Indoor Air Quality Guidelines. 2005. Available online: http://www.euro.who.int/__data./pdf_file/0009/128169/e94535.pdf (accessed on 27 August 2015).

40. World Health Organization. Exposure to Lead: A Major Public Health Concern. 2010. Available online: http://www.who.int/ipcs/assessment/public_health/lead/en/ (accessed on 27 August 2014).

41. European Commission. Air Quality Standards. Available online: http://ec.europa.eu/environment/air/quality/standards.htm (accessed on 27 August 2015).

42. Ministry of Environmental Protection of the People's Republic of China. Ambient Air Quality Standards. 2012. Available online: http://kjs.mep.gov.cn/hjbhbz/bzwb/dqhjbh/dqhjzlbz/201203/t20120302_224165.htm (accessed on 27 August 2015).

43. Environmental Protection Department of Hong Kong. Hong Kong's Air Quality Objectives. Available online: http://www.epd.gov.hk/epd/english/environmentinhk/air/air_quality_objectives/air_quality_objectives.html (accessed on 27 August 2015).

44. United States Environmental Protection Agency. List of Designated Reference and Equivalent Methods. Available online: http://www.epa.gov/ttnamti1/files/ambient/criteria/reference-equivalent-methods-list.pdf (accessed on 8 September 2015).

45. Environmental Protection Department of Hong Kong. Air Quality Monitoring Equipment. Available online: http://www.aqhi.gov.hk/en/monitoring-network/air-quality-monitoring-equipment.html (accessed on 8 September 2015).

46. Aleixandre, M.; Gerboles, M. Review of Small Commercial Sensors for Indicative Monitoring of Ambient Gas. Chem. Eng. Trans. **2012**, 30, 169–174.

47. Bender, F.; Barié, N.; Romoudis, G.; Voigt, A.; Rapp, M. Development of a preconcentration unit for a SAW sensor micro array and its use for indoor air quality monitoring. Sens. Actuators B: Chem. **2003**, 93, 135–141.

48. Lee, Y.J.; Kim, H.B.; Roh, Y.R.; Cho, H.M.; Baik, S. Development of a SAW gas sensor for monitoring SO_2 gas. Sens. Actuators A: Phys. **1998**, 64, 173–178.

49. Fanget, S.; Hentz, S.; Puget, P.; Arcamone, J.; Matheron, M.; Colinet, E.; Andreucci, P.; Duraffourg, L.; Myers, E.; Roukes, M.L. Gas sensors based on gravimetric detection-A review. Sens. Actuators B Chem. **2011**, 160, 804–821.

50. Boussaad, S.; Tao, N.J. Polymer Wire Chemical Sensor Using a Microfabricated Tuning Fork. Nano Lett. **2003**, 3, 1173–1176.

51. Ren, M.; Forzani, E.S.; Tao, N. Chemical Sensor Based on Microfabricated Wristwatch Tuning Forks. Anal. Chem. **2005**, 77, 2700–2707.

52. Shiina, T. LED mini-lidar as minimum setup. In Proceedings of the SPIE 9246, Amsterdam, Netherlands, 22 September 2014; Lidar Technologies, Techniques, and Measurements for Atmospheric Remote Sensing X, 92460F, Edinburgh, UK. 2014; pp. 92460F-1–92460F-6.

53. Chiang, C.W.; Das, S.K.; Chiang, H.W.; Nee, J.B.; Sun, S.H.; Chen, S.W.; Lin, P.H.; Chu, J.C.; Su, C.S.; Su, L.S. A new mobile and portable scanning lidar for profiling the lower troposphere. Geosci. Instrum. Methods Data Syst. **2015**, 4, 35–44.

54. Rionda, A.; Marin, I.; Martinez, D.; Aparicio, F.; Alija, A.; Garcia Allende, A.; Minambres, M.; Paneda, X.G. UrVAMM-A full service for environmental-urban and driving monitoring of professional fleets. In Proeedings of the 2013 International Conference on New Concepts in Smart Cities: Fostering Public and Private Alliances (SmartMILE), Gijon, Spain, 11–13 December 2013; pp. 1–6.

55. Ltd., D.T. DUVAS Series. Available online: http://www. duvastechnologies. com/ (accessed on 18 June 2015).

56. Data Sheet of BAM-1020 Beta Attenuation Monitor. 2013. Available online: http://www.metone.com/documents/BAM-1020_Datasheet.pdf (accessed on 10 September 2014).

57. AEROCET 831 Aerosol Mass Monitor. 2014. Available online: http://www.metone.com/docs/831_datasheet.pdf (accessed on 10 September 2015).

58. OPC-N2 Particle Monitor. 2015. Available online: http://www.alphasense. com/WEB1213/wp-content/uploads/2015/05/OPC-N2.pdf (accessed on 9 August 2015).

59. Sharp DN7C3CA006 PM2.5 Sensor Module. 2014. Available online: http://media.digikey.com/pdf/Data%20Sheets/Sharp%20PDFs/DN7C3CA0 06_Spec.pdf (accessed on 9 August 2015).

60. Data Sheet of Model 602 Beta^{PLUS} Particle Measurement System. 2012. Available online: http://www.teledyne-api.com/pdfs/602_Literature_ RevB.pdf (accessed on 10 September 2015).

61. Sharp GP2Y1010AU Compact Dust Sensor for Air Conditioners. Available online: http://media.digikey.com/pdf/Data%20Sheets/Sharp%20PDFs/GP2Y1010A U.pdf (accessed on 9 August 2015).

62. Data Sheet of ModelT300U Ultra-Sensitive Carbon Monoxide Analyzer. 2011. Available online: http://www.teledyne-ml.com/pdfs/T300U.pdf (accessed on 10 September 2015).

63. Aeroqual Carbon Monoxide Sensor Head 0–25 ppm. 2014. Available online: http://www.aeroqual.com/product/carbon-monoxide-sensor-0-25ppm (accessed on 9 August 2015).

64. CO-B4 Carbon Monoxide Sensor 4-Electrode. 2015. Available online: http://www.alphasense.com/WEB1213/wp-content/uploads/2015/04/COB41.pdf (accessed on 9 August 2015).

65. MQ-7 Gas Sensor. Available online: http://www.ventor.co.in/Datasheet/ MQ-7.pdf (accessed on 9 August 2015).

66. Data Sheet of Model T500U CAPS Nitrogen Dioxide Analyzer. 2014. Available online: http://www.teledyne-api.com/pdfs/T500U_Literature.pdf (accessed on 10 September 2015).

67. Aeroqual Nitrogen Dioxide Sensor Head 0-1 ppm. 2014. Available online: http://www.aeroqual.com/product/nitrogen-dioxide-sensor-0-1ppm (accessed on 9 August 2015).

68. NO2-B42F Nitrogen Dioxide Sensor 4-Electrode. 2015. Available online: http://www.alphasense.com/WEB1213/wp-content/uploads/ 2015/03/ NO2B42F.pdf (accessed on 9 August 2015).

69. The MiCS-2714 is a compact MOS sensor. 2014. Available online: http://www.sgxsensortech.com/content/uploads/2014/08/1107_Datasheet-MiCS-2714.pdf (accessed on 9 August 2015).

70. Data Sheet of Model 265E Chemiluminescence Ozone Analyzer. Available online: http://www.teledyne-api.com/pdfs/265E_Literature_RevC.pdf (accessed on 10 September 2014).

71. Aeroqual Ozone Sensor Head 0-0.15 ppm. 2014. Available online: http://www.aeroqual.com/product/ozone-sensor-ozu (accessed on 9 August 2015).

72. OX-B421 Oxidising Gas Sensor Ozone + Nitrogen Dioxide 4-Electrode. 2015. Available online: http://www.alphasense.com/WEB1213/wp-content/uploads/2015/04/OX-B421.pdf (accessed on 9 August 2015).

73. MQ-131 Semiconductor Sensor for Ozone. Available online: http://www.datasheet-pdf.com/PDF/MQ131-Datasheet-HenanHanwei-770516 (accessed on 9 August 2015).

74. Data Sheet of Model 6400T/6400E Sulfur Dioxide Analyzer. Available online: http://www.teledyne-ai.com/pdf/6400t_Rev-B.pdf (accessed on 10 September 2015).

75. Aeroqual Sulfur Dioxide Sensor Head 0-10 ppm. 2014. Available online: http://www.aeroqual.com/product/sulfur-dioxide-sensor-0-10ppm (accessed on 9 August 2015).

76. SO2-B4 Sulfur Dioxide Sensor 4-Electrode. 2014. Available online: http://www.alphasense.com/WEB1213/wp-content/uploads/2014/08/SO2B4.pdf (accessed on 9 August 2015).

77. MQ-136 Semiconductor Sensor for Sulfur Dioxide. Available online: http://www.china-total.com/Product/meter/gas-sensor/MQ136.pdf (accessed on 9 August 2015).

78. Chou, J. Electrochemical Sensors. In Hazardous Gas Monitors—A Practical Guide to Selection, Operation and Applications; McGraw-Hill and SciTech Publishing: New York, NY, USA, 1999; pp. 27–35.

79. Chou, J. Catalytic Combustible Gas Sensors. In Hazardous Gas Monitors—A Practical Guide to Selection, Operation and Applications; McGraw-Hill and SciTech Publishing: New York, NY, USA, 1999; pp. 37–45.

80. Chou, J. Solid-State Gas Sensors. In Hazardous Gas Monitors—A Practical Guide to Selection, Operation and Applications; McGraw-Hill and SciTech Publishing: New York, NY, USA, 1999; pp. 47–53.

81. Chou, J. Infrared Gas Sensors. In Hazardous Gas Monitors—A Practical Guide to Selection, Operation and Applications; McGraw-Hill and SciTech Publishing: New York, NY, USA, 1999; pp. 55–72.

82. Chou, J. Photoionization Detectors. In Hazardous Gas Monitors—A Practical Guide to Selection, Operation and Applications; McGraw-Hill and SciTech Publishing: New York, NY, US, 1999; pp. 73–81.

83. Williams, R.; Kilaru, V.; Snyder, E.; Kaufman, A.; Dye, T.; Rutter, A.; Russell, A.; Hafner, H. Air Sensor Guidebook; Technical report, United States Environmental Protection Agency; 2004.

84. Chou, J. Sensor Selection Guide. In Hazardous Gas Monitors—A Practical Guide to Selection, Operation and Applications; McGraw-Hill and SciTech Publishing: New York, NY, USA, 1999; Chapter 8; pp. 103–109.

85. Gerboles, M.; Buzica, D. Evaluation of Micro-Sensors to Monitor Ozone in Ambient Air; Technical report, Joint Research Center, Institute for Environment and Sustainability; 2009.

86. Air Quality Expert Group. Methods for Monitoring Particulate Concentrations. In Particulate Matter in the United Kingdom; Department for the Environment, Food and Rural Affairs: London, UK, 2005; pp. 125–129.

87. Grover, B.D.; Kleinman, M.; Eatough, N.L.; Eatough, D.J.; Hopke, P.K.; Long, R.W.; Wilson, W.E.; Meyer, M.B.; Ambs, J.L. Measurement of total PM2.5 mass (nonvolatile plus semivolatile) with the Filter Dynamic Measurement System tapered element oscillating microbalance monitor. J. Geophys. Res: Atmos. **2005**, 110, 148–157.

88. Air Quality Expert Group. Methods for Monitoring Particulate Concentrations. In Particulate Matter in the United Kingdom; Department for the Environment, Food and Rural Affairs: London, UK, 2005; pp. 129–131.

89. Air Quality Expert Group. Methods for Monitoring Particulate Concentrations. In Particulate Matter in the United Kingdom; Department for the Environment, Food and Rural Affairs: London, UK, 2005; pp. 131–133.

90. Air Quality Expert Group. Methods for Monitoring Particulate Concentrations. In Particulate Matter in the United Kingdom; Department for the Environment, Food and Rural Affairs: London, UK, 2005; pp. 134–143.

91. Air Quality Expert Group. Methods for Monitoring Particulate Concentrations. In Particulate Matter in the United Kingdom; Department for the Environment, Food and Rural Affairs: London, UK, 2005; pp. 133–137.

92. United States Environmental Protection Agency. Compact Nephelometer System for On-Line Monitoring of Particulate Matter Emissions. 2004. Available online: http://cfpub.epa.gov// ncer_abstracts/ index.cfm/ fuseaction/display.abstractdetail/abstract/6539 (accessed on 25 September 2015).

93. Choi, S.; Kim, N.; Cha, H.; Ha, R. Micro Sensor Node for Air Pollutant Monitoring: Hardware and Software Issues. Sensors **2009**, 9, 7970–7987.

94. Ikram, J.; Tahir, A.; Kazmi, H.; Khan, Z.; Javed, R.; Masood, U. View: Implementing low cost air quality monitoring solution for urban areas. Environ. Syst. Res. **2012**, 1, 10–15.

95. Hasenfratz, D.; Saukh, O.; Walser, C.; Hueglin, C.; Fierz, M.; Thiele, L. Pushing the Spatio-Temporal Resolution Limit of Urban Air Pollution Maps. In Proceedings of the 12th International Conference on Pervasive Computing and Communications (PerCom 2014), Budapest, Hungary, 24–28 March 2014; pp. 69–77.

96. Burke, J.A.; Estrin, D.; Hansen, M.; Parker, A.; Ramanathan, N.; Reddy, S.; Srivastava, M.B. Participatory sensing. In Proceedings of the 4th ACM Conference on Embedded Network Sensor Systems (SenSys '6), Boulder, CO, USA, 1–3 November 2006; pp. 1124–1127.

97. Mendez, D.; Perez, A.J.; Labrador, M.A.; Marron, J.J. P-Sense: A participatory sensing system for air pollution monitoring and control. In Procedings of the 2011 IEEE International Conference on Pervasive Computing and Communications Workshops (PERCOM Workshops), Seattle, WA, USA, 21–25 March 2011; pp. 344–347.

98. Hu, S.C.; Wang, Y.C.; Huang, C.Y.; Tseng, Y.C. A vehicular wireless sensor network for CO_2 monitoring. IEEE Sens. **2009**, 2009, 1498–1501.

99. Wong, K.J.; Chua, C.C.; Li, Q. Environmental Monitoring Using Wireless Vehicular Sensor Networks. In Proceedings of the 5th International Conference on Wireless Communications, Networking and Mobile Computing, 2009 (WiCom '09), Beijing, China, 24–26 September 2009; pp. 1–4.

100. Hasenfratz, D. Enabling Large-Scale Urban Air Quality Monitoring with Mobile Sensor Nodes. Ph.D. Thesis, ETH-Zürich, ZÄrich, Switzerland, 2015.

101. Murty, R.N.; Mainland, G.; Rose, I.; Chowdhury, A.R.; Gosain, A.; Bers, J.; Welsh, M. CitySense: An Urban-Scale Wireless Sensor Network and

Testbed. In Proceedings of the 2008 IEEE Conference on Technologies for Homeland Security, Waltham, MA, USA, 12–13 May 2008; pp. 583–588.

102. Liu, J.H.; Chen, Y.F.; Lin, T.S.; Lai, D.W.; Wen, T.H.; Sun, C.H.; Juang, J.Y.; Jiang, J.A. Developed urban air quality monitoring system based on wireless sensor networks. In Proceedings of the 2011 Fifth International Conference on Sensing Technology (ICST), Palmerston North, New Zealand, 28 November 2011–1 December 2011; pp. 549–554.

103. Kadri, A.; Yaacoub, E.; Mushtaha, M.; Abu-Dayya, A. Wireless sensor network for real-time air pollution monitoring. In Proceedings of the 2013 1st International Conference on Communications, Signal Processing, and their Applications (ICCSPA), Sharjah, UAE, 12–14 February 2013; pp. 1–5.

104. Khedo, K.K.; Perseedoss, R.; Mungur, A. A Wireless Sensor Network Air Pollution Monitoring System. IJWMN **2010**, 2, 1–15.

105. Mansour, S.; Nasser, N.; Karim, L.; Ali, A. Wireless Sensor Network-based air quality monitoring system. In Proceedings of the 2014 International Conference on Computing, Networking and Communications (ICNC), Honolulu, HI, USA, 3–6 February 2014; pp. 545–550.

106. Libelium. Libelium Waspmote. Available online: http://www. libelium. com/products/waspmote/ (accessed on 27 May 2015).

107. Jelicic, V.; Magno, M.; Brunelli, D.; Paci, G.; Benini, L. Context-Adaptive Multimodal Wireless Sensor Network for Energy-Efficient Gas Monitoring. IEEE Sens. J. **2013**, 13, 328–338.

108. Postolache, O.A.; Pereira, J.M.D.; Girao, P.M.B.S. Smart Sensors Network for Air Quality Monitoring Applications. IEEE Trans. Instrum. Meas. **2009**, 58, 3253–3262.

109. Jiang, Y.; Li, K.; Tian, L.; Piedrahita, R.; Yun, X.; Mansata, O.; Lv, Q.; Dick, R.P.; Hannigan, M.; Shang, L. MAQS: A Personalized Mobile Sensing System for Indoor Air Quality Monitoring. In Proceedings of the 13th International Conference on Ubiquitous Computing (UbiComp '11), Beijing, China, 17–21 September 2011; pp. 271–280.

110. Aberer, K.; Sathe, S.; Chakraborty, D.; Martinoli, A.; Barrenetxea, G.; Faltings, B.; Thiele, L. OpenSense: Open Community Driven Sensing of Environment. In Proceedings of the ACM SIGSPATIAL International Workshop on GeoStreaming (GIS-IWGS), San Jose, CA, USA, 2 November 2010; pp. 39–42.

111. Honicky, R.; Brewer, E.A.; Paulos, E.; White, R. N-smarts: Networked Suite of Mobile Atmospheric Real-time Sensors. In Proceedings of the Second

ACM SIGCOMM Workshop on Networked Systems for Developing Regions (NSDR'08), Seattle, WA, USA, 18 August 2008; pp. 25–30.

112. Maisonneuve, N.; Stevens, M.; Niessen, M.E.; Hanappe, P.; Steels, L. Citizen Noise Pollution Monitoring. In Proceedings of the 10th Annual International Conference on Digital Government Research: Social Networks: Making Connections Between Citizens, Data and Government (dg.o '09), Puebla, Mexico, 17–21 May 2009; pp. 96–103.

113. Tsow, F.; Forzani, E.; Rai, A.; Wang, R.; Tsui, R.; Mastroianni, S.; Knobbe, C.; Gandolfi, A.J.; Tao, N.J. A Wearable and Wireless Sensor System for Real-Time Monitoring of Toxic Environmental Volatile Organic Compounds. IEEE Sens. J. **2009**, 9, 1734–1740.

114. United States Environmental Protection Agency. Buildings and their Impact on the Environment: A Statistical Summary. 2009. Available online: http://www.epa.gov/greenbuilding/pubs/gbstats.pdf (accessed on 16 June 2015).

115. Lo Re, G.; Peri, D.; Vassallo, S. Urban Air Quality Monitoring Using Vehicular Sensor Networks. In Advances onto the Internet of Things; Gaglio, S., Lo Re, G., Eds.; Springer International Publishing: Gewerbestrasse, Switzerland, 2014; pp. 311–323.

116. Hu, S.C.; Wang, Y.C.; Huang, C.Y.; Tseng, Y.C. Measuring air quality in city areas by vehicular wireless sensor networks. Mobile Applications: Status and Trends. J. Syst. Softw. **2011**, 84, 2005–2012.

117. Devarakonda, S.; Sevusu, P.; Liu, H.; Liu, R.; Iftode, L.; Nath, B. Real-time Air Quality Monitoring Through Mobile Sensing in Metropolitan Areas. In Proceedings of the 2nd ACM SIGKDD International Workshop on Urban Computing (UrbComp '13), Chicago, IL, USA, 11 August 2013; pp. 15:1–15:8.

118. Al-Ali, A.R.; Zualkernan, I.; Aloul, F. A Mobile GPRS-Sensors Array for Air Pollution Monitoring. IEEE Sens. J. **2010**, 10, 1666–1671.

119. Liu, J.; Diamond, J. Revolutionizing China's Environmental Protection. Science **2008**, 319, 37–38.

120. Al-Saadi, J.; Szykman, J.; Pierce, R.B.; Kittaka, C.; Neil, D.; Chu, D.A.; Remer, L.; Gumley, L.; Prins, E.; Weinstock, L.; MacDonald, C.; Wayland, R.; Dimmick, F.; Fishman, J. Improving National Air Quality Forecasts with Satellite Aerosol Observations. Bull. Am. Meteorol. Soc. **2005**, 86, 1249–1261.

121. Engel-Cox, J.; Hoff, R.; Weber, S.; Zhang, H.; Prados, A. Three Dimensional Air Quality System (3D-AQS). 2007. Available online: http://alg.umbc.edu/3d-aqs/doc/3daqs_agu_winter2007.pdf (accessed on 17 July 2015).

122. Calpini, B.; Simeonov, V.; Jeanneret, F.; Kuebler, J.; Sathya, V.; van den Bergh, H. Ozone LIDAR as an Analytical Tool in Effective Air Pollution Management: The Geneva 96 Campaign. CHIMIA Int. J. Chem. **1997**, 51, 700–704.

INDEX